Plates and FEM

SOLID MECHANICS AND ITS APPLICATIONS

Volume 171

Aims and Scope of the Series

The fundamental questions arising in mechanics are: *Why?*, *How?*, and *How much?* The aim of this series is to provide lucid accounts written by authoritative researchers giving vision and insight in answering these questions on the subject of mechanics as it relates to solids.

The scope of the series covers the entire spectrum of solid mechanics. Thus it includes the foundation of mechanics; variational formulations; computational mechanics; statics, kinematics and dynamics of rigid and elastic bodies: vibrations of solids and structures; dynamical systems and chaos; the theories of elasticity, plasticity and viscoelasticity; composite materials; rods, beams, shells and membranes; structural control and stability; soils, rocks and geomechanics; fracture; tribology; experimental mechanics; biomechanics and machine design.

The median level of presentation is the first year graduate student. Some texts are monographs defining the current state of the field; others are accessible to final year undergraduates; but essentially the emphasis is on readability and clarity.

For other titles published in this series, go to
www.springer.com/series/6557

Johan Blaauwendraad

Plates and FEM

Surprises and Pitfalls

Springer

Johan Blaauwendraad
Professor Emeritus of Structural Mechanics
Delft University of Technology
Klinkenbergerweg 74
6711 ML Ede
The Netherlands
j.blaauwendraad@tudelft.nl

ISSN 0925-0042
ISBN 978-94-007-3189-9 ISBN 978-90-481-3596-7(eBook)
DOI 10.1007/978-90-481-3596-7
Springer Dordrecht Heidelberg London New York

Softcover reprint of the hardcover 1st edition 2010

Cover design: eStudio Calamar S.L.

Printed on acid-free paper

Springer is part of Springer Science+Business Media (www.springer.com)

To Henny

Contents

Preface

The Finite Element Method, shortly FEM, is a widely used computational tool in structural engineering. For basic design purposes it usually suffices to apply a linear-elastic analysis. Only for special structures and for forensic investigations the analyst need to apply more advanced features like plasticity and cracking to account for material nonlinearities, or nonlinear relations between strains and displacements for geometrical nonlinearity to account for buckling. Advanced analysis techniques may also be necessary if we have to judge the remaining structural capacity of aging structures. In this book we will abstain from such special cases and focus on everyday jobs. Our goal is the worldwide everyday use of linear-elastic analysis, and dimensioning on basis of these elastic computations. We cover steel and concrete structures, though attention to structural concrete prevails.

Structural engineers have access to powerful FEM packages and apply them intensively. Experience makes clear that often they do not understand the software that they are using. This book aims to be a bridge between the software world and structural engineering. Many problems are related to the correct input data and the proper interpretation and handling of output. The book is neither a text on the Finite Element Method, nor a user manual for the software packages. Rather it aims to be a guide to understanding and handling the results gained by such software.

We purposely restrict ourselves to structure types which frequently occur in practise. We consider shear walls and other two-dimensional structures loaded in their plane (*membrane state*), and we deal with floor and bridge structures loaded perpendicular to their plane (*bending state*). The content reflects the subject matter of continuing education courses in European countries, such as Belgium and the Netherlands.

The book is composed of four parts. In the first one we present an overview of the classical theory of plates in the membrane and bending state. This is considered necessary in order to better understand instructions to be given in later parts. For a few relevant examples which play an illustrative role exact solutions are presented. To some extent the presentation for plates in the membrane state diverges from classical text books. There all membrane problems are discussed on basis of the well-known Airy function. This theory can be considered as an application of the force method in which stress function is chosen as the statically dependent quantity. However, this method is completely opposite to the current displacement method (stiffness method) as applied in the finite element method. We consider this an unwanted discrepancy and anachronism, and therefore have chosen a displacement method for the membrane state as well. The degree of difficulty does not become higher, while there is now uniformity with the theory for the bending state. We resort to Airy's theory only for some well-known classical solutions. The problem of discrepancy and anachronism does not occur for plates in bending and transverse shear, because the relevant classical differential equations are already based on the displacement method. For this category another aspect needs attention, more than usually given in classical text books. Nowadays, explicit clarification is needed on the difference between thick plates (Mindlin–Reissner) and thin plates (Kirchhoff), because both options are offered in commercial software and surprising choices appear to have been made as default option or are made by users.

Before moving to the main body of the book we direct attention in Part 2 of the book to some instructive computational models of pre-FEM days. These are re-called for both plates in membrane state and plates in bending. It is believed that they are of great didactical value, particularly for plates in bending.

The contents of Part 3 of the book provide the impetus for publication. We intend to help the structural engineer in handling linear-elastic computational results in daily practice. This part starts with an overview of the stiffness method, as applied in commercial FEM packages. We review element types which are relevant to the subject matter of the book, discuss input options, and review output selection. We address the subject that results of the calculation are highly dependent on the structural engineer who performs the computation, and give hints how to avoid dependency on mesh fineness. We show how FE analysis can support application of the strut-and-tie design model and draw attention to several surprises and pitfalls. We are pleased with the contribution of Professor Paulo Lourenço of Minho University in Portugal on reinforcement design of plates in membrane state, bending-shear state or combined membrane-bending-shear state. We also review a couple of contemporary two-way slab systems, and consider options for calculating the deflection and crack-width in the serviceability state.

The final part of the book (Part 4) is another reason for publication: it deals with orthotropy as it occurs in bridge and floor systems. The focus for these orthotropic properties is on calculating correct input rigidity data, and determining whether computed plate moments and shear forces match code requirements. Noticing odd outcomes we observe a definite need for clarification of this subject matter. Orthotropic properties also render a service when structural engineers want to calculate deflections and crack-widths in the cracked serviceability state.

Acknowledgements

We happily acknowledge the support of Dr. Anton W.M. Kok, the spiritual father of the FE code *Kola*. We are indebted to Dr. Cox W.M. Sitters for brushing up the text and figures on theory, gratefully acknowledge the help of Dr. Pierre C.J. Hoogenboom in calculating examples, and highly value the writing hints of Professor Graham M.L. Gladwell of the University of Waterloo. We very much appreciated the contributions of software providers Nemetschek-Scia, Buildsoft, Technosoft and Matrix. The Structural Mechanics group of Civil Engineering in Delft University of Technology, PAO Foundation for Post Graduate Education, Delft Cluster organization on delta technology and infrastructural research, and Martens Group in the Netherlands have shown stimulating interest, facilitating a productive stay at the University of California in San Diego. We enjoyed the encouraging exchange of thoughts with Professors José Restrepo and Joel Conte. Particular thanks are due to publishing editor Nathalie Jacobs of Springer for her dedicated promotion, to Anneke Pot for assistance, and to Jolanda Karada for her great job in styling the book.

Free Software

For the majority of the Finite Element Analyses in this book we applied the program *KolaLight*, which was developed for teaching purposes by Anton W.M. Kok, former Associate Professor in the Structural Mechanics Group of the Faculty of Civil Engineering and Geosciences (CiTG) in Delft University of Technology in the Netherlands. The reader can freely download the program and documentation from internet. The link is

http://www.mechanics.citg.tudelft.nl/kolalight

We also offer a free program *RCShell* for the application of the theory for reinforcement design on the basis of linear-elastic analysis as discussed in Chapter 16. The author of the program is Paulo B. Lourenço, Professor in the Masonry and Historical Constructions Group, Department of Civil Engineering, University of Minho, Guimarães, Portugal. The links to the menu and fortran files are

http://www.civil.uminho.pt/masonry/Software/RCShell.zip (menu)
http://www.civil.uminho.pt/masonry/Software/Reinforcement.zip (fortran)

Conversion of SI Units to Imperial Units

1 SI unit = C USCS units

$\frac{1}{C}$ SI units = 1 USCS unit

SI Unit	***C***	***1/C***	***Imperial***
Area			
1 m^2	10.76	0.0929	ft^2
1 mm^2	1.55×10^{-3}	645	in^2
Distributed line load			
1 N/m	0.0685	1.46	lbf/ft
1 kN/m	0.0306	32.7	tonf/ft
1 N/m	5.71×10^{-3}	175	lb/in
1 kN/m	2.55×10^{-3}	392	tonf/in
Distributed area load			
1 N/m^2	0.0209	47.9	lbf/ft^2
1 kN/m^2	0.0933	10.7	$tonf/ft^2$
1N/m^2	0.145×10^{-3}	6.89×10^3	lbf/in^2
1kN/m^2	64.7×10^{-6}	15.5×10^3	$tonf/in^2$
Force			
1 N	0.225	4.45	lbf
1 kN	0.100	9.96	tonf
Length			
1 m	3.28	0.305	ft
1 mm	0.0394	25.4	in

SI Unit	C	1/C	Imperial
Moment. Torque			
1 N·m	0.738	1.36	lbf·ft
1 N·m	8.85	0.113	lbf·in
1 kN·m	0.329	3.04	tonf·ft
1 kN·m	3.95	0.253	tonf·in
Moment in plate			
1 N·m/m	0.225	4.45	lbf·ft/ft
1 N·m/m	0.225	4.45	lbf·in/in
1 kN·m/m	0.100	9.96	tonf·ft/ft
1 kN·m/m	0.100	9.96	tonf·in/in
Shear force in plate			
1 N/m	0.0685	14.6	lbf/ft
1 kN/m	0.0305	32.8	tonf/ft
1 N/m	5.71×10^{-3}	175	lbf/in
1 kN/m	2.55×10^{-3}	392	tonf/in
Moment of inertia (second moment of area)			
1 mm^4	2.40×10^{-6}	0.416×10^{6}	in^4
1 m^4	2.40×10^{6}	0.416×10^{-6}	in^4
Pressure, Stress			
1 Pa	0.0209	47.9	lbf/ft^2
1 Pa	0.145×10^{-3}	6890	lbf/in^2
1 MPa	20.9×10^{3}	47.9 lbf/ft^2	
1 MPa	0.145×10^{-3}	7.20	lbf/in^2
1 MPa	93.3	0.0107	tonf/ft^2
1 MPa	0.648	1.54	tonf/in^2
Section modulus. Volume			
1 mm^3	0.0610×10^{-3}	16.4×10^{3}	in^4
1 m^3	0.0610×10^{6}	16.4×10^{-6}	in^4

Conversion of SI Units to US Customary System

1 SI unit = C USCS units

$\frac{1}{C}$ SI units = 1 USCS unit

SI Unit	***C***	***1/C***	***US CS***
Area			
1 m^2	10.76	0.0929	ft^2
1 mm^2	1.55×10^{-3}	645	in^2
Distributed line load			
1 N/m	0.0685	1.46	lb/ft
1 kN/m	0.0685	14.6	k/ft
1 N/m	5.71×10^{-3}	175	lb/in
1 kN/m	5.71×10^{-3}	175	k/in
Distributed area load			
1 N/m^2	0.0209	147.9	lb/ft^2
1 kN/m^2	0.0209	47.9	k/ft^2
1 N/m^2	0.145×10^{-3}	6.89×10^3	lb/in^2
1 kN/m^2	0.145×10^{-3}	6.89×10^3	k/in^2
Force			
1 N	0.225	4.45	lb
1 kN	0.225	4.45	k
Length			
1 m	3.28	0.305	ft
1 mm	0.0394	25.4	in

SI Unit	C	1/C	US CS
Moment. Torque			
1 N·m	0.738	1.36	ft-lb
1 N·m	8.85	0.113	in-lb
1 kN·m	0.738	1.36	ft-k
1 kN·m	8.85	0.113	in-k
Moment in plate			
1 N·m/m	0.225	4.45	ft-lb/ft
1 N·m/m	0.225	4.45	in-lb/in
1 kN·m/m	0.225	4.45	ft-k/ft
1 kN·m/m	0.225	4.45	in-k/in
Shear force in plate			
1 N/m	0.0685	14.6	lb/ft
1 kN/m	0.0685	14.6	k/ft
1 N/m	5.71×10^{-3}	175	lb/in
1 kN/m	5.71×10^{-3}	175	k/in
Moment of inertia (second moment of area)			
1 mm^4	2.40×10^{-6}	0.416×10^{6}	in^4
1 m^4	2.40×10^{6}	0.416×10^{-6}	in^4
Pressure. Stress			
1 Pa	0.0209	47.9	psf
1 Pa	0.145×10^{-3}	6890	psi
1 MPa	20.9	0.0479	ksf
1 MPa	0.145	6.89	ksi
Section Modulus. Volume			
1 mm^3	0.0610×10^{-3}	16.4×10^{3}	in^4
1 m^3	0.0610×10^{6}	16.4×10^{-6}	in^4

Part 1
Theory of Plates

Chapter 1
Plate Membrane Theory

The word plate is a collective term for systems in which transfer of forces occurs in two directions; walls, deep beams, floors and bridge slabs are all plates. We distinguish two main categories, plates that are loaded in their plane, and plates loaded perpendicularly to their plane. For both categories we give an approach with differential equations, such that a basic understanding is provided and for certain characteristic cases an exact solution can be determined. We follow the displacement method, working with differential equations. In plates that are loaded in their plane, the plane stress state is called the *membrane state*. All stress components are parallel to the mid-plane of the plate. In special cases we can simply determine the stresses. A well-known example is a prismatic slender beam of rectangular cross-section, loaded by bending moments and shear forces as shown in Figure 1.1. For this special case we can easily calculate the bending stresses σ_{xx} and the shear stresses σ_{xy} from Euler–Bernoulli beam theory. However, less simple are other cases like a deep beam or wall (Figure 1.2).

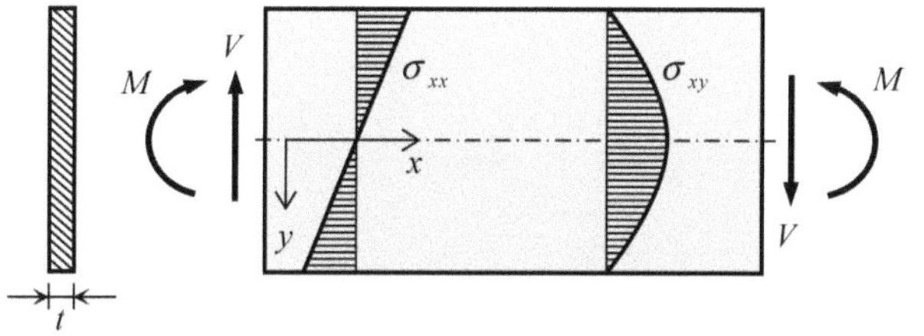

Figure 1.1 Stresses in prismatic beam in classical beam theory.

J. Blaauwendraad, *Plates and FEM: Surprises and Pitfalls*, Solid Mechanics and Its Applications 171, DOI 10.1007/978-90-481-3596-7_1,

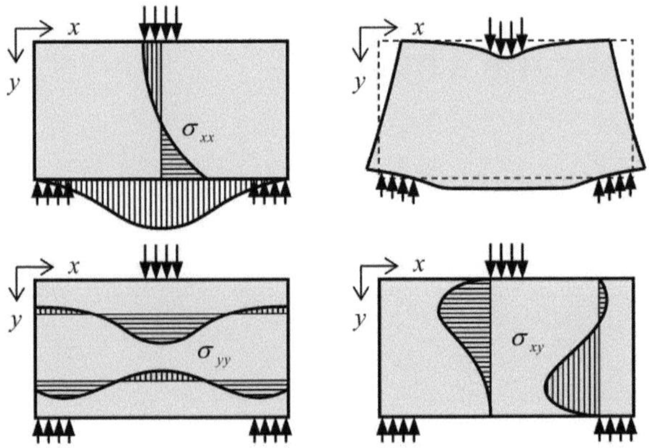

Figure 1.2 Example of the deformation and stress distribution in a deep beam.

In a deep beam, the stress distribution differs from what the classical beam theory predicts. The bending stress σ_{xx} is no longer linear and beam theory does not give any information about the vertical normal stresses σ_{yy}, but they do of course occur. Finally, the shear stress σ_{xy} in a deep beam does not have a parabolic distribution, as in the classic beam theory. Classical text books on the subject refer to Girkmann [2] and Timoshenko [3]. This chapter will offer a solution method for such general problems. We deal with a group of problems that we can consider to be two-dimensional. Plane stress occurs in a thin flat plate, which is loaded in its plane by a perimeter load f or/and a distributed load p over the plate with components p_x and p_y, see Figure 1.3.

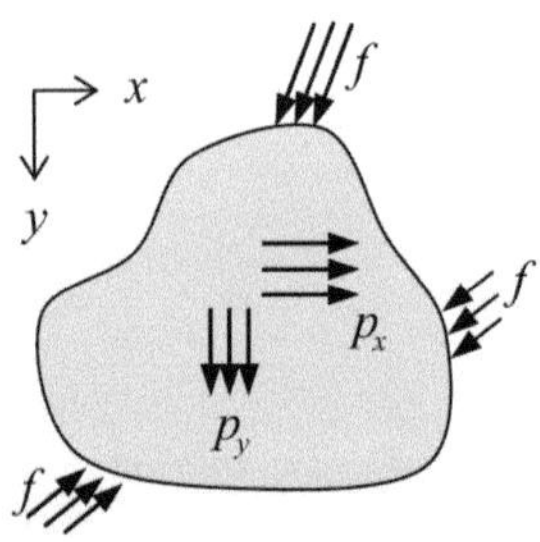

Figure 1.3 Thin flat plate loaded in-plane.

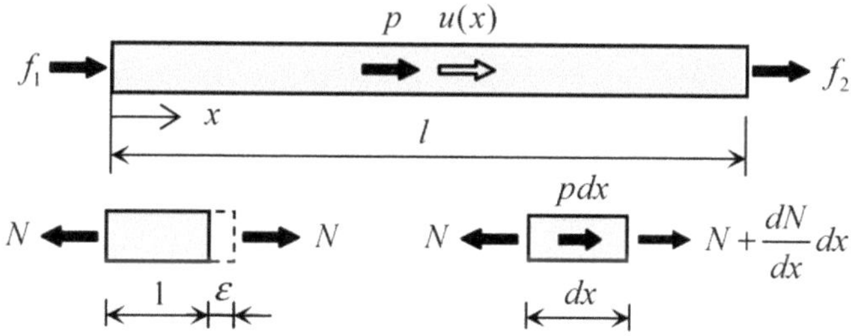

Figure 1.4 Bar subjected to extension with relevant quantities.

1.1 Introduction: Special Case of a Plate, the Truss

Before continuing with the general two-dimensional case, we first discuss the one-dimensional case as a special case. So, we take a plate, on which a load is applied in one direction, the x-axis. Everything in y-direction is constant, and Poisson's ratio is zero. Figure 1.4 shows this uni-axial situation. In fact we can handle it as a truss. The x-axis is chosen from left to right. At the position x, the cross-section displaces $u(x)$ in x-direction after applying the load; the displacement is accompanied by a specific strain $\varepsilon(x)$. The truss with length l is loaded along its length with a distributed load $p(x)$ per unit length, and at both ends with forces f_1 and f_2. The cross-section of the truss has area A. The modulus of elasticity of the material is E. The stress resultant in the cross-section is the normal force $N(x)$. For this problem we can derive three basic equations between the quantities u, ε, N, and p, as is shown in Figure 1.5, and these relationships are the basis for the differential equation.

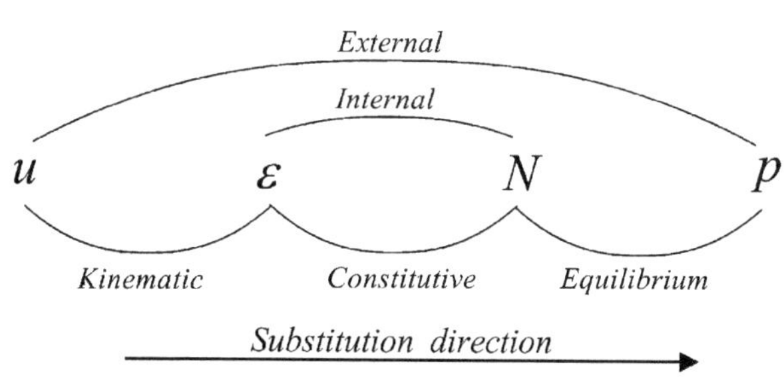

Figure 1.5 Scheme of three basic equations.

Recurring scheme

Hereafter in the book such a scheme will occur each time that we will derive differential equations. The three relationships are the *kinematic equation*, the *constitutive equation* and the *equilibrium equation.* In short, we refer to them hereafter as kinematic, constitutive and equilibrium.

$$\varepsilon = \frac{du}{dx} \quad \textit{Kinematic} \tag{1.1}$$

$$N = EA\varepsilon \quad \textit{Constitutive} \tag{1.2}$$

$$-\frac{dN}{dx} = p \quad \textit{Equilibrium} \tag{1.3}$$

Substitution of the kinematic equation in the constitutive law and after that the changed constitutive in the equilibrium equation transforms the latter into

$$-EA\frac{d^2u}{dx^2} = p \tag{1.4}$$

This is a differential equation in the degree of freedom u. The procedure of subsequent substitution from left to right will occur in this book each time that differential equations are to be derived, and these equations always will be differential equations in degrees of freedom.

The second-order differential equation (1.4) can be solved if two boundary conditions are specified, one at the left end and one at the right end. At position $x = 0$ the boundary condition is either

$$u_1 = u_{1,0} \tag{1.5}$$

in which $u_{1,0}$ is a prescribed value, or

$$-N_1 = f_1 \rightarrow -EA\left(\frac{du}{dx}\right)_1 = f_1 \tag{1.6}$$

where f_1 is a given load. At $x = l$ the boundary condition is either

$$u_2 = u_{2,0} \tag{1.7}$$

or

$$N_2 = f_2 \rightarrow EA\left(\frac{du}{dx}\right)_2 = f_2 \tag{1.8}$$

From the differential equation (1.4) and the boundary conditions in Eq. (1.5) up to Eq. (1.8), we can solve the displacement u. When the solution u has been ob- tained, we can calculate the normal force N from Eq. (1.2) accounting for Eq. (1.1). If the displacement u at an end is a constraint, we can compute the support reaction f. Conversely, if the load f is specified at a free end, we find a solution for the displacement u at that position.

Recurring procedure

The method stated here is known as the *displacement method* or *stiffness method.* In the next sections we will solve the plate problem with a two-way load along similar lines, and throughout the whole book a similar approach will be followed. Our procedure will always be to track down the quantities which govern the problem and should be included in the scheme for the three basic sets of relationships. Each time we must decide which *displacements* are the degrees of freedom and will introduce associate *external loads.* We also choose proper *deformations* and their associated *stress resultants.* The kinematic relationships relate the displacements and deformations to each other, the constitutive relationships (laws) the deformations and stress resultants, and the equilibrium relationships the stress resultants to the external loads.

1.2 Membrane Plate Problem Statement

Every point (x, y) of a plate loaded in its plane goes through a displacement $u_x(x, y)$ in the direction of the x-axis and a displacement $u_y(x, y)$ in the direction of the y-axis, see Figure 1.6. So, the displacement field is defined by two degrees of freedom. That means that distributed external loads p_x and p_y per unit area can be applied in these two directions. Three deformations occur internally in the plate, the strain ε_{xx} in x-direction, the strain ε_{yy} in the y-direction, and a shear strain γ_{xy}. This conjugates with the stresses σ_{xx}, σ_{yy} and σ_{xy}, respectively, see Figure 1.6. As said before, throughout the book we make a clear distinction between displacements and deformations. Structural engineers often use displacements and de-

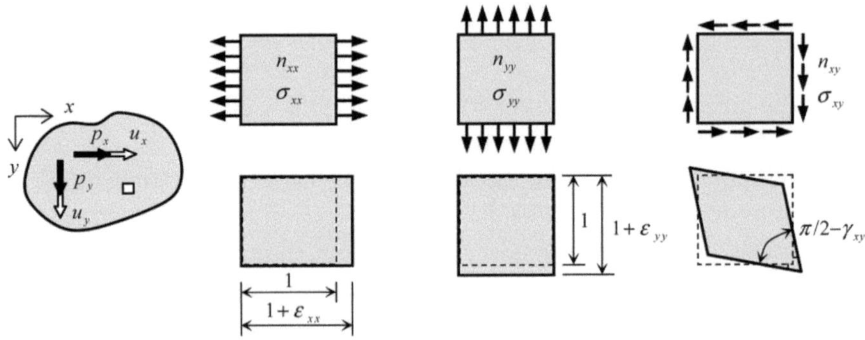

Figure 1.6 Quantities which play a role in a plate loaded in-plane.

formations synonymously; we do not. Displacements indicate a shift of a point relative to a coordinate frame; deformations relate to change of shape.

Notation and sign convention

The *notation* for the stress in the membrane state may require clarification. For each stress we use two indices. The first indicates the face on which the stress acts. It is the direction of the normal of the face. The second index is the direction in which the stress is acting. For instance, the stress σ_{xy} acts on a face with normal in the x-direction and is directed in the y-direction.
The *sign convention* is as follows. A stress is *positive* if it acts in the positive coordinate direction on a plane with the normal vector in the *positive* coordinate direction. Correspondingly, a stress is positive if it acts in the *negative* coordinate direction on a plane with its normal in the *negative* coordinate direction, see Figure 1.6.

The stresses are constant through the thickness. Common practice is to multiply them by the plate thickness t. The resulting *membrane forces* n_{xx}, n_{yy}, and n_{xy} are the stress resultants per unit plate width, see Figure 1.7, having the dimension force per unit length. The sign convention for the membrane forces is the same as for the stresses. While we use the capitol symbol N for a truss element, we now use lower cast n for membrane forces. This is because of difference in units. The normal force in a truss has the unit of force (N), and the membrane forces have the unit force per unit width (N/m).

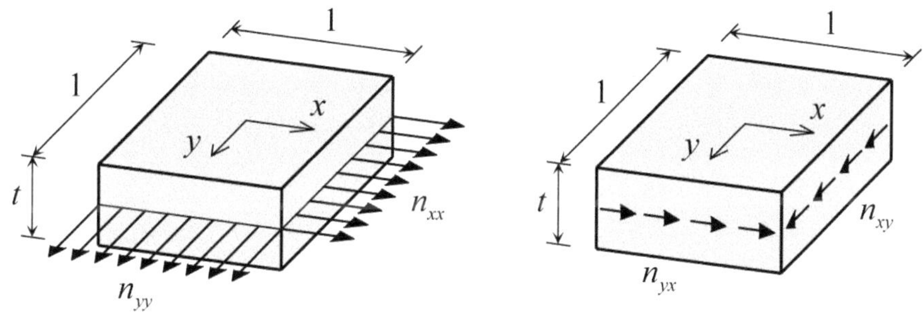

Figure 1.7 Membrane forces used by the designer

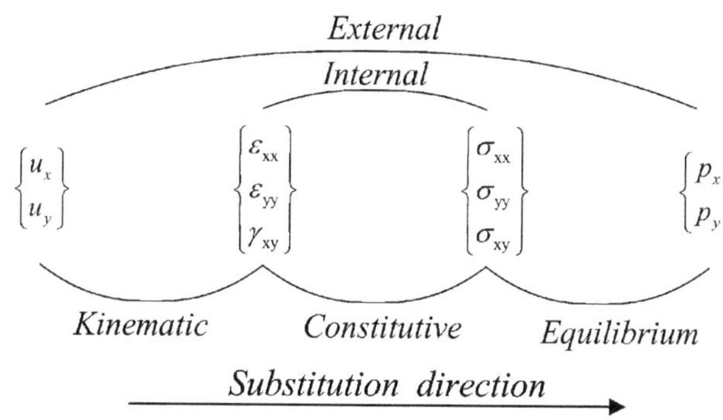

Figure 1.8 Scheme of relationships in a plate in membrane state.

We can define relationships between the degrees of freedom, associated loads, membrane forces and associate strains as is shown in the scheme of Figure 1.8 for plates in membrane state, just as with the truss problem in Section 1.1.

1.2.1 Kinematic Equations

Consider an elementary rectangular plate particle with sides dx and dy in an orthogonal set of axes x, y in an unloaded state. These are the dotted lines in Figure 1.9. After a load is applied, this particle is *displaced* and *deformed.* The new position in the set of axes can be established by three rigid body displacements and three deformations. The three rigid body displacements are

- a translation u_x in x-direction;
- a translation u_y in y-direction;

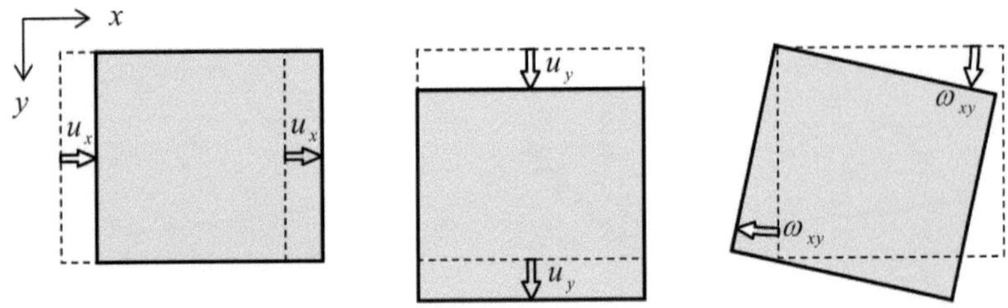

Figure 1.9 Three rigid body displacements.

- a rotation ω_{xy}, positive as shown.

The three deformations are shown in Figure 1.10; they are

- strain ε_{xx} in x-direction; this is positive when elongation is involved;
- strain ε_{yy} in y-direction; this is positive when elongation is involved;
- shear strain γ_{xy}; this deformation changes a square shape into a rhombus, such that the diagonal coinciding with the bisector of the first quadrant becomes larger and the other diagonal shorter; the magnitude of γ_{xy} is the angular deviation of the initially right angle.

The rigid body displacements are strain-less movements and occur without generating any stresses. The deformations are associated with strains and do create stresses. The kinematic equations define the relationship between the displacements and the strains. The effect of the displacement field is shown in Figure 1.11. An elementary plate part $ABCD$ in an unloaded state transforms into the quadrilateral $A'B'C'D'$ after application of the load. According to the definitions of Figure 1.6 the following relationships are valid:

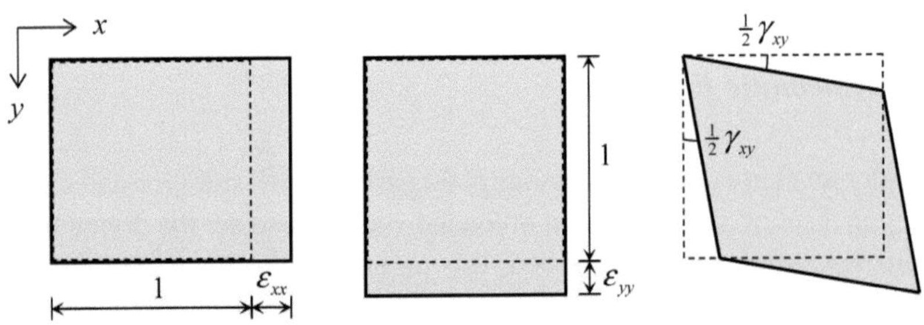

Figure 1.10 Three deformations.

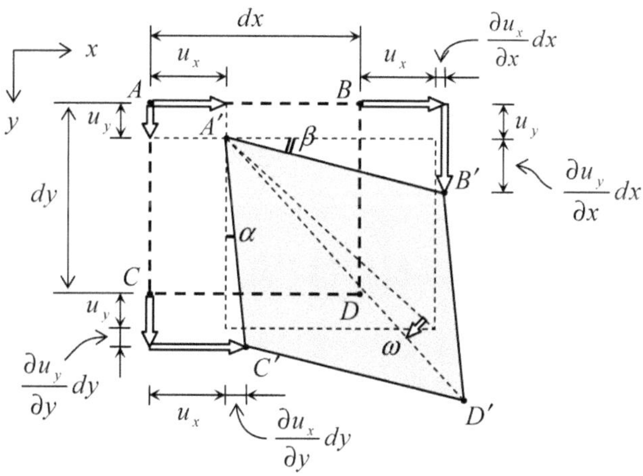

$$\gamma_{xy} = \alpha + \beta \; ; \quad \omega_{xy} = ½(\alpha - \beta) \; ; \quad \alpha = \frac{\partial u_x}{\partial y} \; ; \quad \beta = \frac{\partial u_y}{\partial x}$$

Figure 1.11 Displaced and deformed state of an elementary plate part.

$$\begin{aligned} \varepsilon_{xx} &= \frac{\partial u_x}{\partial x} \\ \varepsilon_{yy} &= \frac{\partial u_y}{\partial y} \qquad\qquad \textit{Kinematic} \\ \gamma_{xy} &= \frac{\partial u_x}{\partial y} + \frac{\partial u_y}{\partial x} \end{aligned} \tag{1.9}$$

Equation (1.9) is a generalization of Eq. (1.1) for the truss. From Figure 1.10 we also deduce the determination of the rotation ω_{xy} from the displacements u_x and u_y

$$\omega_{xy} = \left(-\frac{\partial u_x}{\partial y} + \frac{\partial u_y}{\partial x} \right) \tag{1.10}$$

1.2.2 Constitutive Equations

The constitutive equations give us information about the material behaviour, by providing the relation between the stresses and the strains. Hooke's law is considered in its most general form for linear-elastic materials. Figure 1.12 displays the deformation states due to a normal stress σ_{xx}, a normal stress σ_{yy} and a shear stress σ_{xy}. Normal stresses cause an elongation in the direc-

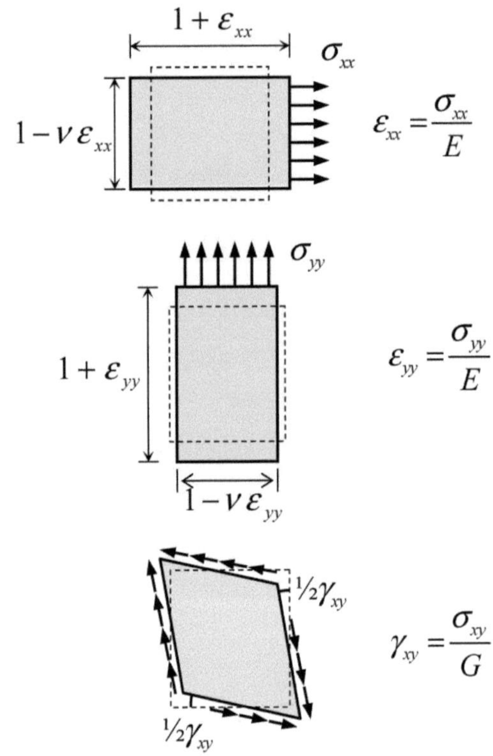

Figure 1.12 Stress-strain relations for non-zero Poisson's ratio.

tion they act and a lateral contraction due to Poisson's ratio. Shear stresses cause a square to deform to a rhombus. The mathematical relation between deformations and stress resultants become

$$\begin{aligned}\varepsilon_{xx} &= \frac{1}{Et}(n_{xx} - \nu n_{yy})\\ \varepsilon_{yy} &= \frac{1}{Et}(n_{yy} - \nu n_{xx})\\ \gamma_{xy} &= \frac{n_{yy}}{Gt}\end{aligned} \tag{1.11}$$

where $G = E/2(1+\nu)$. These constitutive relations presuppose that the plate carries no stresses in the direction z perpendicular to the x, y-plane. In matrix notation the three relations read

$$\begin{Bmatrix}\varepsilon_{xx}\\ \varepsilon_{yy}\\ \gamma_{xy}\end{Bmatrix} = \frac{1}{Et}\begin{bmatrix}1 & -\nu & 0\\ -\nu & 1 & 0\\ 0 & 0 & 2(1+\nu)\end{bmatrix}\begin{Bmatrix}n_{xx}\\ n_{yy}\\ n_{xy}\end{Bmatrix} \tag{1.12}$$

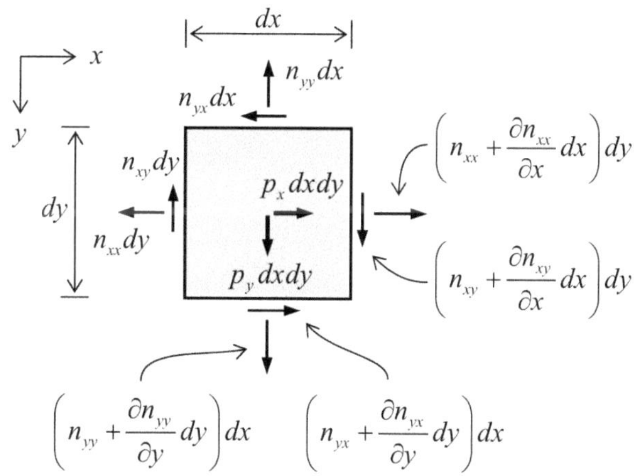

Figure 1.13 Equilibrium of an elementary plate part.

This is the flexibility formulation of the constitutive equations. By inverting Eq. (1.12) we obtain the stiffness formulation

$$\begin{Bmatrix} n_{xx} \\ n_{yy} \\ n_{xy} \end{Bmatrix} = \frac{Et}{1-\nu^2} \begin{bmatrix} 1 & \nu & 0 \\ \nu & 1 & 0 \\ 0 & 0 & \frac{1}{2}(1-\nu) \end{bmatrix} \begin{Bmatrix} \varepsilon_{xx} \\ \varepsilon_{yy} \\ \gamma_{xy} \end{Bmatrix} \qquad \textit{Constitutive} \tag{1.13}$$

Eq. (1.13) is a generalization of Eq. (1.2) for the truss.

1.2.3 Equilibrium Equations

Equilibrium equations give the relations between the loads and the membrane forces. An equilibrium equation can be formulated in the direction of both degrees of freedom u_x and u_y (see Figure 1.13). In the x-direction, the equation for an elementary plate part with sizes dx and dy is

$$\begin{aligned} -n_{xx}\,dy + \left(n_{xx} + \frac{\partial n_{xx}}{\partial x}\,dx\right) dy - n_{yx}\,dx \\ + \left(n_{yx} + \frac{\partial n_{yx}}{\partial y}\,dy\right) dx + p_x\,dx\,dy = 0 \end{aligned}$$

In the y-direction a similar equation is valid that is obtained by simply interchanging all x and y. Some terms will cancel out. Then division through $dx\,dy$ leads to

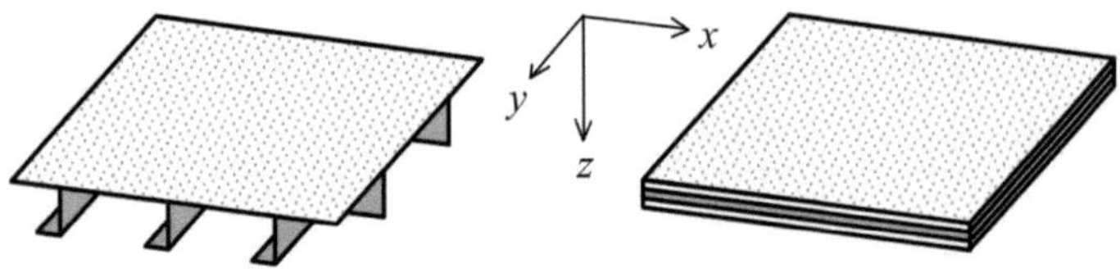

Figure 1.14 Shape-orthotropic plates.

$$\left.\begin{array}{l} -\left(\dfrac{\partial n_{xx}}{\partial x}+\dfrac{\partial n_{yx}}{\partial y}\right)=p_x \\ -\left(\dfrac{\partial n_{xy}}{\partial y}+\dfrac{\partial n_{yy}}{\partial x}\right)=p_y \end{array}\right\} \quad \textit{Equilibrium} \tag{1.14}$$

Equation (1.14) is a generalization of Eq. (1.3) for the truss. All basic equations have been determined now.

Remark

The derivations of the basic equations are valid for a homogenous isotropic plate. Plate theory is also used for homogenous orthotropic plates or structures of shape-orthotropy (Figure 1.14). Then more generally the constitutive relationship of Eq. (1.15) holds

$$\begin{Bmatrix} n_{xx} \\ n_{yy} \\ n_{xy} \end{Bmatrix} = \begin{bmatrix} d_{xx} & d_v & 0 \\ d_v & d_{yy} & 0 \\ 0 & 0 & d_{xy} \end{bmatrix} \begin{Bmatrix} \varepsilon_{xx} \\ \varepsilon_{yy} \\ \gamma_{xy} \end{Bmatrix} \tag{1.15}$$

The rigidity terms in Eq. (1.15) will have to be determined separately for each case, depending on the structure of the plate field. This is the subject of Chapter 20.

1.2.4 The Displacement Method

The procedure in the displacement method is to substitute the kinematic equations and the constitutive equations into the equilibrium equations. First the kinematic relationships in Eq. (1.9) will be introduced into the constitutive relationships in Eq. (1.13), so the membrane forces are expressed in the displacements

$$
\begin{aligned}
n_{xx} &= \frac{Et}{1-\nu^2}\left(\frac{\partial u_x}{\partial x} + \nu\frac{\partial u_y}{\partial y}\right) \\
n_{yy} &= \frac{Et}{1-\nu^2}\left(\frac{\partial u_y}{\partial y} + \nu\frac{\partial u_x}{\partial x}\right) \\
n_{xy} &= \frac{Et}{2(1+\nu)}\left(\frac{\partial u_x}{\partial y} + \frac{\partial u_y}{\partial x}\right)
\end{aligned} \tag{1.16}
$$

The second step is substituting this intermediate result into the equilibrium relationships of Eq. (1.14), which leads to two partial differential equations in u_x and u_y, the Navier equations

$$
\begin{aligned}
-\frac{Et}{1-\nu^2}\left(\frac{\partial^2 u_x}{\partial x^2} + \frac{1-\nu}{2}\frac{\partial^2 u_x}{\partial y^2} + \frac{1+\nu}{2}\frac{\partial^2 u_y}{\partial x \partial y}\right) &= p_x \\
-\frac{Et}{1-\nu^2}\left(\frac{\partial^2 u_y}{\partial y^2} + \frac{1-\nu}{2}\frac{\partial^2 u_y}{\partial x^2} + \frac{1+\nu}{2}\frac{\partial^2 u_x}{\partial x \partial y}\right) &= p_y
\end{aligned} \tag{1.17}
$$

These equations are the generalization of the differential equation in Eq. (1.4) for the truss problem. If there are no variations in the y-direction, the first differential equation becomes equal to Eq. (1.4) if one substitutes $t = A$ and $\nu = 0$. To be complete, we will write Eq. (1.17) in a matrix operator formulation:

$$
-\frac{Et}{1-\nu^2}\begin{bmatrix} \frac{\partial^2}{\partial x^2} + \frac{1-\nu}{2}\frac{\partial^2}{\partial y^2} & \frac{1+\nu}{2}\frac{\partial^2}{\partial x \partial y} \\ \frac{1+\nu}{2}\frac{\partial^2}{\partial x \partial y} & \frac{\partial^2}{\partial y^2} + \frac{1-\nu}{2}\frac{\partial^2}{\partial x^2} \end{bmatrix}\begin{Bmatrix} u_x \\ u_y \end{Bmatrix} = \begin{Bmatrix} p_x \\ p_y \end{Bmatrix} \tag{1.18}
$$

In Eq. (1.18) we have derived two coupled partial differential equations in two unknown displacements u_x and u_y, which have to be solved simultaneously. We can replace the set of second-order differential equations by one of the fourth order, by eliminating one of the displacements. If we choose to eliminate u_y we must perform the lower right differential operation of the matrix on both members of the first equation in (1.18) and the upper right operation with a minus sign on the second equation. If we then sum the two equations, the displacement u_y will disappear and we find a fourth-order differential equation for u_x

$$
\begin{aligned}
&-\frac{Et}{2(1+\nu)}\left(\frac{\partial^4}{\partial x^4} + 2\frac{\partial^4}{\partial x^2 \partial y^2} + \frac{\partial^4}{\partial y^4}\right) u_x \\
&= \left(\frac{\partial^2}{\partial y^2} + \frac{1-\nu}{2}\frac{\partial^2}{\partial x^2}\right) p_x - \left(\frac{1+\nu}{2}\frac{\partial^2}{\partial x \partial y}\right) p_y
\end{aligned} \tag{1.19}
$$

Introducing the harmonic Laplace-operator ∇^2 (pronounce: nabla squared)

$$\nabla^2 = \frac{\partial^2}{\partial x^2} + \frac{\partial^2}{\partial y^2} \tag{1.20}$$

we can rewrite Eq. (1.19) as the bi-harmonic equation

$$-\frac{Et}{2(1+\nu)}\nabla^2\nabla^2 u_x = \left(\frac{\partial^2}{\partial y^2} + \frac{1-\nu}{2}\frac{\partial^2}{\partial x^2}\right)p_x - \left(\frac{1+\nu}{2}\frac{\partial^2}{\partial x \partial y}\right)p_y \tag{1.21}$$

If the plate is loaded only by edge line forces f and no distributed surface loads p_x and p_y occur, the bi-harmonic equation takes the simple form

$$\nabla^2\nabla^2 u_x = 0 \tag{1.22}$$

We could also have chosen to single out the displacement u_x; we would have obtained a bi-harmonic equation in the displacement u_y.

1.3 Boundary Conditions

We must solve the differential equations (1.21) taking into account the boundary conditions. We name the part of the edge where displacements u have been specified S_u and the part where the load f is prescribed S_f. Together S_u and S_f form the total perimeter. On S_u the prescribed displacements are indicated by $u_{x,\text{o}}$ and $u_{y,\text{o}}$. The prescribed perimeter load generally consists of two components f_x and f_y. Both these distributed edge loads have the dimension force per unit of length. If $u_{x,\text{o}}$ is specified, f_x cannot be prescribed on that same part of the edge and vice-versa. The same is true for $u_{y,\text{o}}$ and f_y. However, it is possible for $u_{x,\text{o}}$ and f_y to be prescribed on the same part of the edge, as goes for $u_{y,\text{o}}$ and f_x simultaneously. A formal way of writing is

$$\left.\begin{aligned} u_x &= u_{x,\text{o}} \\ u_y &= u_{y,\text{o}} \end{aligned}\right\} \quad \text{on } S_u \tag{1.23}$$

These are the *kinematic* boundary conditions. Where the load is given, we speak of *dynamic* boundary conditions. Prescribed values of the load are basically a condition for the stresses on the edge, for the following applies:

$$\left.\begin{aligned} \sigma_{xx}e_x + \sigma_{yx}e_y &= f_x \\ \sigma_{xy}e_x + \sigma_{yy}e_y &= f_y \end{aligned}\right\} \quad \text{on } S_f \tag{1.24}$$

Here e_x and e_y are the components of the unit normal outward-pointing vector on the edge.

1.4 Message of the Chapter

- Differential equations have been derived in the framework of the displacement method (stiffness method). Throughout the book we will distinguish the following four quantities: displacements, their associate external loading, deformations and their associate internal stress resultants.

- The plane stress state in the membrane plate theory is defined by two degrees of freedom (displacements) and two corresponding load components. The three deformations are strains, and the associate stress resultants are membrane forces.

- Three basic sets of equations must be derived: kinematic, constitutive and equilibrium equations. The kinematic equations relate the deformations (strains) to the displacements. The constitutive equations define the relationship between deformations and stress resultants (membrane forces). The equilibrium relations relate the stress resultants to the external loads.

- Subsequent substitution of the one in the other leads to the required differential equations, which are equilibrium equations, expressed in displacements. In the membrane plate theory we obtain two simultaneous partial differential equations of the second-order, which can be replaced by one fourth-order bi-harmonic equation.

- In the membrane plate theory two boundary conditions can be specified per plate edge, either the displacement normal to the edge (kinematic condition) or the edge load in that direction (dynamic condition). In the same way either the displacement parallel to the edge or the load in that direction.

Chapter 2
Applications of the Plate Membrane Theory

In this chapter we will give solutions for plates, which are loaded only on their edges. This implies that no distributed forces p_x and p_y occur, and the fourth-order bi-harmonic equation (1.23) reduces to the simple form

$$\nabla^2\nabla^2 u_x = 0 \tag{2.1}$$

When a general solution has been found for u_x, the solution for u_y can be derived from the relation between u_x and u_y as given in Eq. (1.17). If we choose the first equation, the relation is ($p_x = p_y = 0$)

$$\left(\frac{\partial^2}{\partial x^2} + \frac{1-\nu}{2}\frac{\partial^2}{\partial y^2}\right) u_x + \left(\frac{1+\nu}{2}\frac{\partial^2}{\partial x \partial y}\right) u_y = 0 \tag{2.2}$$

We will demonstrate two types of solution. In the first type, solutions for the displacements u_x and u_y will be tried, which are polynomials in x and y. We will see that interesting problems can be solved through this 'inverse method'. The second type of solution is found by assuming a periodic distribution (sine or cosine) in one direction. Then in the other direction an ordinary differential equation has to be solved. This approach is suitable for deep beams or walls.

2.1 Trial Solutions in the Form of Polynomials

In this section we consider problems of which we know the stress state. For this stress state we want to determine the displacement field. For this purpose

J. Blaauwendraad, *Plates and FEM: Surprises and Pitfalls*, Solid Mechanics and Its Applications 171, DOI 10.1007/978-90-481-3596-7_2,

appropriate trial functions for the displacements u_x and u_y will be chosen, in which a number of coefficients occur, yet to be determined. As a trial solution in its most general form we can choose

$$\begin{aligned} u_x(x, y) &= a_1 + a_2 x + a_3 y + a_4 x^2 + a_5 xy + a_6 y^2 + a_7 x^3 \\ &\quad + a_8 x^2 y + a_9 x y^2 + a_{10} y^3 + a_{11} x^3 y + a_{12} x y^3 \end{aligned} \tag{2.3}$$

$$\begin{aligned} u_y(x, y) &= b_1 + b_2 x + b_3 y + b_4 x^2 + b_5 xy + b_6 y^2 + b_7 x^3 \\ &\quad + b_8 x^2 y + b_9 x y^2 + b_{10} y^3 + b_{11} x^3 y + b_{12} x y^3 \end{aligned} \tag{2.4}$$

All 12 polynomial terms in Eq. (2.3) for u_x are independent solutions of the differential equation (2.1), so the 12 coefficients a_i are independent of each other. In the same way is Eq. (2.4) a general solution for the bi-harmonic equation for u_y.

In this section we start with the simple case that only constant and linear polynomial terms are chosen. After that a problem is solved for which we have to consider quadratic terms. Finally a problem will be solved for which we also have to include cubic terms.

2.1.1 Homogeneous Stress States

We consider the constant and linear terms with coefficients a_1, a_2, a_3, b_1, b_2, and b_3

$$u_x(x, y) = a_1 + a_2 x + a_3 y; \quad u_y(x, y) = b_1 + b_2 x + b_3 y \tag{2.5}$$

Together the six terms determine all possible states of homogeneous strains and all possible rigid body displacements, as can easily be shown. Applying the kinematic relations (1.9) we find the strains

$$\left.\begin{aligned} \varepsilon_{xx} &= a_2 \\ \varepsilon_{yy} &= b_3 \\ \gamma_{xy} &= a_3 + b_2 \end{aligned}\right\} \quad \textit{Homogeneous strains}$$

These strains are constant over the plate domain. The constants a_1 and b_1 do not appear in the strains at all. Those represent the rigid body translations. Of the constants a_3 and b_2, only the sum appears in the strains. The difference of these constants defines a rigid body rotation. The three rigid body motions are

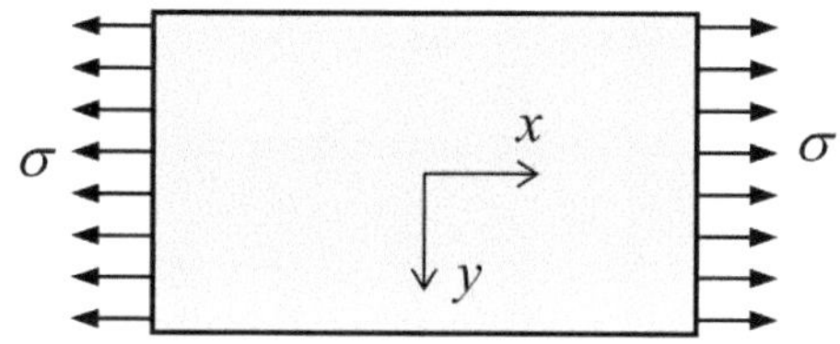

Figure 2.1 Constant tensile stress.

$$\left.\begin{aligned} u_x &= a_1 \\ u_y &= b_1 \\ \omega_{xy} &= \tfrac{1}{2}(-a_3 + b_2) \end{aligned}\right\} \quad \textit{Rigid body motions}$$

The homogenous strain state of Eq. (2.6) defines the stresses. From the constitutive law in Eq. (1.13) we find for $t = 1$

$$\begin{aligned} \sigma_{xx} &= \frac{E}{1-\nu^2}\,(a_2 + \nu b_3) \\ \sigma_{yy} &= \frac{E}{1-\nu^2}\,(b_3 + \nu a_2) \\ \sigma_{xy} &= \frac{E}{2(1+\nu)}\,(a_3 + b_2) \end{aligned} \tag{2.8}$$

In the two following examples we will determine the three coefficients a_i and three coefficients b_i for some special cases.

Case 1: Constant Normal Stress

A plate of unit thickness will be analyzed; it is subjected to a constant (uniaxial) tensile stress σ in the x-direction (see Figure 2.1). The loads p_x and p_y are zero. We need six conditions to find the coefficients a_i and b_i. We know $\sigma_{xx} = \sigma$, $\sigma_{yy} = 0$, and $\sigma_{xx} = 0$, and we prescribe that no translations or rotations occur at the origin of the coordinate system. The stresses satisfy the equilibrium conditions in (1.14). Accounting for Eq. (2.8), we obtain three conditions for stresses

$$\frac{E}{1-\nu^2}(a_2 + \nu b_3) = \sigma; \quad b_3 + \nu a_2 = 0; \quad a_3 + b_2 = 0 \tag{2.9}$$

Equation (2.7) for the rigid body motions specifies three other conditions:

$$a_1 = 0; \quad b_1 = 0; \quad -a_3 + b_2 = 0 \tag{2.10}$$

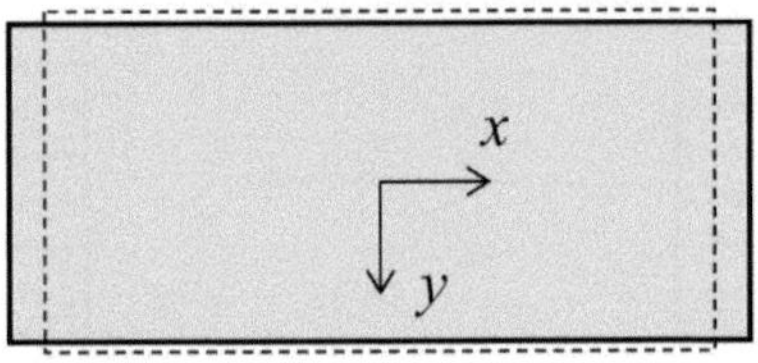

Figure 2.2 Deformation without zero rigid body motion.

These six equations imply

$$a_1 = 0; \quad a_2 = \frac{\sigma}{E}; \quad a_3 = 0$$

$$b_1 = 0; \quad b_2 = 0; \quad b_3 = -\nu\frac{\sigma}{E} \tag{2.11}$$

The displacement field equation (2.5) then becomes

$$u_x = \frac{\sigma}{E}x; \quad u_y = -\nu\frac{\sigma}{E}y \tag{2.12}$$

The middle of the plate does not translate or rotate. So the left-hand side of the plate moves towards the left, and the right-hand side towards the right (see Figure 2.2). In the lateral direction contraction takes place; this yields a negative displacement u_y for positive values y, and a positive displacement for negative values.

As an alternative we could have required the left-hand side not to move. In that case a rigid body displacement u_o has to be added (a displacement of the plate towards the right as shown in Figure 2.3, and instead of $a_1 = 0$ we should choose $a_1 = u_o$.

Note

We have two states of equal stresses, but different displacement fields. The difference consists of rigid body motions.

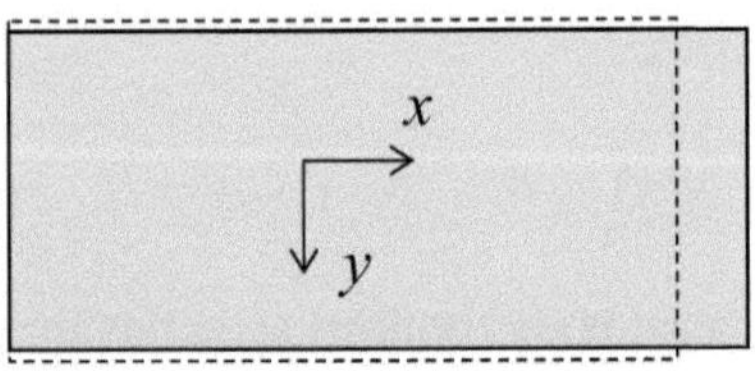

Figure 2.3 Deformation with rigid body motion in x-direction.

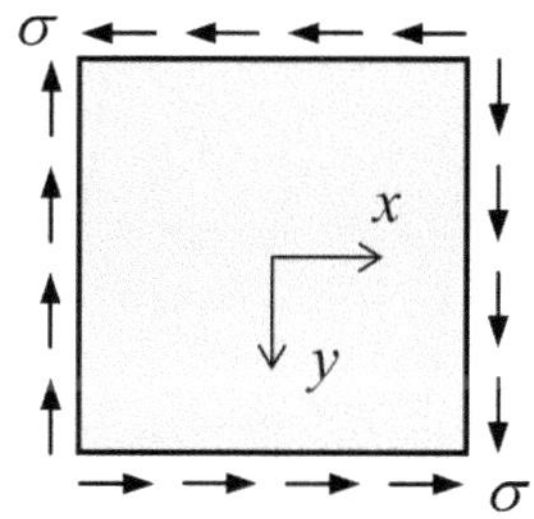

Figure 2.4 Constant shear stress.

Case 2: Constant Shear Stress

Consider a plate undergoing pure shear σ (see Figure 2.4). This stress state satisfies the equilibrium conditions in Eq. (1.14). We do not permit rigid body displacements a_1 and b_1. However, we take into account a rigid body rotation ω. The stresses are, see Eq. (2.8)

$$a_2 + \nu b_3 = 0; \quad b_3 + \nu a_2 = 0; \quad \frac{E}{2(1+\nu)}(a_3 + b_2) = \sigma \tag{2.13}$$

The rigid body displacements are, see Eq. (2.76)

$$a_1 = 0; \quad b_1 = 0; \quad \frac{1}{2}(-a_3 + b_2) = \omega \tag{2.14}$$

From Eqs. (2.13) and (2.14) we find four zero values

$$a_1 = 0; \quad a_2 = 0; \quad b_1 = 0; \quad b_3 = 0 \tag{2.15}$$

Only two equations remain with coefficients a_3 and b_2

$$\sigma = \frac{E}{2(1+\nu)}(a_3 + b_2); \quad \omega = \frac{1}{2}(-a_3 + b_2) \tag{2.16}$$

Therefore the displacements will be

$$u_x = a_3 y; \quad u_y = b_2 x \tag{2.17}$$

Now, we will consider three subcases.

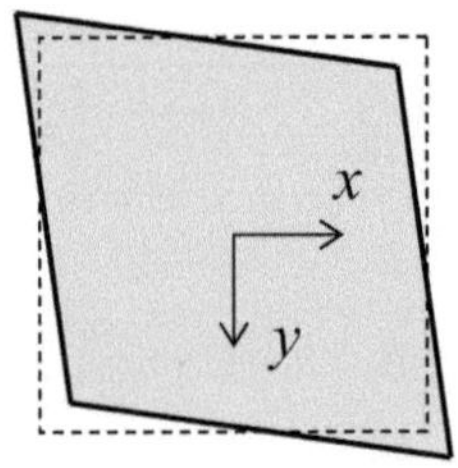

Figure 2.5 Deformation without rigid body rotation.

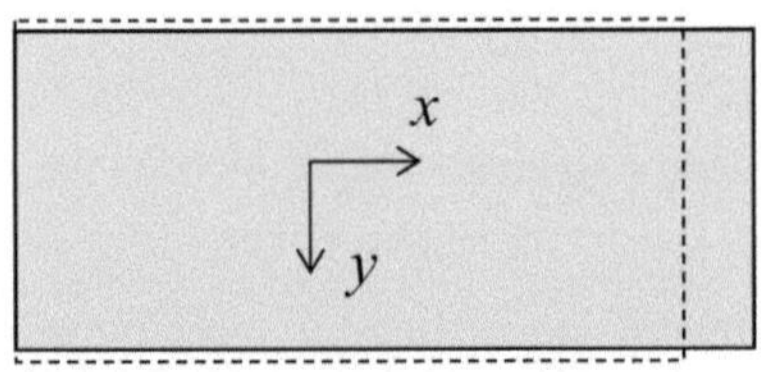

Figure 2.6 Deformation with rigid body motion in x-direction.

Subcase 2.1

No rigid body rotation (see Figure 2.5) occurs. We choose

$$\omega = 0 \;\rightarrow\; a_3 = b_2 = \frac{(1+\nu)\sigma}{E} \tag{2.18}$$

and therefore

$$u_x = \frac{(1+\nu)\sigma}{E} y; \qquad u_y = \frac{(1+\nu)\,\sigma}{E} x \tag{2.19}$$

The linear distribution of u_x in y-direction and of u_x in x-direction is confirmed by the deformation as shown in Figure 2.5.

Subcase 2.2

No displacement in x-direction (see Figure 2.6) takes place. The plate has vertical edges after a rigid body rotation. Now

$$a_3 = 0 \;\rightarrow\; b_2 = \frac{2(1+\nu)\sigma}{E}; \qquad \omega = \frac{(1+\nu)\sigma}{E} \tag{2.20}$$

and therefore

$$u_x = 0; \qquad u_y = \frac{2(1+\nu)\sigma}{E} x \tag{2.21}$$

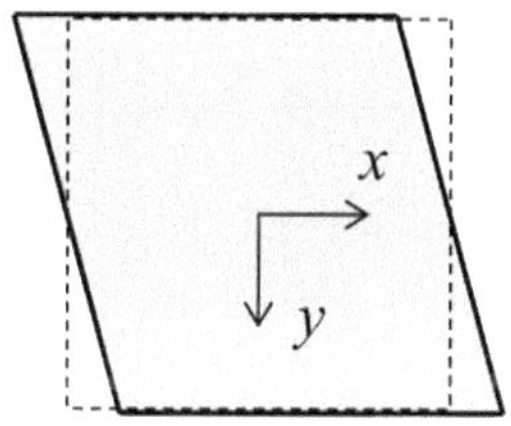

Figure 2.7 Deformation with zero displacement in y-direction.

Subcase 2.3

There is no displacement in the y-direction (see Figure 2.7). The plate has horizontal edges after a rigid body rotation. Now

$$b_2 = 0 \rightarrow a_3 = \frac{2(1+\nu)\sigma}{E}; \quad \omega = -\frac{(1+\nu)\sigma}{E} \tag{2.22}$$

and therefore

$$u_x = \frac{2(1+\nu)\sigma}{E} y; \quad u_y = 0 \tag{2.23}$$

Same Stress, Different Displacements

In all three cases the same shear stress occurs, however the displacement fields are different. The difference is related to the magnitude and sign of the rigid body rotation.

Case 3: Rigid Body Displacements

There is a field of displacements that consists only of rigid body displacements

$$\begin{aligned} u_x(x, y) &= C_1 - C_3 y \\ u_y(x, y) &= C_2 + C_3 x \end{aligned} \tag{2.24}$$

Substitution into the kinematic equations (1.9) shows that the three strains are zero. Therefore the three stresses will be zero too. The constants C_1 and C_2 relate to translations; the constant C_3 to rotation.

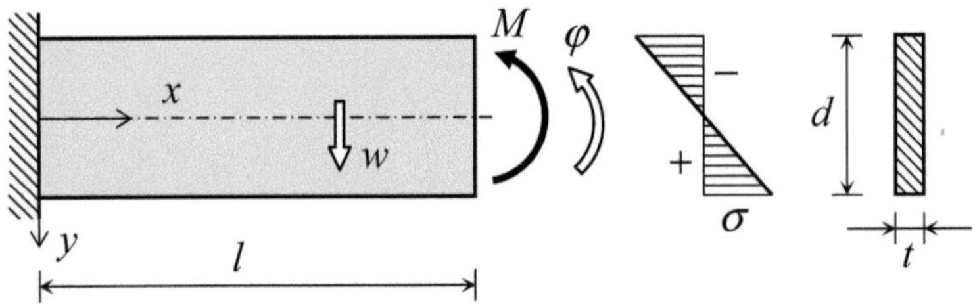

Figure 2.8 Cantilever beam subjected to pure bending.

2.1.2 Constant Bending Moment in Beam

Consider the classic Euler–Bernoulli beam theory [1] of a cantilever beam loaded by a moment at the free end, see Figure 2.8. In this case, no shear force V occurs, and the bending moment M is constant (and positive) over the length of the beam. In the beam theory the stresses in the beam are

$$\sigma_{xx} = \frac{M}{I} y; \quad \sigma_{yy} = 0; \quad \sigma_{xy} = 0 \tag{2.25}$$

Here $I = d\, t^3/12$ is the second moment of the cross-sectional area, where d is the width and t the depth of the beam. This stress distribution has been derived in classical beam theory on the assumption that a plane cross-section remains plane after applying the load. The stress state equation (2.25) satisfies the equilibrium conditions in (1.14) and therefore is a set of equilibrating stresses. In the stresses of Eq. (2.25) a term occurs which is linear in y, which means that we also can expect linear terms in the strains. Because strains are first derivatives of displacements, we therefore must consider quadratic displacement terms. We start with the most general form of all quadratic terms

$$u_x(x, y) = a_4 x^2 + a_5 xy + a_6 y^2; \quad u_y(x, y) = b_4 x^2 + b_5 xy + b_6 y^2 \tag{2.26}$$

The strains and stresses are now

$$\begin{aligned} \varepsilon_{xx} &= 2a_4 x + a_5 y \\ \varepsilon_{yy} &= b_5 x + 2b_6 y \\ \gamma_{xy} &= (a_5 + 2b_4)\, x + (2a_6 + b_5)\, y \end{aligned} \tag{2.27}$$

$$\sigma_{xx} = \frac{E}{1-\nu^2}\{(2a_4 + \nu\, b_5)x + (a_5 + 2\nu b_6)y\}$$
$$\sigma_{yy} = \frac{E}{1-\nu^2}\{(2\nu a_4 + b_5)x + (\nu\, a_5 + 2b_6)y\} \qquad (2.28)$$
$$\sigma_{xy} = \frac{E}{2(1+\nu)}\{(a_5 + 2b_4)x + (2a_6 + b_5)y\}$$

A comparison of these stresses with the actual stresses leads to six conditions

$$2a_4 + \nu b_5 = 0, \quad \nu a_5 + 2b_6 = 0$$
$$a_5 + 2b_4 = 0, \quad \frac{E}{1-\nu^2}(a_5 + 2\nu b_6) = \frac{M}{I} \qquad (2.29)$$
$$2\nu a_4 + b_5 = 0, \quad 2a_6 + b_5 = 0$$

The solution of these six equations is

$$a_4 = 0; \quad a_5 = \frac{M}{EI}; \quad a_6 = 0$$
$$b_4 = -\frac{M}{2EI}; \quad b_5 = 0; \quad b_6 = -\frac{\nu M}{2EI} \qquad (2.30)$$

Therefore the displacements are

$$u_x(x, y) = \frac{M}{EI}xy; \quad u_y(x, y) = -\frac{M}{2EI}(x^2 + \nu y^2) \qquad (2.31)$$

For a homogenous moment distribution, the classical assumption that a plane cross-section remains plane after loading is correct, as appears from Eq. (2.31), because u_x has a linear dependence on y for each value of x. The stress state does not change when a rigid body displacement is added, as defined in Eq. (2.24). In total we get

$$u_x(x, y) = \frac{M}{EI}xy + c_1 - c_3 y$$
$$u_y(x, y) = -\frac{M}{2EI}(x^2 + \nu y^2) + c_2 + c_3 x \qquad (2.32)$$

The three constants c_1, c_2 and c_3 have to be found from the boundary conditions. In the example we have a support in $x = 0$. We interpret this support as conditions that hold for $x = 0$, $y = 0$. The axis of the beam at $x = 0$ cannot translate and rotate, the bar axis remains horizontal.

$$\left.\begin{aligned} &u_x = 0; \quad u_y = 0 \\ &\frac{\partial u_y}{\partial x} = 0 \end{aligned}\right\} \quad \text{for} \quad x = 0, y = 0 \tag{2.33}$$

Substitution of Eq. (2.32) into Eq. (2.33) leads to

$$c_1 = 0; \quad c_2 = 0; \quad c_3 = 0 \tag{2.34}$$

Apparently the displacements in Eq. (2.32) already fully meet the boundary conditions. To interpret these results, we move over to the deflection w and the rotation φ of the cross-section. Because the section is plane after deformation we can write

$$u_x = y\varphi, \quad u_y = w \tag{2.35}$$

This changes Eq. (2.32) into

$$\varphi = \frac{M}{EI}x, \quad w = -\frac{1}{2}\frac{M}{EI}(x^2 + \nu y^2) \tag{2.36}$$

At the free end of the beam, at the position of the axis ($x = l$, $y = 0$), we find

$$\varphi = \frac{M\,l}{EI}, \quad w = -\frac{1}{2}\frac{M\,l^2}{EI} \tag{2.37}$$

These results are well known from elementary beam theory. The rotation φ is both the inclination of the beam axis and the tilt of the cross-section plane.

Check on Euler–Bernoulli beam theory

We conclude that the well known results of Euler–Bernoulli beam theory are confirmed by plate theory. From Eq. (2.36) it follows, that the rotation φ increases linearly with x and the vertical deflection w is square in x. In one way the results of the plane stress theory differ from Euler–Bernoulli beam theory. The predicted deflections are the same only along the axis of the beam, where $y = 0$. Outside the beam axis a small correction factor is needed when $\nu \neq 0$. So, strictly speaking, the assumption of the deflection on all points along the height of the beam being the same is incorrect. However, for slender beams the correction term is an order $\nu(d/l)^2$ smaller than the main term. This is of the order of 1% or less, so the assumption in the beam theory is acceptable.

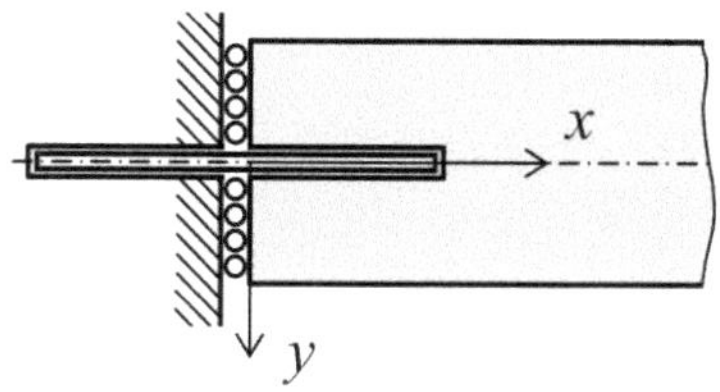

Figure 2.9 Detail of the boundary condition at the restrained end.

Remark

The boundary condition in $x = 0$, $y = 0$ in fact means that the horizontal displacement u_x is obstructed in the complete vertical cross-section in $x = 0$, but that the vertical displacement u_y could occur freely in this section, except for $y = 0$, see Figure 2.9. The bar axis is horizontal at the clamped end.

2.1.3 Constant Shear Force in Beam

We increase the complexity of the cantilever beam by replacing the moment at the free end by a downward vertical force F as shown in Figure 2.10. Now there is a constant shear force V (positive) and the bending moment M varies linearly along the beam axis (negative). The expressions for M and V are

$$M = F(x - l), \quad V = F \tag{2.38}$$

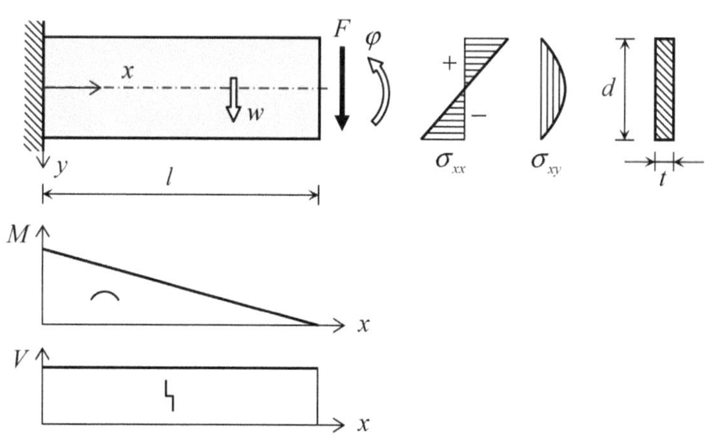

Figure 2.10 Cantilever beam loaded by point load.

The stresses are

$$\begin{aligned}\sigma_{xx} &= \frac{My}{I} = \frac{F}{I}(xy - ly)\\ \sigma_{yy} &= 0\\ \sigma_{xy} &= \frac{3}{2}\left(1-\frac{4y^2}{d^2}\right)\frac{V}{A} = \frac{3F}{2A}\left(1-\frac{4y^2}{d^2}\right)\end{aligned} \tag{2.39}$$

where

$$A = td; \quad I = \frac{1}{12}td^3 \tag{2.40}$$

This set of equations satisfies the equilibrium equations in (1.14). The boundary conditions in the left end of the beam axis ($x = 0$, $y = 0$) are chosen in the same fashion as in the previous example with the moment load (horizontal bar axis)

$$u_x = 0; \quad u_y = 0; \quad \frac{\partial u_y}{\partial x} = 0 \tag{2.41}$$

In the expression for the stress σ_{xx}, a term $-Fly/I$ is present which we recognize as the distribution of a constant moment $M = -Fl$. For such a stress state we already found

$$u_x = \frac{M}{EI}xy = -\frac{Fl}{EI}xy; \quad u_y = \frac{1}{2}\frac{M}{EI}(x^2+\nu y^2) = \frac{Fl}{2EI}(x^2+\nu y^2) \tag{2.42}$$

In the stress σ_{xy}, a constant part $3F/2A$ is also present. Taking into account the boundary conditions, subcase 3 of case 2 (Section 2.1.1) is applicable. We substitute $G = E/2(1+\nu)$

$$u_x = \frac{\sigma}{G}y = \frac{3F}{2GA}y, \quad u_y = 0 \tag{2.43}$$

The residual part of the stresses is

$$\sigma_{xx} = \frac{F}{I}xy; \quad \sigma_{yy} = 0; \quad \sigma_{xy} = -\frac{6F}{Ad^2}y^2 \tag{2.44}$$

The displacement field corresponding with these stresses still needs to be determined. Quadratic stress polynomials imply quadratic strain polynomials and cubic displacement polynomials, because strains are the first derivative of the displacements. So, we start from the most general cubic terms

$$\begin{aligned}u_x(x,y) &= a_7x^3 + a_8x^2y + a_9xy^2 + a_{10}y^3\\ u_y(x,y) &= b_7x^3 + b_8x^2y + b_9xy^2 + b_{10}y^3\end{aligned} \tag{2.45}$$

The corresponding strains are

$$\begin{aligned}\varepsilon_{xx} &= 3a_7x^2 + 2a_8xy + a_9y^2\\ \varepsilon_{yy} &= b_8x^2 + 2b_9xy + 3b_{10}y^2\\ \gamma_{xy} &= (a_8 + 3b_7)\,x^2 + 2\,(a_9 + b_8)\,xy + (3a_{10} + b_9)\,y^2\end{aligned} \tag{2.46}$$

and the stresses

$$\begin{aligned}\sigma_{xx} &= \frac{E}{1-\nu^2}\left\{(3a_7 + \nu b_8)\,x^2 + 2\,(a_8 + \nu b_9)\,xy + (a_9 + 3\nu b_{10})\,y^2\right\}\\ \sigma_{yy} &= \frac{E}{1-\nu^2}\left\{(3\nu a_7 + b_8)\,x^2 + 2\,(\nu a_8 + b_9)\,xy + (\nu a_9 + 3b_{10})\,y^2\right\}\\ \sigma_{xy} &= \frac{E}{2(1+\nu)}\left\{(a_8 + 3b_7)\,x^2 + 2\,(a_9 + b_8)\,xy + (3a_{10} + b_9)\,y^2\right\}\end{aligned} \tag{2.47}$$

A comparison with Eq. (2.44) leads to the conditions

$$\begin{aligned}&3a_7 + \nu b_8 = 0, \quad &&\frac{2E}{1-\nu^2}(a_8 + \nu b_9) = \frac{F}{I}, \quad &&a_9 + 3\nu b_{10} = 0\\ &3\nu a_7 + b_8 = 0, &&\nu a_8 + b_9 = 0, &&\nu a_9 + 3b_{10} = 0\\ &a_8 + 3b_7 = 0, &&a_9 + b_8 = 0, &&G(3a_{10} + b_9) = -\frac{6F}{Ad^2}\end{aligned} \tag{2.48}$$

The solution of these nine equations for eight unknown coefficients produces only four no-zero coefficients

$$a_8 = \frac{F}{2EI}, \quad a_{10} = \frac{\nu F}{6EI} - \frac{2F}{GAd^2}, \quad b_7 = -\frac{F}{6EI}, \quad b_9 = -\frac{\nu F}{2EI} \tag{2.49}$$

The displacements in this case have become

$$u_x = a_8x^2y + a_{10}y^3, \quad u_y = b_7x^3 + b_9x\,y^2 \tag{2.50}$$

Substitution of Eq. (2.49) leads to

$$\begin{aligned}u_x &= \frac{F}{2EI}x^2y + \left(\frac{\nu F}{6EI} - \frac{2F}{GAd^2}\right)y^3\\ u_y &= -\frac{F}{6EI}x^3 - \frac{\nu F}{2EI}xy^2\end{aligned} \tag{2.51}$$

The total displacement field is found by adding Eqs. (2.42), (2.43) and (2.51), and the addition of a rigid body displacement. In this final result we assemble the terms with EI and the terms with GA and find

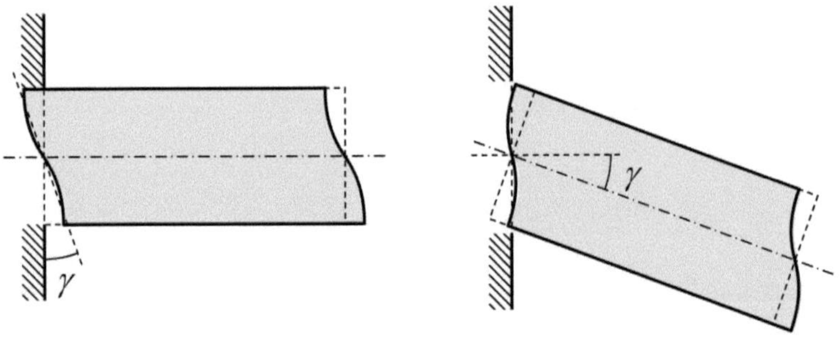

Figure 2.11 Rotation due to 'mean' shear deformation.

$$u_x = \frac{F}{EI}\left\{x\left(\frac{1}{2}x - l\right)y + \frac{\nu}{6}y^3\right\} + \frac{F}{GAd^2}\left(-2y^3 + \frac{3}{2}d^2y\right) + c_1 - c_3y$$
$$u_y = \frac{F}{EI}\left\{x^2\left(-\frac{1}{6}x + \frac{1}{2}l\right) + \nu\left(-\frac{1}{2}x - \frac{1}{2}l\right)y^2\right\} + c_2 + c_3x \qquad (2.52)$$

The boundary conditions (2.41) are met for $c_1 = 0$, $c_2 = 0$, $c_3 = 0$. If we define the rotation φ as the inclination of the beam axis

$$\varphi = -\frac{\partial u_y}{\partial x} \qquad (2.53)$$

and the displacement w as the vertical displacement u_y of the beam axis, the rotation φ and deflection w of the free beam end (in the axis of the beam) are

$$\varphi = -\frac{1}{2}\frac{Fl^2}{EI}, \qquad w = \frac{1}{3}\frac{Fl^3}{EI} \qquad (2.54)$$

Again, these are equal to the well-known results of classical beam theory. However, the cross-sections no longer remain plane. In u_x, not only linear terms in y are present, but also terms y^3, even when $\nu = 0$. Nonetheless, the bending stress develops linearly over the height of the beam. So, an erroneous assumption in classical beam theory has led in the past to correct solutions for the stresses!

Now we want to have a closer look at the shape of the deformed beam at the restrained end (see the left figure of Figure 2.11). We see that a horizontal beam axis does not imply that the cross-section takes up a vertical position. First, the cross-section is distorted. In addition, the 'mean' cross-section is tilted. The distortion and the tilt are the result of lateral contraction (Poisson's ratio) and shear deformation (angle γ), though primarily by the latter.

The shear deformation is recognizable by the term in GA. A rigid body rotation over an angle γ is necessary to eliminate the tilt caused by the shear deformation. This rigid body rotation generates an additional displacement at the free end of the beam. This is the contribution of the shear deformation to the deflection. The value of γ is

$$\gamma = \eta \frac{F}{GA}. \tag{2.55}$$

The shape factor η has a value of 1 if the shear stress is constant over the cross-section. For the parabolic variation over a rectangular cross-section the value is 6/5. At the free end of the beam we obtain

$$\varphi = -\frac{1}{2}\frac{Fl^2}{EI} - \gamma, \quad w = \frac{1}{3}\frac{Fl^3}{EI} + \gamma l \tag{2.56}$$

Introduction of γ from Eq. (2.55) and accounting for $A = td$, $I = td^3/12$ and the shear modulus $G = E/2(1+\nu)$ leads to

$$\varphi = -\frac{1}{2}\frac{Fl}{EI}\left(1 + \frac{\eta\,(1+\nu)}{3}\frac{d^2}{l^2}\right), \quad w = \frac{1}{3}\frac{Fl^3}{EI}\left(1 + \frac{\eta\,(1+\nu)}{2}\frac{d^2}{l^2}\right) \tag{2.57}$$

The term d^2/l^2 mirrors the influence of slenderness of the beam on the end rotation and deflection. When l/d is larger than five, this term may be neglected. The shear force or shear deformation is not of any importance for slender beams.

Assumption of plane sections

When deriving the classic beam theory, people like Euler and Bernoulli and after them Navier [1] started from the supposition that a plane section normal to the beam axis remains plane and stays normal to the axis. Supposing this, they made no distinction between a constant and linear moment, and found a linear distribution of bending stresses over the depth of the beam. Their finding of the stress distribution is correct, but the membrane theory shows that their supposition holds true only for a constant moment, and that it is at best a good approximation for linear moments in case of slender beams. They were just lucky!

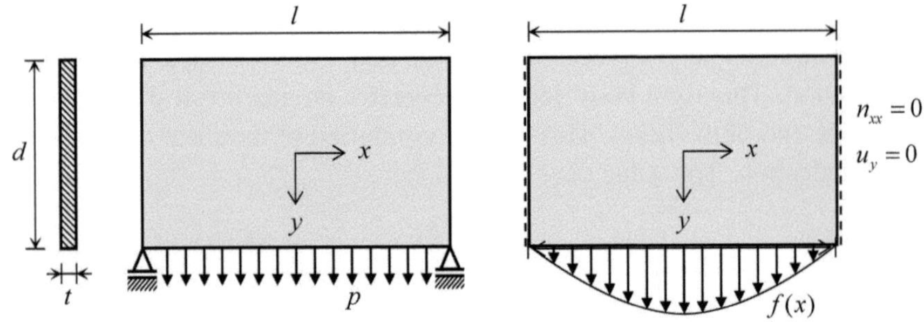

Figure 2.12 Deep beam with edge load.

2.2 Solution for a Wall

Consider a wall on two simple supports for ratios of depth and span ranging from tall wall to slender beam. The wall is loaded along its lower edge by a homogeneously distributed load f, as shown in the left-hand part of Figure 2.12. We want to determine the distribution of the bending stresses σ_{xx} in the vertical axis of symmetry $x = 0$. We replace the structure and load by the problem stated in the right-hand part of Figure 2.12. The supports in the two lower corners have been replaced by boundary conditions for both vertical edges. These edges can freely move horizontally, but prohibit vertical displacements. In the figure this is indicated by the dotted lines. This means that the reaction force will be distributed along the vertical edge. This can be done without changing the bending moment in the vertical cross-section mid-span ($x = 0$). The homogeneously distributed line load f is replaced by a varying load $f(x)$, which has a cosine distribution

$$f(x) = f_m \cos(\alpha x) \tag{2.58}$$

in which $\alpha = \pi/l$ and f_m is the maximum load value at mid-span. This cosine load is the first term in a Fourier series development of load f, so the value of f_m is

$$f_m = \frac{4}{\pi} f \tag{2.59}$$

2.2.1 Beam Intermezzo

We will show that the value of the bending moment M in the mid-span cross-section is practically the same for the actual load p and the replacing load

$f(x)$. The differential equation for beams in bending is

$$EI\frac{d^4w}{dx^4} = f(x) \tag{2.60}$$

in which EI is the bending stiffness and $f((x)$ is a distributed load. The bending moment M is computed by

$$M = -EI\frac{d^2w}{dx^2} \tag{2.61}$$

Load f and displacement w are positive if pointing downwards. The bending moment M is positive when tensile stresses are generated in the lower part of the beam. For the homogeneously distributed load, $f(x) = f$. This is a classical case with a well-known solution

$$\begin{aligned} w_{\max} &= \frac{5}{384}\frac{fl^4}{EI}\left(= 0.0130\frac{fl^4}{EI}\right) \\ M_{\max} &= \frac{1}{8}fl^2(= 0.125fl^2) \end{aligned} \tag{2.62}$$

The solution for the cosine load is easily found by substitution of the trial displacement function

$$w(x) = w_m \cos\alpha x \tag{2.63}$$

in the differential equation in combination with the cosine load of Eq. (2.58). This gives us a particular solution

$$\begin{aligned} w(x) &= \frac{f_m l^4}{\pi^4 EI}\cos\alpha x \\ M(x) &= \frac{f_m l^2}{\pi^2}\cos\alpha x \end{aligned} \tag{2.64}$$

Substitution of $f_m = 4f/\pi$ leads to maximum values

$$\begin{aligned} w_m &= \frac{(4f/\pi)l^4}{\pi^4 EI} = 0.0131\frac{fl^4}{EI} \\ M_m &= \frac{(4f/\pi)l^2}{\pi^2} = 0.129fl^2 \end{aligned} \tag{2.65}$$

which are close to the correct values shown above. For the cosine load, the proposed shape of the deflection $w(x)$ is the exact one for a beam on simple supports. At the supports, the boundary conditions are $w = 0$ and $M = 0$. These conditions are satisfied, so the particular solution we have found is the real solution. No homogeneous solution needs to be added.

2.2.2 Solution for the Wall

Encouraged by the good result for a beam subjected to a cosine load, we propose a similar cosine displacement field in x-direction for $u_y(x, y)$. This choice meets the conditions that the vertical displacement must be zero at the vertical edges and maximum at mid-span. The horizontal displacement u_x must be zero in the vertical line of symmetry ($x = 0$) and can have values that are not zero (equal, but with an opposite sign) at the two vertical edges. Therefore, we use a sine distribution for u_x. So our expectation for the displacements is

$$\begin{aligned} u_x(x, y) &= u_{xm}(y) \sin \alpha x \\ u_y(x, y) &= u_{ym}(y) \cos \alpha x \end{aligned} \tag{2.66}$$

Here $u_{xm}(y)$ is the distribution of the horizontal displacement along the vertical edges and $u_{ym}(y)$ is the distribution of the vertical displacement along the line of symmetry at mid span. We can choose to work with either $u_{xm}(y)$ or $u_{ym}(y)$. Choosing the former, we substitute the expectation for $u_x(x, y)$ into the bi-harmonic differential equation (2.1), which leads to a normal differential equation for $u_{ym}(y)$

$$\alpha^4 u_{xm} - 2\alpha^2 \frac{d^2 u_{xm}}{dy^2} + \frac{d^4 u_{xm}}{dy^4} = 0 \tag{2.67}$$

We suppose a solution of the form

$$u_{xm} = A e^{ry} \tag{2.68}$$

Substitution in Eq. (2.67) leads to a characteristic equation for the roots r

$$\alpha^4 - 2\alpha^2 r^2 + r^4 = 0 \tag{2.69}$$

which can be rearranged to

$$(r - \alpha)^2 (r + \alpha)^2 = 0 \tag{2.70}$$

There are two equal roots α and two equal roots $-\alpha$. For equal roots r the general solution has a term with e^{ry} and a term with ye^{ry}. So the solution for $u_{xm}(y)$ becomes

$$u_{xm}(y) = A_1 e^{\alpha y} + A_2 \alpha y e^{\alpha y} + A_3 e^{-\alpha y} + A_4 \alpha y e^{-\alpha y} \tag{2.71}$$

We added a constant α in the second and fourth term in order to give the coefficient A_1 up to and included A_4 equal dimensions. This can be done without loss of generality. Now Eq. (2.2) is used to determine $u_{ym}(y)$. From here on we choose without loss of understanding $v = 0$. Accounting for Eq. (2.2) and after integration, we find

$$u_{ym}(y) = (-A_1+3A_2)e^{\alpha y} - A_2\alpha y e^{\alpha y} + (A_3+3A_4)e^{-\alpha y} + A_4\alpha y e^{-\alpha y} \quad (2.72)$$

Based on Eqs. (1.9), (2.71) and (2.72) the strains can be expressed in terms of the constants too, and therefore also the membrane forces n_{xx}, n_{yy} and n_{xy}. The four constants then can be determined from four boundary conditions

$$\begin{aligned} y = -d/2 &\rightarrow n_{yy} = 0, & n_{xy} = 0 \\ y = d/2 &\rightarrow n_{yy} = f_m \cos\alpha x, & n_{xy} = 0 \end{aligned} \quad (2.73)$$

The elaboration is skipped here. For n_{xx} in the line of symmetry ($x = 0$) we find

$$n_{xx} = \alpha(A_1 e^{\alpha y} + A_2\alpha y e^{\alpha y} + A_3 e^{-\alpha y} + A_4\alpha y e^{-\alpha y}) \quad (2.74)$$

Case 1

We consider the case $d/l \gg 1$. This occurs for a tall wall. The stress distribution is highly nonlinear over the depth; in the upper part of the wall the influence of the load on the lower edge is not noticeable, see the left part of Figure 2.13.

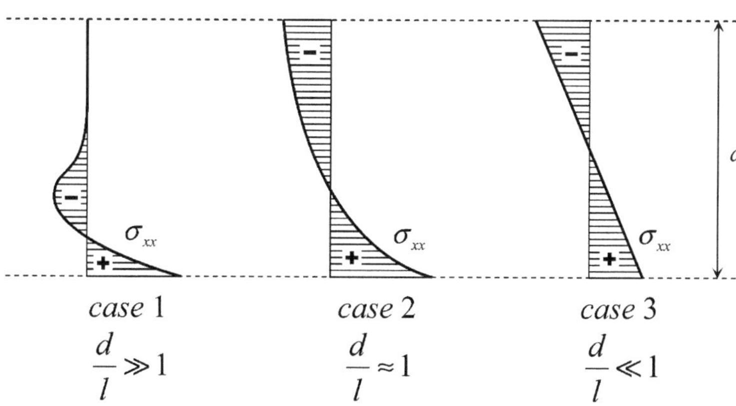

Figure 2.13 Deep beam results for several depth-span ratios.

Case 2

We now consider the case $d/l \approx 1$. This occurs for a wall or beam of which the height and length are nearly equal. The bending stress distribution is again nonlinear over the height, but approaches to classical beam theory. The middle part of Figure 2.13 displays the result.

Case 3

Finally we consider the case $d/l \ll 1$. This is the case for the slender beam and we expect to find the solution for Euler beam theory as drawn in the right-hand part of Figure 2.13. If $d/l \ll 1$, then $\alpha y \ll 1$. For these arguments αy, all the exponential functions can be expanded in a Taylor series around $y = 0$. It appears that the contributions to n_{xx} of powers of αy larger than 1, are negligibly small, so a linear distribution remains. This is the classical solution.

2.2.3 *Practical Application*

The discussed case of a high wall ($d/l \gg 1$) can be used to estimate the stress distribution in practical structures. An example of this is a silo wall on columns, loaded by a uniformly distributed load, shown in Figure 2.14. This may be its own weight, and wall friction forces due to the bulk material in the silo. To estimate the horizontal stress σ_{xx} in the wall halfway between the columns, we adopt the following approach. The load can be split up into

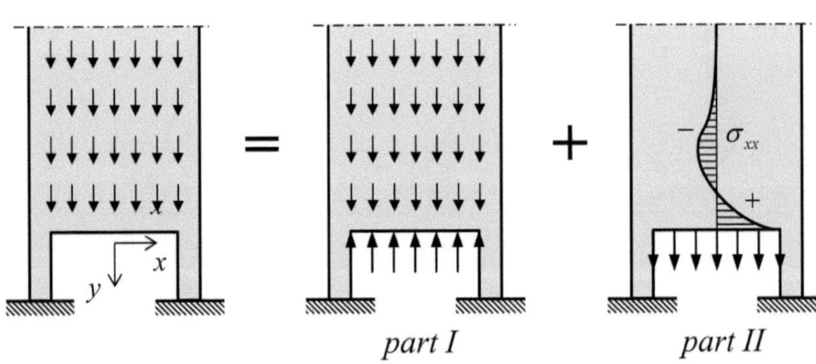

Figure 2.14 Silo wall on columns.

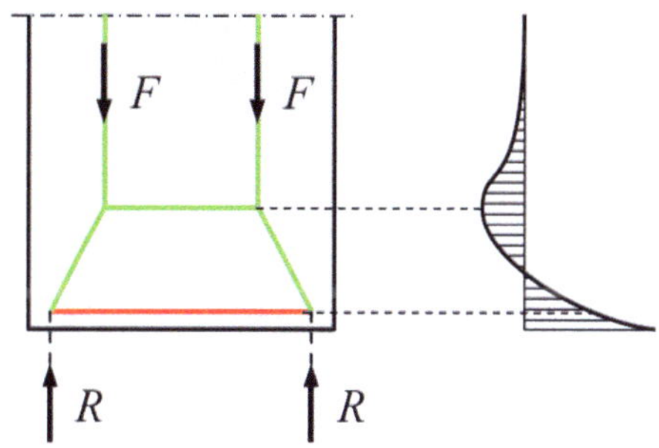

Figure 2.15 Strut-and-tie model for silo wall.

two parts. Part one is a simple stress state in which only vertical stresses σ_{yy} are present and no stresses σ_{xx} occur. We are not interested in this part. The second part is the load case in which the solution for the high wall in Section 2.2.2 can be applied.

Structural engineers who must design reinforced concrete walls often apply truss models for the determination of the reinforcement. For the silo wall they may concentrate the total distributed load in two forces F as shown in Figure 2.15. Each support reaction R is equal to F. The green lines carry compressive forces and the red line the tensile force. The structural engineer wants to know where to place the horizontal compressive strut and the tensile tie, because the distance between them influences the magnitude of the forces in the strut and tie. Knowledge about the elastic solution will be a great help.

2.3 Stresses, Transformations and Principal Stresses

The stresses we have discussed until now have been chosen to be in directions parallel to the x-axis or the y-axis. Sometimes it is useful to know the stresses σ_{nn}, σ_{tt} and σ_{nt} in the directions n and t that make an angle α with the x-axis and y-axis (Figure 2.16). With the help of simple transformation rules, such stresses can be calculated if σ_{xx}, σ_{yy} and σ_{xy} are known

$$\begin{aligned} \sigma_{nn} &= \sigma_{xx}\cos^2\alpha + \sigma_{yy}\sin^2\alpha + \sigma_{xy}\sin 2\alpha \\ \sigma_{tt} &= \sigma_{xx}\sin^2\alpha + \sigma_{yy}\cos^2\alpha - \sigma_{xy}\sin 2\alpha \\ \sigma_{nt} &= -\frac{1}{2}\sigma_{xx}\sin 2\alpha + \frac{1}{2}\sigma_{yy}\sin 2\alpha + \sigma_{xy}\cos 2\alpha \end{aligned} \tag{2.75}$$

Written in another way

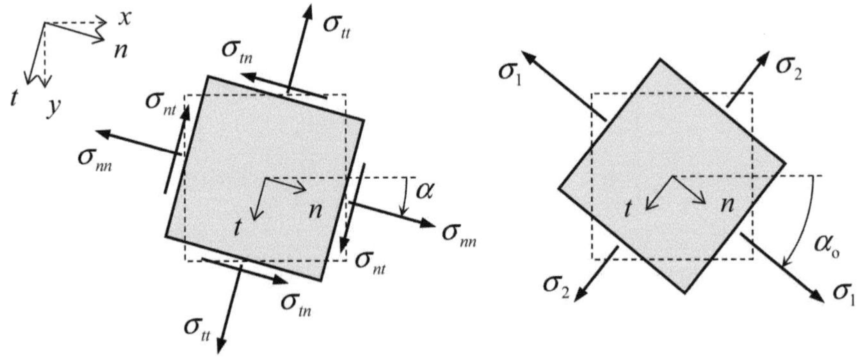

Figure 2.16 Transformation of stresses. Principle stresses and direction.

$$\begin{bmatrix} \sigma_{nn} & \sigma_{nt} \\ \sigma_{nt} & \sigma_{tt} \end{bmatrix} = \begin{bmatrix} \cos\alpha & \sin\alpha \\ -\sin\alpha & \cos\alpha \end{bmatrix} \begin{bmatrix} \sigma_{xx} & \sigma_{xy} \\ \sigma_{xy} & \sigma_{yy} \end{bmatrix} \begin{bmatrix} \cos\alpha & -\sin\alpha \\ \sin\alpha & \cos\alpha \end{bmatrix} \tag{2.76}$$

We see on the basis of Eq. (2.75) that $\sigma_{nn} + \sigma_{tt} = \sigma_{xx} + \sigma_{yy}$. The sum of the normal stresses is invariant under rotations of the axes. An alternative for this transformation is the graphic determination using the Mohr's circle. There is one special value for α that leads to a shear stress value of zero. Then the two normal stresses reach an extreme value. These stresses are called *principal stresses* σ_1 and σ_2 and have the direction α_0, which is called the *principal stress direction* (Figure 2.16). The principal stresses are

$$\sigma_{1,2} = \frac{\sigma_{xx} + \sigma_{yy}}{2} \pm \sqrt{\left(\frac{\sigma_{xx} - \sigma_{yy}}{2}\right)^2 + \sigma_{xy}^2} \tag{2.77}$$

The direction α_o belonging to Eq. (2.77) is computed from

$$\tan 2\alpha_o = \frac{2\sigma_{xy}}{\sigma_{xx} - \sigma_{yy}} \tag{2.78}$$

FE codes may offer the option to show this direction of the principle stresses and refer to it as *trajectories*.

2.4 Other Applications

Consider a circular hole in a plate subjected to a homogenously distributed stress state in which the (normal) stress σ is equal in all directions.

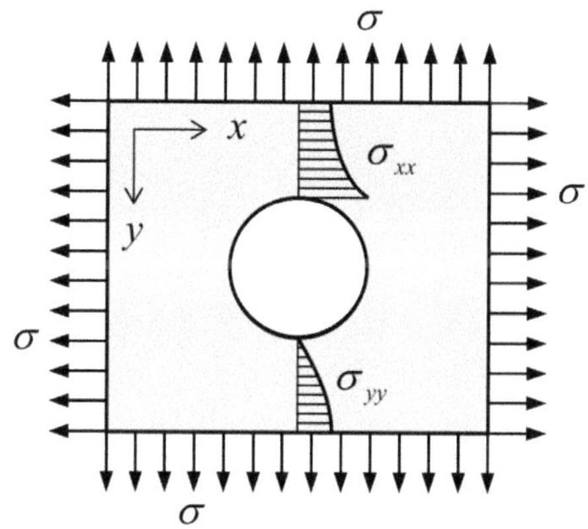

Figure 2.17 Plate with circular hole subjected to a biaxial homogeneous stress state.

The hole causes a disturbance in this homogenously distributed stress field. Figure 2.17 shows the variation of the stresses σ_{xx} and σ_{yy} along the vertical through the centre of the hole. On the edge of the hole, the stress is: $\sigma_{xx} = 2\sigma$. This means a doubling of the stress of the homogenously distributed stress state. The factor 2 is called the *stress concentration factor*. We refer to Section 6.1.2 for the derivation.

A higher stress concentration factor occurs at a circular hole in a plate in a uni-axial stress state (see Figure 2.18). At the edge of the hole, a stress of magnitude $\sigma_{xx} = 3\sigma$ can be found.

Another example is a curved beam (see Figure 2.19). The bending stresses in a cross-section no longer vary linearly. In the direction towards the centre of curvature they strongly increase and the maximum stress on the inside may be much larger than can be expected on basis of the elementary bending theory for a straight beam. This is the subject of Section 6.1.3.

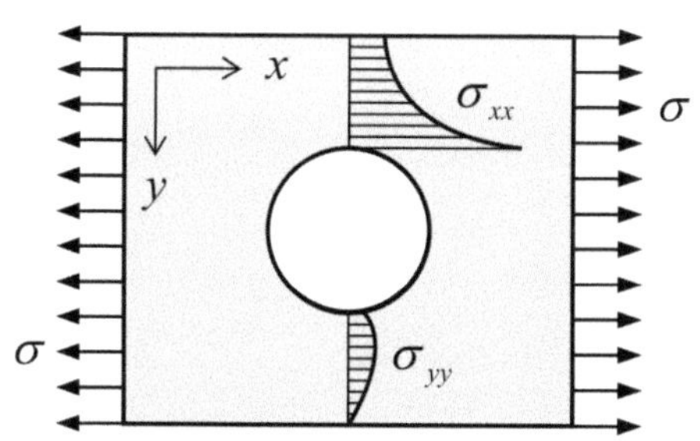

Figure 2.18 Plate with circular hole subjected to a uniaxial homogeneous stress state.

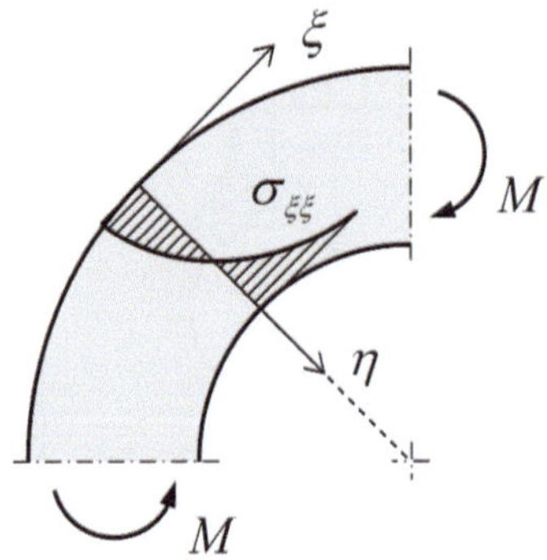

Figure 2.19 Curved bar subjected to a bending moment.

Many interesting stress states can be described with analytical solutions, but many others cannot, for example because the boundary conditions cannot be met or because the contour of the plane stress state cannot be simply described. In such cases numerical methods like the Finite Difference Method or Finite Element Method offer a solution.

We want to give some more examples of stress states that have been determined numerically. First we show another high wall with a load in the middle and restraints along the bottom edge as shown in Figure 2.20. A foundation block can be modeled in this way. The normal stresses σ_{xx} do not vary linearly. The maximum stress at the bottom is noticeably higher than elementary bending theory would have calculated. The moment of these stresses of course should be equal to the total moment in the considered cross-section. Figure 2.20 shows the strut-and-tie model. Green is compression, red is tension.

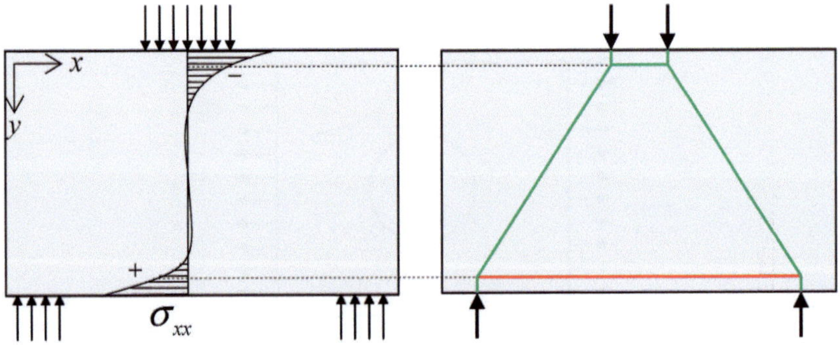

Figure 2.20 Foundation block. Stresses and strut-and-tie scheme.

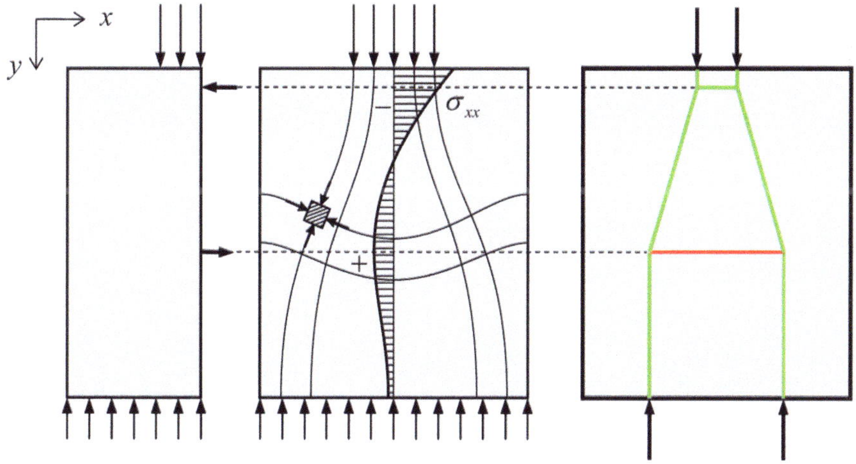

Figure 2.21 Load spreading (for example the anchorage of a pre-stressed cable in a beam).

Another example addresses the load distribution in a beam for the anchorage of a post-tensioned cable. At some distance from the end of the section the forces are distributed uniformly. If we make a vertical cut in the middle and consider one of the halves, then it follows from equilibrium that, in this cutting plane, horizontal stresses σ_{xx} should be present, which are compression stresses at the top and tensile stresses at some distance from the top. The distribution shows the attenuated character again. Practice is not ordinarily prepared for these tensile stresses. They can lead to cracks in the plane of the beam axis; a concrete beam will require reinforcement in the form of stirrups or spiral reinforcement. Figure 2.21 shows some principal stress trajectories. The corresponding strut-and-tie model is also shown.

A similar example of load spreading is the foundation footing, as found under buildings with brick walls (see Figure 2.22). If we make another vertical cut and consider the equilibrium of one of the halves, it will show the presence of horizontal tensile stresses σ_{xx} at the bottom. To determine the magnitude, the stress problem has to be fully solved. The broader the base of the foundation, the lower is the pressure on the soil. However, the tensile stresses in the brickwork will increase, and the poor tensile strength of this material will soon lead to cracks.

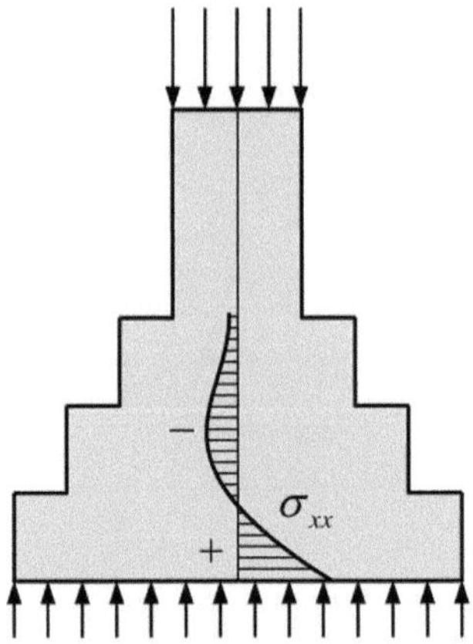

Figure 2.22 Foundation foot. Stresses and strut-and-tie scheme.

Set-back corners (window, door or other openings in a wall) deserve special attention. Figures 2.23 and 2.24 give two more examples. If there is no rounding in the corner the stresses are theoretically infinitely large. In this relation we speak of *notch stresses*. Many cracks are the result of this, and many accidents have occurred (e.g. airplane industry). These corners need special attention from the designer. Often the corners have to be rounded off (plane windows) or strengthened in another way. Concrete structures need special detailed designs for the reinforcement in such corners.

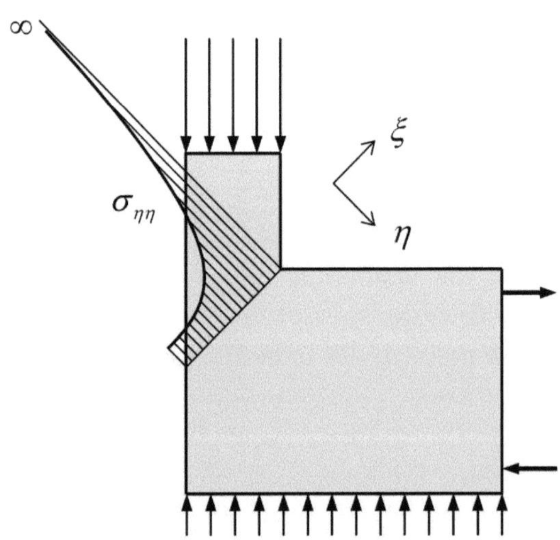

Figure 2.23 Set-back corner.

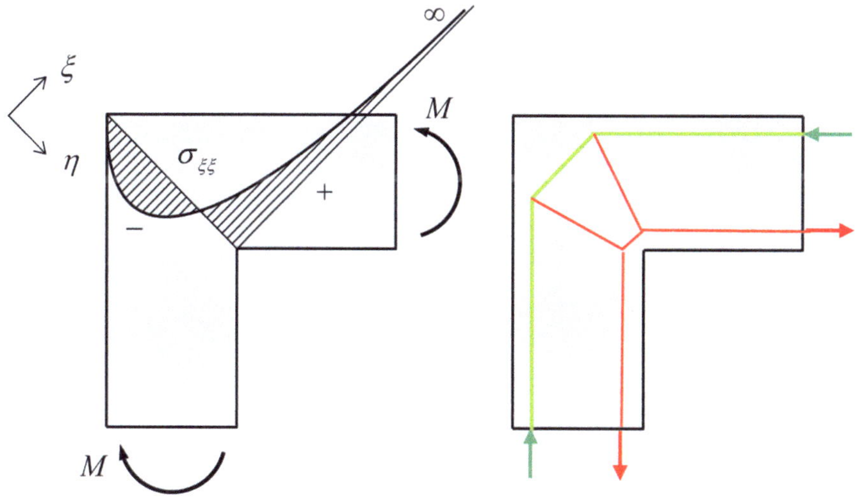

Figure 2.24 Beam-column connection. Stresses and strut-and-tie scheme.

2.5 Message of the Chapter

- One stress state can correspond to more than one displacement field; the difference between the fields are rigid body motions, displacement fields with zero strains. A rigid body motion leaves the structure stress-less.

- Displacements due to a constant bending moment in classical bending theory for a beam of thin cross-section are confirmed by the plane stress membrane theory. Plane sections before loading remain plane after loading. The well-known simple formulas for the deflection and rotation in basic standard cases are confirmed.

- The membrane solution for a constant shear force, in combination with a linearly varying bending moment, deviates from classical beam theory. Plane sections are no longer plane after loading. A linear distribution of bending stresses over the depth of the beam is accompanied by a distorted cross-section. The simple formulas for deflection and rotation in classical beam theory must be amended for shear deformation. This amendment is negligible if the cantilever length is over five times the beam depth.

- The distribution of bending stresses in a shear wall is dependent on the ratio of the wall depth and span. Three aspect ratios are considered. For a high ratio (tall wall) the bending stress distribution is highly nonlinear, and the top part of the wall does not contribute to the load transfer. For a ratio in the order of unity (square wall) the distribution is still nonlinear, but the full cross-section participates in the transfer. For a low ratio (slender beam) the stress distribution approaches to the linear distribution of bending stress in classical beam theory.

- From the computed stress state we can compute two principal stresses and their direction. Trajectories are an instructive and insight-providing aid to structural designers.

Chapter 3
Thick Plates in Bending and Shear

A plate subjected to a load perpendicular to its plane is in a state of bending and transverse shear. If the plate is of concrete, it is called a *slab*. Plates are generalizations of beams. A beam spans one direction, but a plate is able to carry loads in two directions. Figure 3.1 shows an example of a plate on four supports under a point load F_z. The mid-plane of the plate is in the x–y plane and F_z is acting in z-direction perpendicular to the plate. The plate will undergo deflections, and moments and shear forces can be expected. The aim of

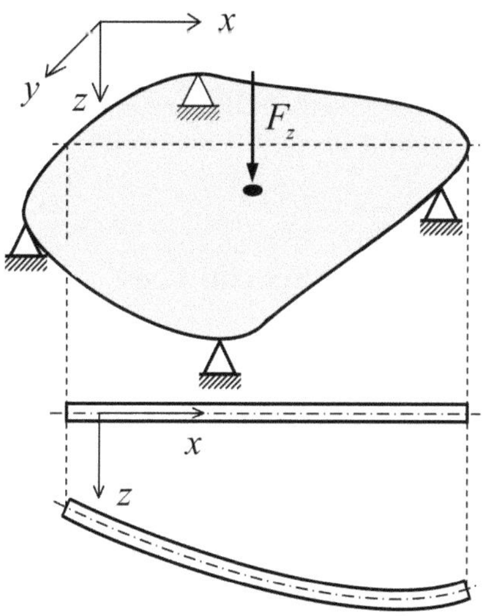

Figure 3.1 Plate with transverse load.

J. Blaauwendraad, *Plates and FEM: Surprises and Pitfalls*, Solid Mechanics and Its Applications 171, DOI 10.1007/978-90-481-3596-7_3,

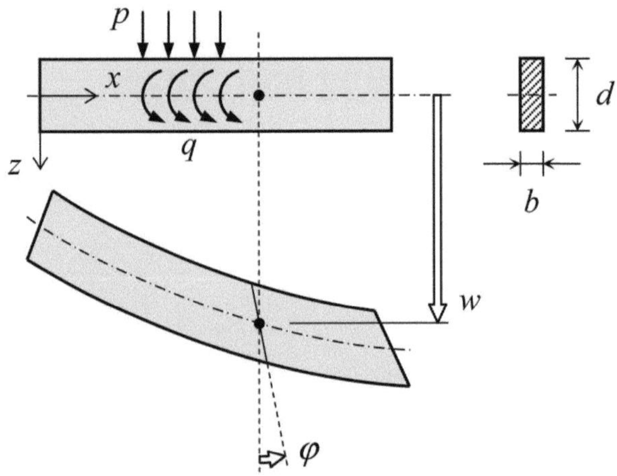

Figure 3.2 Beam with degrees of freedom w and φ and loading p and q.

of this chapter is to explain how these stress resultants can be determined. If both bending moments and shear forces occur, in general bending deformations and shear deformations have to be accounted for. For beams, it is known that shear deformation can be neglected only if the beam is slender. Similarly we must distinguish between *thin plates* and *thick plates*. We will start so that the theory applies for both categories. In Chapter 4 we reduce the complexity and restrict ourselves to the theory for thin plates and its application. An important reason for starting in a general way, including thick plates is, that many computer programs also offer options for thick plates.

3.1 Introduction – Beam as Special Case

As in Chapter 2 for plates loaded in their plane, we will start with the simple case of a plate that spans one direction. We will not consider the effect of Poisson's ratio yet, and leave that for later. Thus, we can consider a strip of width b and depth d as shown in Figure 3.2. We choose a beam axis halfway through the depth d. This axis coincides with the x-axis of a chosen set of axes x and z. The z-axis is pointing downward and is perpendicular to the beam axis. The displacement of the beam axis in the z-direction is called w. In beam theory, it is assumed that no normal force will occur due to constrained supports. This will be true if the deflections are small compared to the depth of the beam ($w \ll d$). Then the beam axis has no axial strain

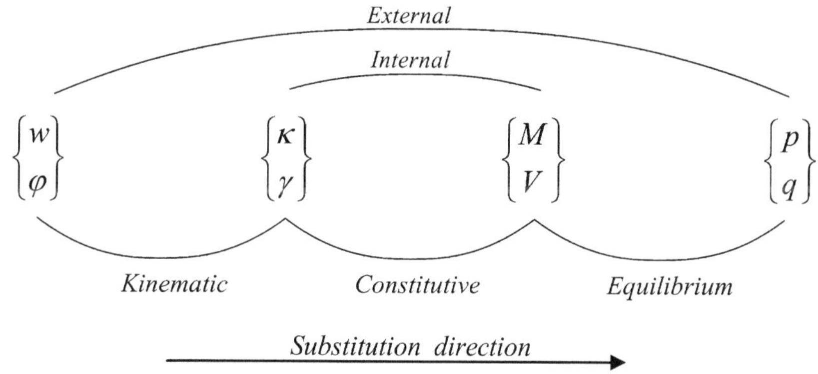

Figure 3.3 Relation scheme for bending and shear in beams.

and will not stretch. This means that after application of the load a cross-section will rotate through an angle φ with the vertical, see Figure 3.2. The rotation φ is an independent degree of freedom, as is the displacement w. The rotation is positive if it leads to positive displacements in x-direction for positive z. Figure 3.2 suggests that an unloaded plane section remains plane after the load is applied. From Figure 2.11 in Chapter 2 we know that this is not generally true. In reality the cross-section is distorted due to shear. We replace the distorted cross-section by a straight plane as shown in Figure 3.2. The straight plane is equivalent to the distorted plane in the sense that they involve an equal amount of shear strain energy.

To have two independent degrees of freedom, means that we can apply two independent load components. In the direction w a distributed line load p can be applied and in the direction of φ a distributed torque q. The stresses in a cross-section have resultants M, the bending moment, and V, the shear force. These stress resultants cause a curvature κ and an (average) shear deformation γ, respectively. The total scheme of quantities for the beam problem is shown in Figure 3.3. The curvature κ is the change of the rotation φ per unit length of the beam, and the shear deformation γ is the change of the right angle between beam axis and cross-section, see Figure 3.4. We introduce the symbol D_b for the *flexural rigidity* EI, and D_s for the *shear rigidity* GA_s, in which A_s is the cross-sectional area A divided by the shape factor η in order to account for the distorted shape due to the inhomogeneous distribution of the shear stresses over the cross-section, see Section 2.1.3. The factor η is 6/5 for a rectangular shape. Equilibrium can be formulated in the two directions w and φ on the basis of the forces acting on a part of the beam of length dx as depicted in Figure 3.4. The three basic sets of equations are

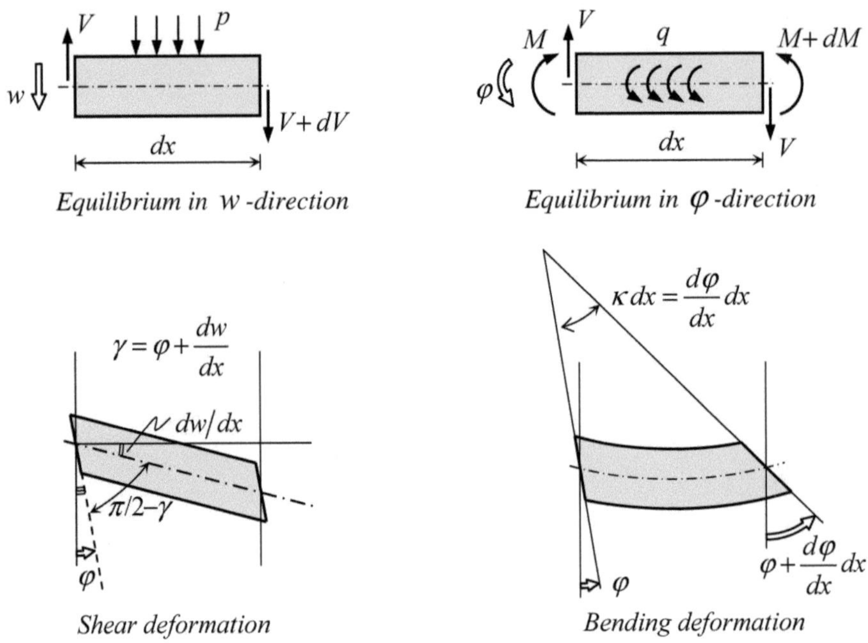

Figure 3.4 Equilibrium in w- and φ-direction; definition of deformations γ and κ.

$$\left.\begin{aligned} \kappa &= \frac{d\varphi}{dx} && \textit{(curvature)} \\ \gamma &= \varphi + \frac{dw}{dx} && \textit{(shear angle)} \end{aligned}\right\} \quad \textit{Kinematic} \tag{3.1}$$

$$\left.\begin{aligned} M &= D_b\kappa && \textit{(bending)} \\ V &= D_s\gamma && \textit{(shear)} \end{aligned}\right\} \quad \textit{Constitutive} \tag{3.2}$$

$$\left.\begin{aligned} &\frac{dV}{dx} + p = 0 && \textit{(w-direction)} \\ &\frac{dM}{dx} - V + q = 0 && \textit{(}\varphi\textit{-direction)} \end{aligned}\right\} \quad \textit{Equilibrium} \tag{3.3}$$

Special attention is drawn to the second equilibrium equation. If the external torque load q is zero, the equation reads that the shear force is the derivative of the bending moment, which is well known to engineers. Note that the shear force is no longer equal to the derivative of the moment if a load q is applied. Substitution of the kinematic equations (3.1) into the constitutive equations (3.2) leads to

$$M = D_b \frac{d\varphi}{dx}$$
$$V = D_s \left(\varphi + \frac{dw}{dx} \right) \tag{3.4}$$

Substitution of these expressions into the equilibrium equations (3.3) transforms these equations into two simultaneous differential equations for w and φ

$$- D_s \frac{d^2 w}{dx^2} - D_s \frac{d\varphi}{dx} = p$$
$$D_s \frac{dw}{dx} + \left(-D_b \frac{d^2}{dx^2} + D_s \right) \varphi = q \tag{3.5}$$

For these two second-order differential equations we need four boundary conditions, two at each end. Per beam end this can be either w or V and either φ or M.

Remark 1

The two differential equations (3.5) in w and φ can be replaced by two equations in w and M if the load q is zero. If we eliminate V from Eq. (3.3), we obtain

$$-\frac{d^2 M}{dx^2} = p \tag{3.6}$$

Combining Eqs. (3.1) and (3.2) we obtain the relations

$$M = D_b \frac{d\varphi}{dx},$$
$$V = D_s \left(\varphi + \frac{dw}{dx} \right) \tag{3.7}$$

Using this information, we can write the first equation of (3.5) as

$$-D_s \left(\frac{d^2 w}{dx^2} + \frac{M}{D_b} \right) = p \tag{3.8}$$

Equations (3.6) and (3.8) now replace Eq. (3.5).

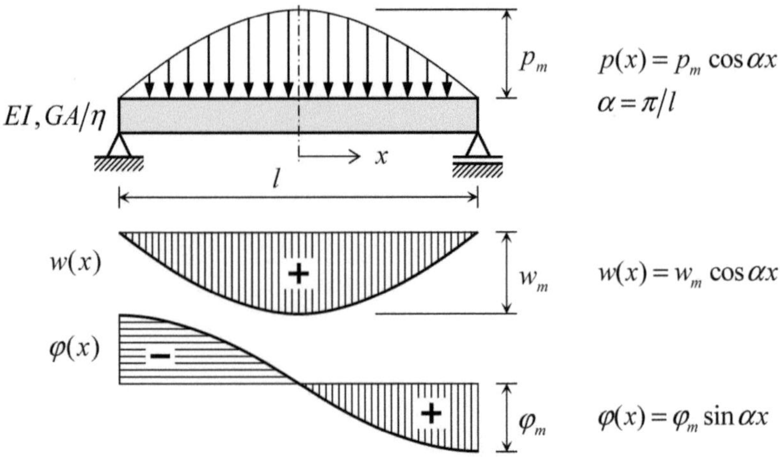

Figure 3.5 Simply-supported beam under cosine load.

Remark 2

In the literature, beams with both flexural and shear deformation are known as Timoshenko beams since Timoshenko discussed such beams in [4].

3.1.1 Illustration

To illustrate this theory we choose the problem of a simply-supported beam of length l, which is subjected to a cosine load p and zero load q, see Figure 3.5. The origin of the axis x is at mid-span. The maximum value of p is p_m. We assume a cosine distribution for w with maximum w_m and a sine distribution for φ with maximum φ_m. For easy writing, we introduce a parameter β, which is defined by

$$\beta^2 = \frac{D_s}{D_b} = \frac{GA/\eta}{EI} \tag{3.9}$$

For a rectangular cross-section with depth d and $\nu = 0.2$ the value of β is with engineering accuracy equal to $2/d$. We see that β has the same dimension as α, which is equal to π/l. The squared quotient $(\alpha/\beta)^2$ is practically equal to $2.5(d/l)^2$ and therefore is apparently a measure for the slenderness of a beam. Substitution of the expected shapes for w and φ into the two differential equations results in two algebraic equations in w_m and φ_m

$$EI\begin{bmatrix} \beta^2\alpha^2 & \beta^2\alpha \\ \beta^2\alpha & \beta^2+\alpha^2 \end{bmatrix}\begin{Bmatrix} w_m \\ \varphi_m \end{Bmatrix}=\begin{Bmatrix} p_m \\ 0 \end{Bmatrix} \tag{3.10}$$

The solution to these equations is

$$\begin{aligned} w_m &= \left\{1+\left(\frac{\alpha}{\beta}\right)^2\right\}\frac{p_m}{\alpha^4 EI} \\ \varphi_m &= -\frac{p_m}{\alpha^3 EI} \end{aligned} \tag{3.11}$$

In this example, the rotation φ is not influenced by the shear rigidity. The displacement w, however, does depend on β, and so on the shear rigidity, but this influence diminishes for slender beams, because $(\alpha/\beta)^2$ approaches zero in that case. If d/l is 1/5, the contribution to w_m due to shear is 10%. If d/l reduces to 1/10, the shear contribution is only 2.5%. For $d/l = 1/20$ the contribution is less than 1%. The chosen simple case could have been solved without the use of the differential equations. This is not done here because we want to demonstrate a general approach.

3.1.2 Simplification for Slender Beam

For slender beams, the theory can be drastically simplified. Let us return to the basic equations in (3.1), (3.2) and (3.3). If the shear deformation can be neglected, then we can state that γ is zero. From the second equation in (3.1) we then conclude

$$\varphi = -\frac{dw}{dx} \tag{3.12}$$

Now the rotation φ is no longer an independent degree of freedom, but is related to the displacement w. Therefore the first equation in (3.1) transforms due to Eq. (3.6) into

$$\kappa = -\frac{d^2w}{dx^2} \quad \textit{Kinematic} \tag{3.13}$$

The second constitutive equation in (3.2) does not make sense any longer. The shear force V still has a value, but the shear rigidity D_s is infinitely large and the shear deformation γ is zero. We just skip this equation, so the only constitutive equation is

$$M = D_b\kappa \quad \textit{Constitutive} \tag{3.14}$$

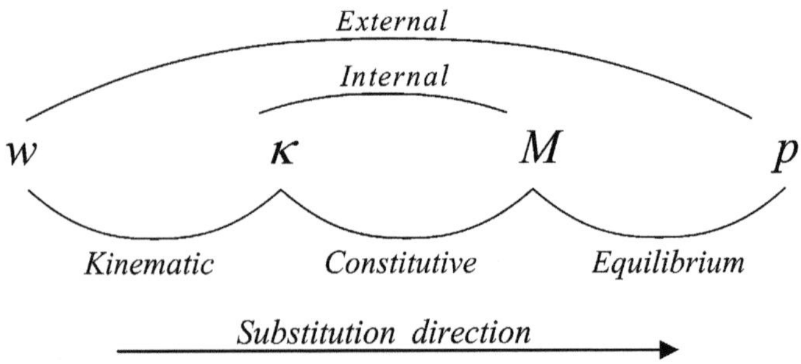

Figure 3.6 Relation scheme for slender beams (bending only).

Finally also the equilibrium equations (3.3) must be inspected. If φ is no longer a degree of freedom, we cannot apply a load q. So it is set to zero. Then the shear force V is the derivative of the bending moment M

$$V = \frac{dM}{dx} \tag{3.15}$$

We now substitute this result into the first equilibrium equation of (3.3), which leads to

$$-\frac{d^2M}{dx^2} = p \quad \textit{Equilibrium} \tag{3.16}$$

Summarizing, the exclusion of the shear deformation reduces the six equations in (3.1), (3.2) and (3.3) to (3.13), (3.14) and (3.16). Figure 3.6 shows the relation scheme for beams subjected to bending if the shear deformation is neglected. The new equations provide, after successive substitution, the classical relation between M and w and between V and w (with $D_s = EI$) given by

$$\begin{aligned} M &= -EI\frac{d^2w}{dx^2} \\ V &= -EI\frac{d^3w}{dx^3} \end{aligned} \tag{3.17}$$

and the fourth-order differential equation then is

$$EI\frac{d^4w}{dx^4} = p \tag{3.18}$$

This differential equation can be solved taking into account four boundary conditions, two at each end. These are either w or V and either dw/dx or M. The reader should be familiar with the application of this classical beam theory.

3.1.3 Suppositions of Timoshenko Beam in Hindsight

In the derivation of the Timoshenko beam theory, we made use of a number of suppositions. It is good to summarize them here:

1. The member is prismatic, and the material is homogeneous and isotropic.
2. No extensional forces occur, so the bar axis is a neutral line. This is valid if $w \ll d$.
3. A plane section in the unloaded state remains plane after the load is applied. In fact, the cross-section will distort, but we can work with an 'average' plane, as explained in Section 2.1.3. The non-homogeneous shear distribution is accounted for by introducing a shape factor η to reduce the shear stiffness.
4. Without saying, it is assumed that σ_{zz} is zero. Due to a non-zero Poisson's ratio, strains ε_{zz} will occur that are not zero. Strictly speaking, the vertical displacement will therefore vary a little bit over the depth of the beam. We have neglected this.
5. At the end of the discussion of theory, the shear deformation has been set to zero, which limits the theory to slender beams and simplifies it noticeably. This last assumption means that a plane section not only remains plane, but also that it will remain normal to the beam axis.

Remark

If we neglect the shear deformations, we find

$$-\frac{d^2M}{dx^2} = p \tag{3.19}$$

$$M = -D_b\frac{d^2w}{dx^2} \tag{3.20}$$

We can compare these equations with (3.6) and (3.8) for the case that shear deformation does occur (special case in which the distributed load $q = 0$). Equation (3.9) is exactly the same as Eq. (3.6). The equivalence between Equations (3.20) and (3.8) is also easily shown if we divide the latter by D_s

$$-\left(\frac{d^2w}{dx^2} + \frac{M}{D_b}\right) = \frac{p}{D_s} \tag{3.21}$$

The case of no shear deformation is achieved when D_s is chosen infinitely large. Then the right-hand member of Eq. (3.8) becomes zero and the equation is equal to (3.20).

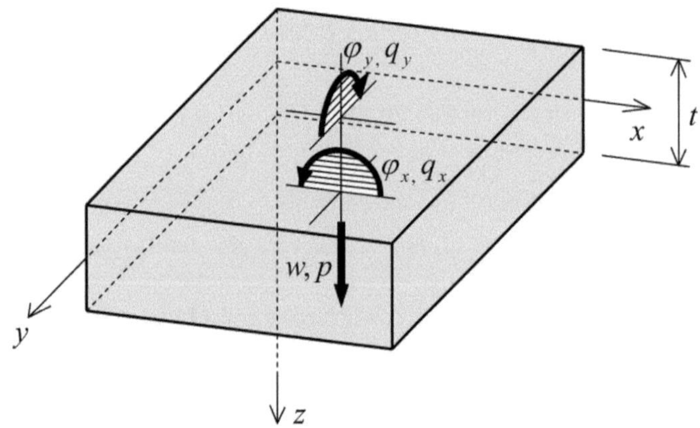

Figure 3.7 Three load components on plate subjected to bending and shear.

3.2 Outline of Thick Plates

In this section we will give the derivation of the differential equations of the homogeneous isotropic plate subjected to bending and shear. For this purpose we will consider a plate with a constant thickness t as shown in Figure 3.7. The mid-plane coincides with the x–y plane of a right-handed orthogonal coordinate system x, y, z. The z-axis is perpendicular to the unloaded plate. After we have discussed plane stress theory in Chapter 1 and the beam theory in Section 3.1 we can be brief in introducing the main symbols for thick plate theory. Hereafter it will be shown that a plate in bending and transverse shear can be considered as a layered system of plane stress states. If we consider a vertical line over the thickness of the plate and normal to the mid-plane, this is subjected to a displacement w in z-direction, a rotation φ_x and rotation φ_y, as shown in Figure 3.7.

Sign convention

Displacement w is positive in the z-direction, the rotation φ_x is positive if it leads to positive displacements u_x at positive z-side of the mid-plane, and the rotation φ_y is positive if it leads to positive displacements u_y on the positive z-side. It is stressed here that the definition of the rotations and their sign convention are different from what is normal in FE codes; this is discussed in Chapter 10. The special choice we

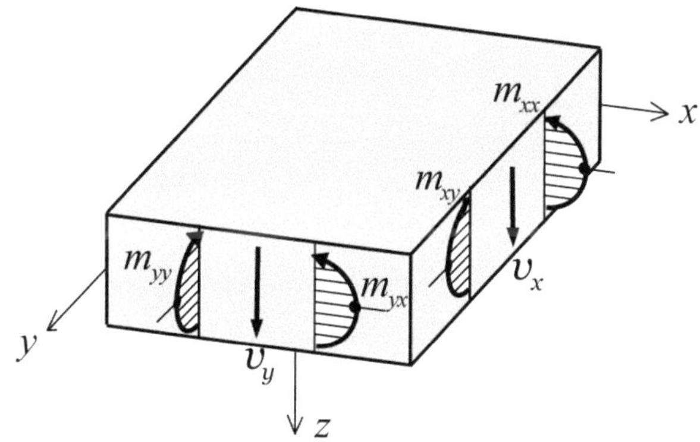

Figure 3.8 Stress resultants in the plate.

have made in the present chapter has advantage in the derivation of the differential equations.

The load contains a *distributed load* p, associated with *displacement* w, that is positive in the positive z-direction, and *distributed couples* q_x and q_y, associated with the *rotations* φ_x and φ_y, see Figure 3.7. Therefore, the couple q_x acts in a plane parallel to the x–z plane, and q_y parallel to the y–z plane. The loads are positive if they act in the positive direction of the degrees of freedom w, φ_x and φ_y.

In each plate layer through the thickness we will find plane stresses σ_{xx}, σ_{yy} and σ_{xy}. Integration over thickness will lead to *bending moments* m_{xx} and m_{yy} and a *twisting moment* m_{xy}. They are positive if they lead to a positive stress in layers on the positive z-side of the mid-plane. For the meaning of the subscripts we refer to Section 1.2. In the same way as we defined strains ε_{xx}, ε_{yy} and γ_{xy} for plane stress, we will now introduce associate *curvatures* κ_{xx}, κ_{yy} and *torsional deformation* ρ_{xy}. Further, we have *transverse shear forces* both in the x-direction and the y-direction, v_x and v_y respectively. They are positive if directed in the positive z-direction on a section with positive normal. These shear forces are associated with *transverse shear angles* γ_x and γ_y. All introduced symbols are shown in Figure 3.9. There we see which symbols will occur in the kinematic, constitutive and equilibrium relationships.

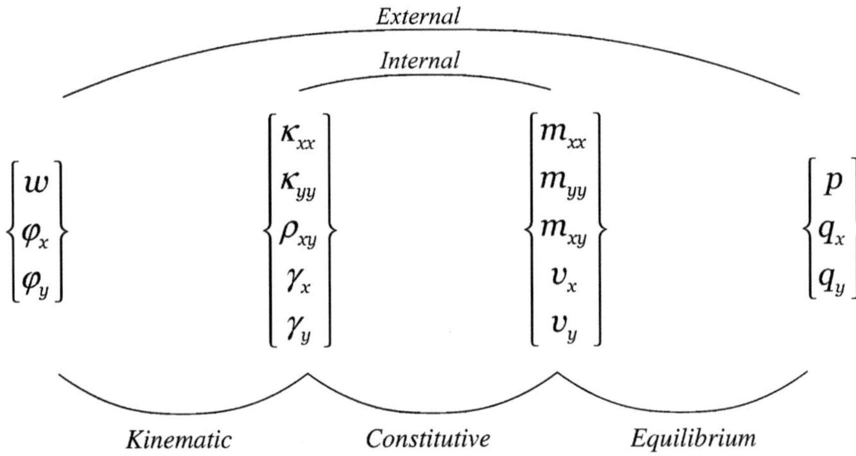

Figure 3.9 Relation scheme for plate with both bending and shear deformation.

> ***Notation***
>
> In this book the shear forces in a beam are expressed by a capital V, and the shear forces in a plate by a letter v. This is done to differentiate between the dimensions of the two quantities, [N] and [N/m], respectively. The dimension of the bending and twisting moments is moment per unit length, so [Nm/m or N]. For plates we use small letters m to emphasize the difference in dimension of the moments in beams for which we use capital letters M.

3.2.1 Suppositions

The analysis of plates is based on suppositions, which are comparable to the suppositions made in beam theory:

1. The plate has a constant thickness, and the material is homogeneous and isotropic.
2. No membrane forces will occur due to support constraints or large deflections. The mid-plane of the plate will remain strain-less after applying the load. This is correct if $w \ll t$.

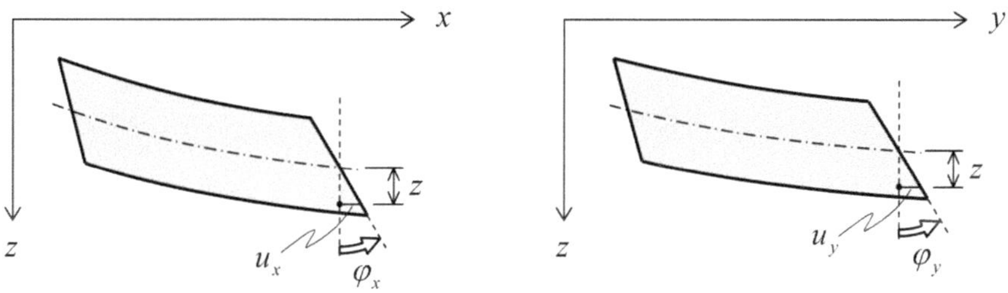

Figure 3.10 Determination of u_x and u_y as function of φ_x and φ_y.

3. It is assumed that a straight line normal to the mid-plane of the plate in an unloaded state remains a straight line after application of the load, however it needs not be normal to the mid-plane of the plate. If one could bring in a needle perpendicular to the mid-plane of the unloaded plate, this needle could freely tilt after loading in the x-direction (angle φ_x) and the y-direction (angle φ_y), but it would remain a straight needle. This *needle hypothesis* is the generalization of the plane section hypothesis in the beam theory. As said, the sign convention for φ_x and φ_y in this chapter is that these rotations are positive if they cause a positive horizontal displacement for plate particles with positive z.
4. The stress σ_{zz} in the direction normal to the mid-plane is negligibly small compared to the bending stresses σ_{xx} and σ_{yy} and is set to zero. The strain ε_{zz} in z-direction is also set to zero. Possible small differences in the displacement w over the thickness of the slab are neglected.
5. The above-mentioned four suppositions are valid for thin and thick plates. For thin plates an additional assumption will be made regarding the shear deformation, which will be neglected if compared to the flexural deformation.

The needle hypothesis implies that the in-plane displacements u_x and u_y vary linearly over the thickness t. Because there are no strains in the mid-plane these displacements will be zero there. From Figure 3.11 it then follows that u_x and u_y can be expressed in terms of the rotations φ_x and φ_y:

$$\begin{aligned} u_x &= z\,\varphi_x \\ u_y &= z\,\varphi_y \end{aligned} \tag{3.22}$$

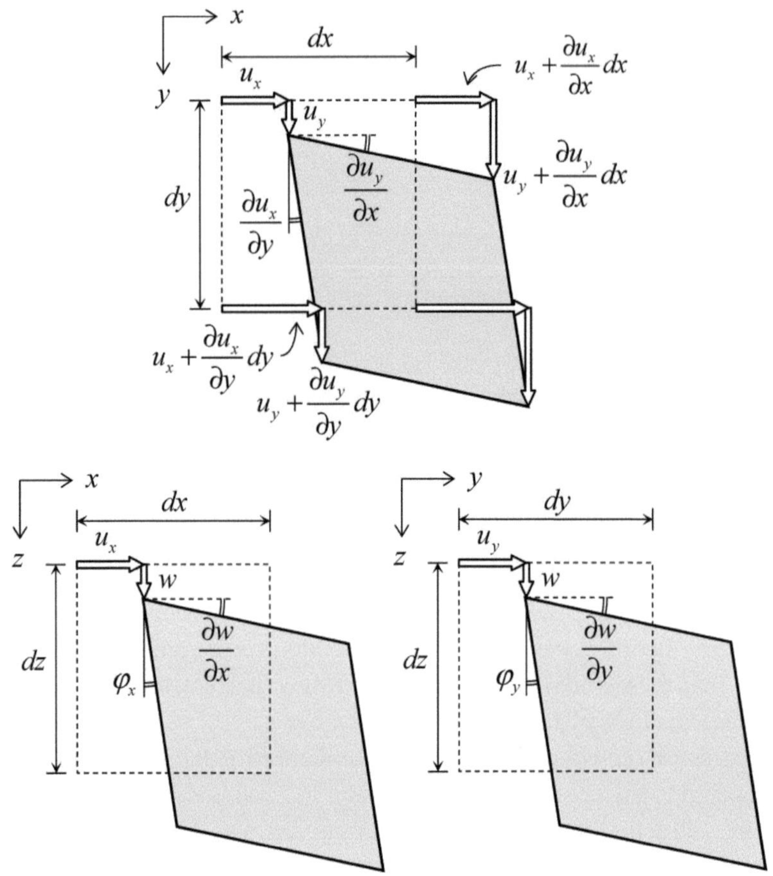

Figure 3.11 Displacements and deformations, observed in three orthogonal planes.

3.3 Basic Equations

We will formulate the three basic equations in the following order: *kinematic equations, constitutive equations* and *equilibrium equations.*

3.3.1 *Kinematic Equations*

The kinematic equations describe the relation between the displacements and the deformations. Figure 3.11 shows three views of an elementary block with the dimensions d_x, d_y and d_z. The dashed line represents the view in an unloaded state and the solid line in the deformed state. From Figure 3.12

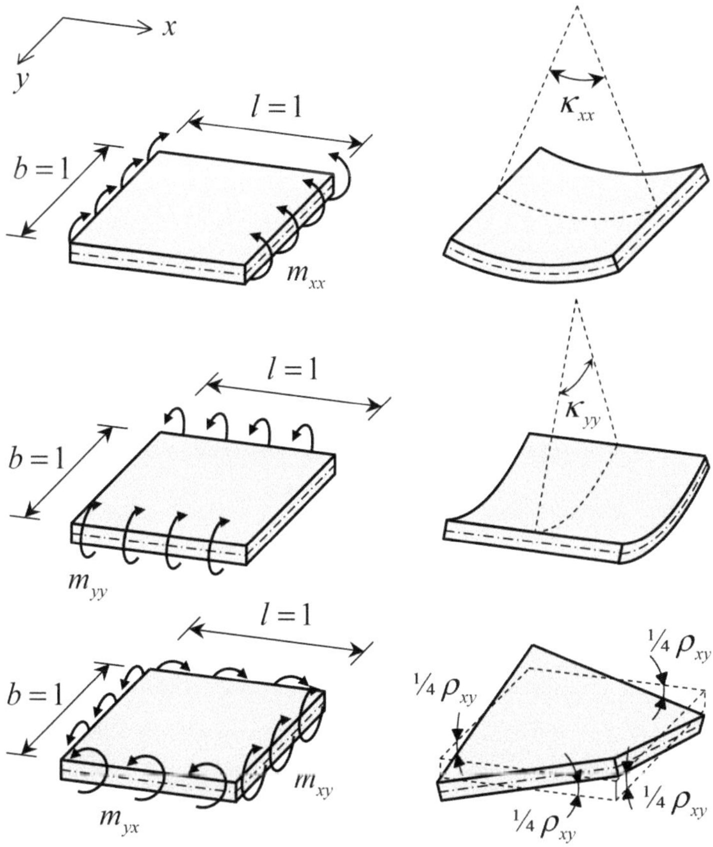

Figure 3.12 Moments and curvatures for zero Poisson's ratio.

as a definition for the strains and shear angles it follows

$$\begin{aligned} \varepsilon_{xx} &= \frac{\partial u_x}{\partial x}, \quad \varepsilon_{yy} = \frac{\partial u_y}{\partial y}, \quad \gamma_{xy} = \frac{\partial u_x}{\partial y} + \frac{\partial u_y}{\partial x} \\ \gamma_{xz} &= \frac{\partial u_x}{\partial z} + \frac{\partial w}{\partial x}, \quad \gamma_{yz} = \frac{\partial u_y}{\partial z} + \frac{\partial w}{\partial y} \end{aligned} \tag{3.23}$$

The strain ε_{zz} has not been included here because it does not play a role in the plate theory. All the five strains can be expressed in terms of the three quantities φ_x, φ_y and w. These degrees of freedom are functions of x and y, i.e. of the position in the mid-plane of the plate. Then the strains γ_{xz} and γ_{yz} in Eq. (3.23), which are associated with the stresses due to the shear forces, are independent of z. They are constant over the thickness as a consequence of the needle hypothesis. In reality the distribution will be parabolic over thickness, for which a correction will be made later when we discuss the constitutive equation. We skip the index z in γ_{xz} and γ_{yz}, and continue using γ_x and γ_y for the shear angles associated with the transverse shear forces v_x and v_y. The other three strains ε_{xx}, ε_{xx} and γ_{xy} are not only dependent on x and y (the position in the mid-plane) but also on z. We find that they vary linearly over the thickness and are zero in the mid-plane. Substitution of Eq. (3.22) into Eq. (3.23) provides

$$\varepsilon_{xx} = z\frac{\partial \varphi_x}{\partial x}, \quad \varepsilon_{yy} = z\frac{\partial \varphi_y}{\partial y}, \quad \gamma_{xy} = z\left(\frac{\partial \varphi_x}{\partial y} + \frac{\partial \varphi_y}{\partial x}\right) \tag{3.24}$$

$$\gamma_x = \varphi_x + \frac{\partial w}{\partial x}, \quad \gamma_y = \varphi_y + \frac{\partial w}{\partial y} \tag{3.25}$$

Introducing here the three curvatures κ_{xx}, κ_y and ρ_{xy} we redefine the three strains in the horizontal planes

$$\varepsilon_{xx} = z\kappa_{xx}, \quad \varepsilon_{yy} = z\kappa_{yy}, \quad \gamma_{xy} = z\rho_{xy} \tag{3.26}$$

The three curvatures in Eq. (3.26) and the two shear deformations in Eq. (3.25) together are five deformations that in general govern the plate problem. Their relations to the three degrees of freedom are the kinematic equations

$$\left.\begin{array}{l} \kappa_{xx} = \dfrac{\partial \varphi_x}{\partial x}, \quad \kappa_{yy} = \dfrac{\partial \varphi_y}{\partial y}, \quad \rho_{xy} = \dfrac{\partial \varphi_x}{\partial y} + \dfrac{\partial \varphi_y}{\partial x} \\ \gamma_x = \varphi_x + \dfrac{\partial w}{\partial x}, \quad \gamma_y = \varphi_y + \dfrac{\partial w}{\partial y} \end{array}\right\} \quad \textit{Kinematic} \tag{3.27}$$

3.3.2 Constitutive Equations

For each horizontal layer of the plate at a distance z from the mid-plane and with thickness d_z we have supposed that the normal stress σ_{zz} is zero. This implies that all layers are in a state of plane stress, so we can apply the constitutive relations (1.13)

$$\begin{aligned} \sigma_{xx} &= \frac{E}{1-\nu^2}(\varepsilon_{xx} + \nu\varepsilon_{yy}) \\ \sigma_{yy} &= \frac{E}{1-\nu^2}(\varepsilon_{yy} + \nu\varepsilon_{xx}) \\ \sigma_{xy} &= \frac{E}{2(1+\nu)}\gamma_{xy} \end{aligned} \tag{3.28}$$

As the normal stresses are functions of the z-coordinate, they provide bending moments per unit length. Also the shear stress σ_{xy} is a function of the z-coordinate, so there is now a resulting twisting moment:

$$m_{xx} = \int_t z\sigma_{xx}\,dz, \quad m_{yy} = \int_t z\sigma_{yy}\,dz, \quad m_{xy} = \int_t z\sigma_{xy}\,dz \tag{3.29}$$

$\sigma_{xy} = \sigma_{yx}$ implies $m_{xy} = m_{yx}$. Positive values of the moments discussed above are shown in Figure 3.8. Substitution of Eq. (3.28) and accounting for Eqs. (3.24) and (3.26) changes these equations into

$$\begin{aligned} m_{xx} &= D_b(\kappa_{xx} + \nu\kappa_{yy}) \\ m_{yy} &= D_b(\kappa_{yy} + \nu\kappa_{xx}) \\ m_{xy} &= \frac{1}{2}\,D_b(1-\nu)\rho_{xy} \end{aligned} \tag{3.30}$$

in which

$$D_b = \frac{E\,t^3}{12(1-\nu^2)} \tag{3.31}$$

is called the *plate flexural rigidity*. If we make use of the matrix notation, Eq. (3.20) becomes

$$\begin{Bmatrix} m_{xx} \\ m_{yy} \\ m_{xy} \end{Bmatrix} = D_b \begin{bmatrix} 1 & \nu & 0 \\ \nu & 1 & 0 \\ 0 & 0 & \frac{1}{2}(1-\nu) \end{bmatrix} \begin{Bmatrix} \kappa_{xx} \\ \kappa_{yy} \\ \rho_{xy} \end{Bmatrix} \tag{3.32}$$

In Figure 3.12 the relation between curvatures and moments is visualized for zero Poisson's ratio. The *flexural rigidity matrix* in Eq. (3.32) then becomes a diagonal matrix, which means that bending in x-direction, bending in y-direction and torsion are uncoupled phenomena. In Figure 3.13 the deformations are shown for a non-zero Poisson's ratio.

The correspondence of Eq. (3.32) with Eq. (1.13) for a plate in plane stress is obvious. If we consider layers over the thickness of the plate, each layer is in a state of plane stress as discussed in Chapter 1. Integrated over thickness the stresses lead to moments.

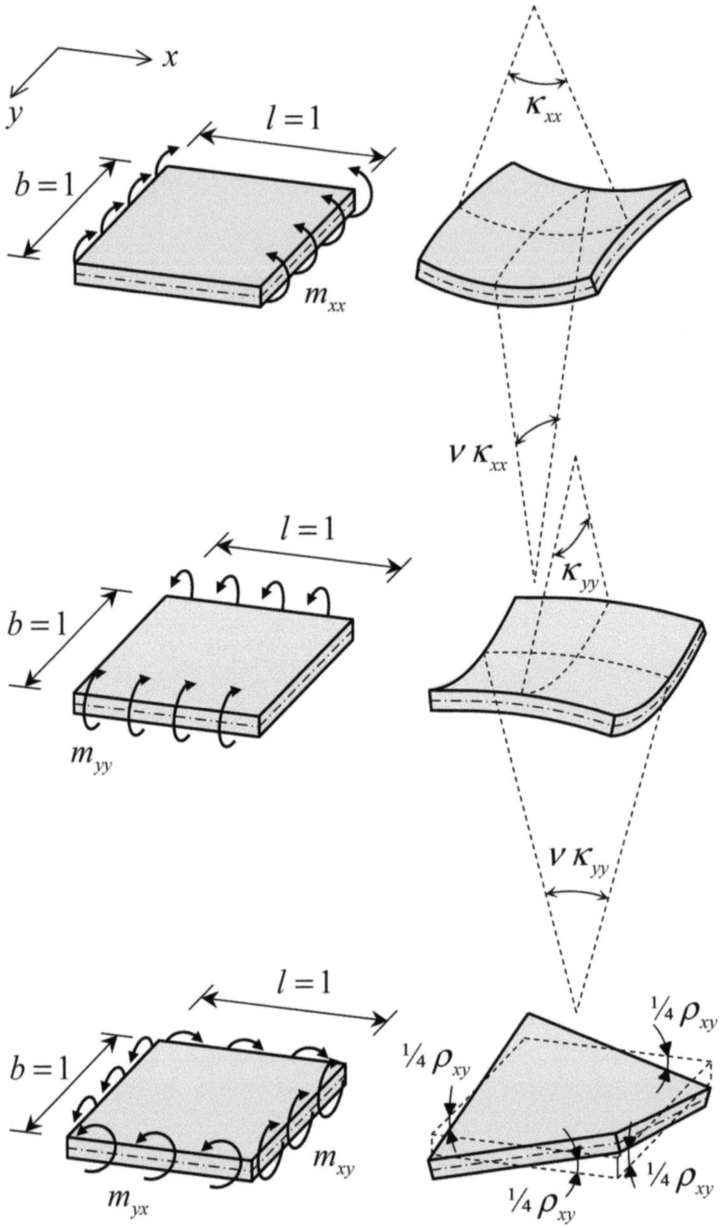

Figure 3.13 Moments and curvatures for non-zero Poisson's ratio.

Warning

At this stage in the derivation a warning is appropriate. In books and papers one may find two different ways of writing the relation between the twisting moment and the associated deformation. The difference regards the definition of the deformation. Authors use either ρ_{xy} or κ_{xy}, which leads to different relationships

$$\text{either } m_{xy} = D_b(1-\nu)\kappa_{xy} \quad \text{or} \quad m_{xy} = \frac{1}{2}D_b(1-\nu)\rho_{xy} \tag{3.33}$$

It is clear that the torsion curvature κ_{xy} is half ρ_{xy}. The second definition in (3.33) has been used in the derivation here, and is common in theory and user manuals of commercial Finite Element software. The first definition is a typical pre-FE notation, and is found in classical books on plate theory. This can cause confusion when users have to determine the torsional rigidity themselves, as may be the case for orthotropic plates. In Chapter 21 we will return to this subject.

Now we proceed to the vertical shear stresses σ_{xz} and σ_{yz}. From beam theory we know that they have a parabolic distribution over the thickness and integration over the thickness yields the shear forces v_x and v_y per unit length. Also from the beam theory we know (see Chapter 2) that a relation exits between the shear forces v_x and v_y and the shear deformations γ_x and γ_y

$$v_x = D_s\gamma_x, \quad v_y = D_s\gamma_y \tag{3.34}$$

in which D_s is the *plate shear rigidity*; $D_s = G_t/\eta$, where G is the shear modulus and $\eta = 6/5$ as explained in Section 2.1.3.

The equations (3.32) and (3.34) together form the constitutive equations for a plate with bending and shear deformation

$$\left.\begin{aligned} m_{xx} &= D_b(\kappa_{xx} + \nu\kappa_{yy}) \\ m_{yy} &= D_b(\kappa_{yy} + \nu\kappa_{xx}) \\ m_{xy} &= \tfrac{1}{2}\, D_b(1-\nu)\rho_{xy} \\ v_x &= D_s\gamma_x \\ v_y &= D_s\gamma_y \end{aligned}\right\} \quad \textit{Constitutive} \tag{3.35}$$

Herein $D_b = Et^3/\{12(1-\nu^2)\}$, the stiffness term $\frac{1}{2}D_b(1-\nu)$ for torsion can be written as $Gt^3/12$ and D_s as Gt/η.

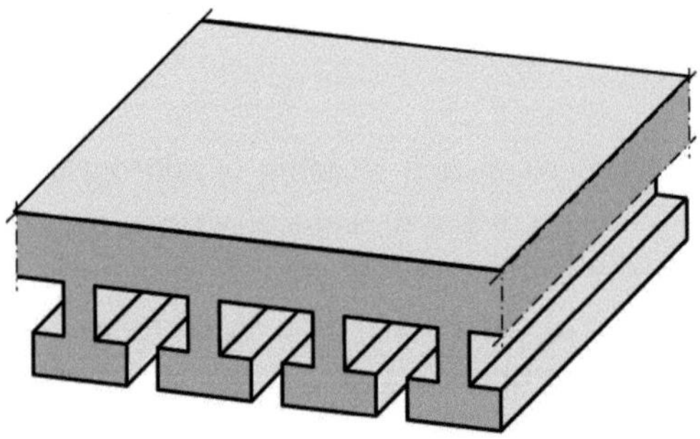

Figure 3.14 Structure that can be considered as an orthotropic plate.

3.3.3 Equilibrium Equations

In the preceding sections we determined the kinematic relations between three degrees of freedom (w, φ_x and φ_y) and five deformations (κ_{xx}, κ_{xx}, ρ_{xy}, γ_x, γ_y) and the constitutive equations between these deformations and five stress resultants (m_{xx}, m_{yy}, m_{xy}, v_x, v_y). Still to be determined are the equilibrium equations between the five stress resultants and three load com-

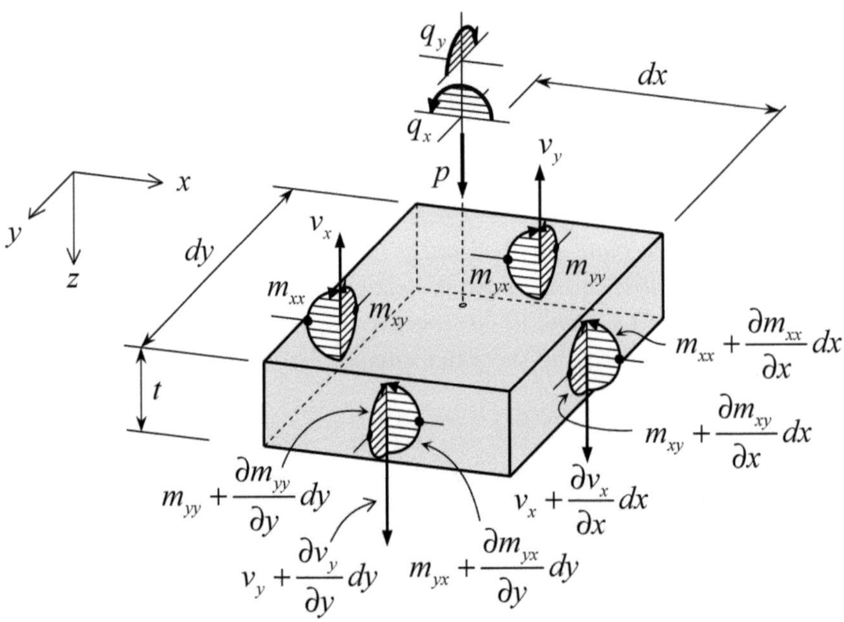

Figure 3.15 Positive loads and stress resultants.

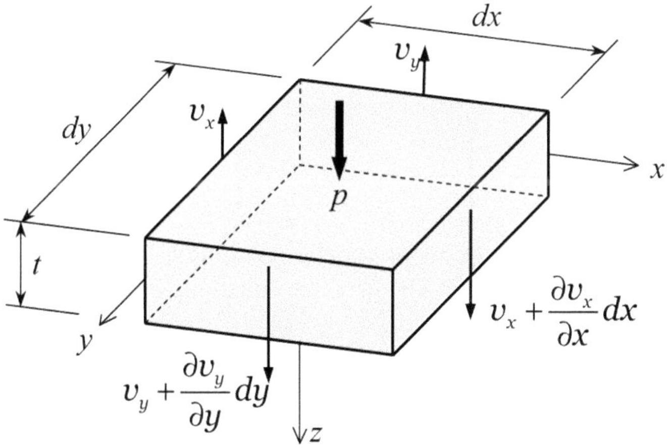

Figure 3.16 Plate equilibrium in z-direction.

ponents p, q_x, q_y. For this purpose we consider equilibrium in the directions w, φ_x and φ_y respectively, see Figure 3.15.

For the equilibrium in w-direction, consider the infinitesimal small plate part with edges d_x and d_y as shown in Figure 3.16. The equilibrium equation contains three terms, i.e. the increase of v_x over the distance d_x, the increase of v_y over the distance d_y, and the load p. Remember that v_x and v_y are forces per unit length, so the increases must be multiplied by d_y and d_x respectively, and p is defined per unit area, and so must be multiplied by $dx\, dy$

$$\left(\frac{\partial v_x}{\partial x} dx\right) dy + \left(\frac{\partial v_y}{\partial y} dy\right) dx + p\, dx\, dy = 0 \tag{3.36}$$

Similarly the moment equilibrium in φ_x- and φ_y-direction requires (see Figures 3.17 and 3.18).

$$\begin{aligned} \left(\frac{\partial m_{xx}}{\partial x} dx\right) dy + \left(\frac{\partial m_{yx}}{\partial y} dy\right) dx - (v_x\, dy)\, dx + q_x\, dx\, dy = 0 \\ \left(\frac{\partial m_{yy}}{\partial y} dy\right) dx + \left(\frac{\partial m_{xy}}{\partial x} dx\right) dy - (v_y\, dx)\, dy + q_y\, dx\, dy = 0 \end{aligned} \tag{3.37}$$

Note that the increase of m_{yx} in the y-direction enters in the equilibrium equation for the φ_x-direction and so does the increase of m_{xy} in x-direction in the equation for the φ_y-direction. In all three equations the product $dx\, dy$ occurs, so it can be left out, which yields the desired equilibrium relations

$$\left.\begin{aligned}
&\frac{\partial v_x}{\partial x}+\frac{\partial v_y}{\partial y}+p=0\\
&\frac{\partial m_{xx}}{\partial x}+\frac{\partial m_{yx}}{\partial y}-v_x+q_x=0\\
&\frac{\partial m_{yy}}{\partial y}+\frac{\partial m_{xy}}{\partial x}-v_y+q_y=0
\end{aligned}\right\}\quad \textit{Equilibrium} \qquad (3.38)$$

It is instructive to compare these equations with the two equilibrium equations for a beam in Eq. (3.3). The first two equations in (3.38) transform into

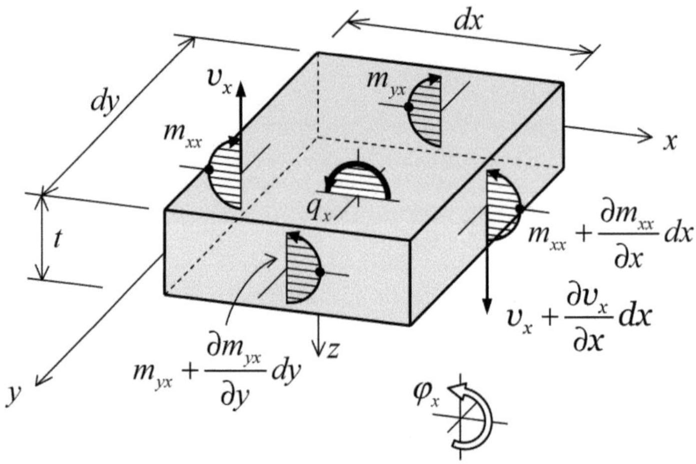

Figure 3.17 Plate equilibrium in φ_x-direction.

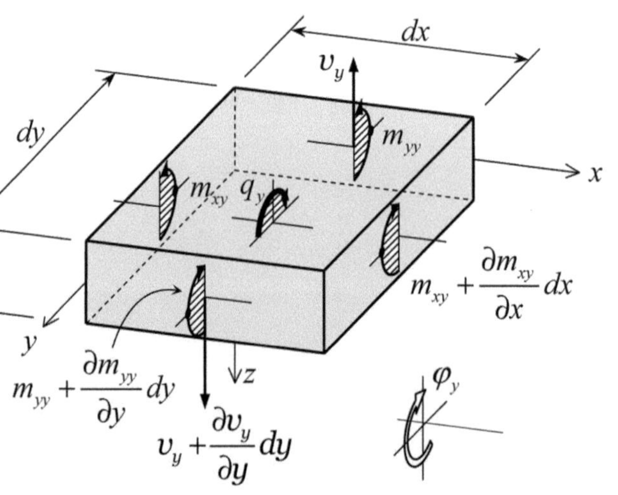

Figure 3.18 Plate equilibrium in φ_y-direction.

(3.3) if w and φ_x are constant in the y-direction and φ_y is zero. Then $\partial v_y/\partial_y$ is cancelled out in the first equation and $\partial m_{yx}/\partial y$ in the second one. Further we see from Eq. (3.38) that for $q_x = 0$ the shear force v_x is no longer the derivative of the bending moment m_{xx}, as we know for beams. An additional derivative in the lateral direction of the twisting moment m_{yx} appears.

3.4 Differential Equations for Thick Plates

As we did earlier for a beam, we substitute the kinematic relations (3.27) into the constitutive equations (3.35) and subsequently substitute the result in the equilibrium equations (3.38). This transforms the latter into three simultaneous partial differential equations expressed in terms of w, φ_x and φ_y

$$
\begin{aligned}
&-D_s\left(\frac{\partial^2}{\partial x^2}+\frac{\partial^2}{\partial y^2}\right)w - D_s\frac{\partial \varphi_x}{\partial x} - D_s\frac{\partial \varphi_y}{\partial y} = p\\
&D_s\frac{\partial w}{\partial x} + \left(D_s - D_b\frac{\partial^2}{\partial x^2} - \frac{1}{2}(1-\nu)\,D_b\frac{\partial^2}{\partial y^2}\right)\varphi_x - \frac{1}{2}(1+\nu)\,D_b\frac{\partial^2 \varphi_y}{\partial x\partial y} = q_x\\
&D_s\frac{\partial w}{\partial y} - \frac{1}{2}(1+\nu)\,D_b\frac{\partial^2}{\partial x\partial y}\varphi_x + \left(D_s - \frac{1}{2}(1-\nu)\,D_b\frac{\partial^2}{\partial x^2} - D_b\frac{\partial^2}{\partial y^2}\right)\varphi_y = q_y
\end{aligned} \tag{3.39}
$$

When written in matrix operator form, they are

$$
\begin{bmatrix}
-D_s\left(\frac{\partial^2}{\partial x^2}+\frac{\partial^2}{\partial y^2}\right) & -D_s\frac{\partial}{\partial x} & -D_s\frac{\partial}{\partial y}\\
D_s\frac{\partial}{\partial x} & D_s-D_b\left(\frac{\partial^2}{\partial x^2}+\frac{1}{2}(1-\nu)\frac{\partial^2}{\partial y^2}\right) & -\frac{1}{2}(1+\nu)D_b\frac{\partial^2}{\partial x\partial y}\\
D_s\frac{\partial}{\partial y} & -\frac{1}{2}(1+\nu)\,D_b\frac{\partial^2}{\partial x\partial y} & D_s-D_b\left(\frac{1}{2}(1-\nu)\frac{\partial^2}{\partial x^2}+\frac{\partial^2}{\partial y^2}\right)
\end{bmatrix}
\begin{Bmatrix} w\\ \varphi_x\\ \varphi_y \end{Bmatrix}
=
\begin{Bmatrix} p\\ q_x\\ q_y \end{Bmatrix} \tag{3.40}
$$

These three differential equations are valid for both thick and thin plates, and follow from the first four suppositions mentioned in Section 3.2.1.

Remark 1

If we set q_x and q_y to zero, the last two equations of (3.28) become

$$v_x = \frac{\partial m_{xx}}{\partial x} + \frac{\partial m_{yx}}{\partial y}, \quad v_y = \frac{\partial m_{yy}}{\partial y} + \frac{\partial m_{xy}}{\partial x} \tag{3.41}$$

These relations between shear forces and moments are a generalization of the relation in (3.15) for beams. Substitution of this result in the first equation of (3.38) leads to

$$-\left(\frac{\partial^2 m_{xx}}{\partial x^2} + 2\frac{\partial^2 m_{xy}}{\partial x \partial y} + \frac{\partial^2 m_{yy}}{\partial y^2}\right) = p \tag{3.42}$$

Making use of the kinematic equations (3.27) and the constitutive equations (3.32) for bending, we can transform Eq. (3.42) into

$$-\nabla^2 m = p \tag{3.43}$$

where ∇^2 is the Laplace operator and m is the weighted sum with respect to Poisson's ratio of the two bending moments

$$m = \frac{m_{xx} + m_{yy}}{1 + \nu} \tag{3.44}$$

This sum of moments is an invariant under rotation of the x–y coordinate system about its origin. Equation (3.43) is the plate generalization of Eq. (3.16) for beams.

We can derive another interesting equation. From the kinematic equations in (3.27) and the constitutive relations in (3.32) we know $m = D_b(\partial\varphi/\partial x + \partial\varphi_y/\partial y)$. With this result, the first equation in (3.39) becomes

$$-D_s\left(\nabla^2 w + \frac{m}{D_b}\right) = p \tag{3.45}$$

We conclude that, if only a load p occurs, we can formulate the problem of a plate (thick or thin) with the two differential equations (3.43) and (3.45). Written in this form we see the correspondence with the beam theory, for which we derived Eqs. (3.6) and (3.8).

Spiritual fathers of thick plate theory

The theory for thick plates was derived independently by E. Reissner (1945) [5] and R.D. Mindlin (1951) [6], with small differences in their theories. We have described Mindlin's theory. FE codes refer to the theory of thick plates as Mindlin theory; we do the same in this book. Reissner did not start from an assumed displacement field but from an assumed stress field, and additionally he took into account the influence of vertical stresses σ_{zz} not equal to zero.

For zero Poisson's ratio, the Reissner and Mindlin theories are the same. For $\nu = 0.3$ and a span to thickness ratio 5 the values of maximum bending moment may differ by 10% and twisting moment by 20% [7].

3.5 Orthotropic Plate

In many cases in structural engineering, particularly in bridge engineering, the slabs are not homogeneous and isotropic, for instance a viaduct cross-section, which is built up of I- or T-sections on top of which a thin concrete deck layer is cast in-situ, see Figure 3.15. Such shape-orthotropic plates can be handled as plates of homogeneous thickness with orthotropic rigidities. The kinematic and equilibrium relationships do not change, but the constitutive equations do change from the homogeneous isotropic case. Now five different flexural rigidities must be determined, being D_{xx}, D_{yy}, D_{υ}, D_{xy} and D_{yx}.

$$\begin{aligned} m_{xx} &= D_{xx}\kappa_{xx} + D_{\nu}\kappa_{yy} \\ m_{yy} &= D_{yy}\kappa_{yy} + D_{\nu}\kappa_{xx} \\ m_{xy} &= D_{xy}\rho_{xy} \\ m_{yx} &= D_{yx}\rho_{xy} \end{aligned} \tag{3.46}$$

It requires experience in structural mechanics to properly estimate the off-diagonal term D_{ν}. In many cases one of the rigidities D_{xx} and D_{yy} is small compared to the other, and then a simple relation exists. If we suppose that D_{yy} is smaller than D_{xx}, we can write

$$D_{\nu} = \nu D_{yy} \tag{3.47}$$

The torsional deformation ρ_{xy} is common for the x- and y-direction, but the moments m_{xy} and m_{yx} are different now. From a virtual work consideration we conclude that it is convenient to work with the average m_{av} of the moments m_{xy} and m_{yx} and with the average rigidity D_{av} of D_{xy} and D_{yx}.

$$\begin{aligned} m_{\text{av}} &= \frac{1}{2}(m_{xy} + m_{yx}) \\ D_{\text{av}} &= \frac{1}{2}(D_{xy} + D_{yx}) \end{aligned} \tag{3.48}$$

We just have to replace Eq. (3.32) by

$$\begin{Bmatrix} m_{xx} \\ m_{yy} \\ m_{\text{av}} \end{Bmatrix} = \begin{bmatrix} D_{xx} & D_v & 0 \\ D_v & D_{yy} & 0 \\ 0 & 0 & D_{\text{av}} \end{bmatrix} \begin{Bmatrix} \kappa_{xx} \\ \kappa_{yy} \\ \rho_{xy} \end{Bmatrix} \tag{3.49}$$

After we have solved the plate problem and therefore found m_{av}, we can calculate the value of the two unequal twisting moments m_{xy} and m_{yx} from

$$m_{xy} = \frac{2D_{xy}}{D_{xy} + D_{yx}} m_{\text{av}}, \quad m_{yx} = \frac{2D_{yx}}{D_{xy} + D_{yx}} m_{\text{av}} \tag{3.50}$$

For the transverse shear forces a small change is necessary. Whereas for isotropic plate material the same shear rigidity D_s can be used in both x- and y-direction, we now must distinguish between D_{sx} and D_{sy}.

$$v_x = D_{sx}\, \gamma_x, \quad v_y = D_{sy}\, \gamma_y \tag{3.51}$$

3.6 Twisted Plate Strip

Following Reissner in his seminal 1945 paper [5], we study the stress distribution in a plate strip of length $2l$ and width a as shown in Figure 3.19. The length l is large compared to a. The thickness of the strip need not be small, however is not expected to exceed $a/2$. The strip is in a state of pure torsion of unrestrained warping. A set of axes x, y, z is chosen, with x in length direction of the strip and y and z in the cross-section, see Figure 3.19. Note the position of the origin of the axes in the very centre of the strip. Hereafter we call the y-direction horizontal and the z-direction vertical. The long strip is twisted about the x-axis by torques M_t. The exact solution for this torsion problem is known from theory of elasticity, thanks to early work of St. Venant [8]. Characteristics of this classic solution are as follows:

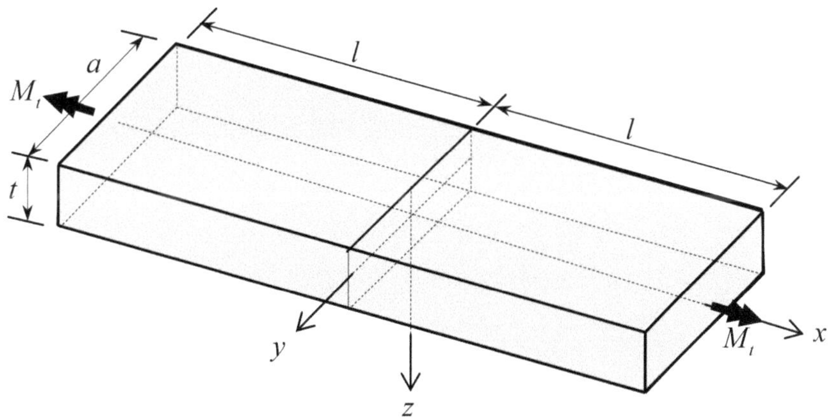

Figure 3.19 Thick plate strip due to torque load.

- The rectangular cross-section before loading remains rectangular in the rotated state after applying torques at both ends.
- Warping of the cross-section occurs, and is the same in all individual cross-sections, independent of x.
- Consequently, all three normal stresses σ_{xx}, σ_{yy} and σ_{zz} are zero, and so is the shear stress σ_{yz}. Only shear stresses σ_{xy} and σ_{xz} in the strip cross-section are present.
- In a cross-section, the horizontal shear stresses σ_{xy} account for one half of the twisting moment, and the vertical shear stresses σ_{xz} for the other half.

In this section we will determine the stress state in the strip by considering it as a thick plate. The plate has dimensions $2l$ and a in the x–y plane, and thickness t in z-direction. Distributed loads p, q_x and q_y are zero. The load consists of torques M_t at the two strip ends. Throughout the length $2l$, the distribution of the shear stresses is equal in all cross-sections. We will interpret the consequences of the classical assumptions for a thick plate. The zero value of normal stress σ_{zz} in the thickness direction is in agreement with the assumption in plate theory. The absence of normal stresses σ_{xx} and σ_{yy} implies that there are no bending moments m_{xx} and m_{yy}. The zero shear stress σ_{yz} means that the shear force v_y is zero all over the plate. The two non-zero shear stresses σ_{xy} and σ_{xz} represent the twisting moment m_{xy} and shear force v_x in the cross-section normal to the x-axis, respectively. Therefore, the distribution of the shear stream in the classical theory of elasticity is replaced by a distribution of twisting moments m_{xy} and shear forces v_x. This is depicted in Figure 3.20. The twisting moment is shown for the left edge zone of the strip, and the shear force for the right edge zone.

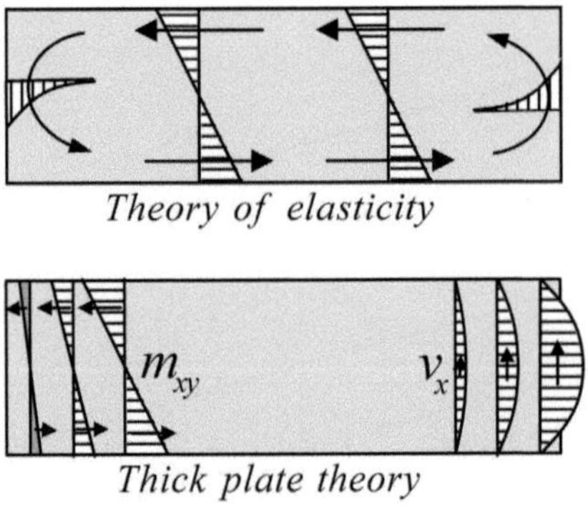

Figure 3.20 Stress resultants in strip due to torque load. Distribution of twisting moment shown left and for shear force right.

The twisting moment must decrease in the direction of the strip edge and the shear force increase. The distribution of the horizontal shear stress of the twisting moment is linear over the thickness, and that of the transverse shear force parabolic.

At the plate edge $x = -l$ a rotation of the cross-section is imposed, such that the deflection is $w = -\theta yl$. Here θ is the *twist*, the rotation per unit length. In the plate edge $x = l$ the rotation is in the opposite direction. The deflection at this end is $w = \theta yl$. Linear interpolation in the x-direction between these two ends leads to the general expression for the deflection $w = \theta xy$. Because the cross-section keeps its shape in the y–z plane, and rotates about the x-axis as a rigid body, the curvature κ_{yy} is zero. Therefore, the rotation φ_y is independent of y and depends on x only. The deflection w and rotation φ_y are coupled due to the rigid body rotation of the cross-section. The zero moment m_{xx} implies a zero curvature κ_{xx}, so a rotation φ_x independent of x. Summing up, we choose the displacement field

$$w(x, y) = \theta xy, \quad \varphi_x(y), \quad \varphi_y = -\theta x \tag{3.52}$$

In fact, only φ_x is unknown. In general, we must satisfy three boundary conditions at each edge in thick plate theory. At the edges $y = \pm\frac{1}{2}a$ the conditions are $m_{yy} = 0$, $m_{yx} = 0$, $v_y = 0$. Because the moment m_{yy} and shear force v_y are zero all over the plate, it remains to require that the twisting moment is zero at the edges $y = \pm a/2$

$$m_{yx} = 0 \tag{3.53}$$

At the edges $x = \pm l$ we must in general consider the edge tractions m_{xx}, m_{xy} and v_x. The moment m_{xx} is zero all over the plate, and the twisting moment m_{xy} and shear force v_x are applied loads. Together they deliver the torque M_t. The distribution of m_{xy} and v_x and the value of the resulting torque are aims of the computation.

Substituting the trial solution (3.52) into the three governing differential equations of (3.39), we find that the first and third equation are satisfied identically. We need not consider them any more. The second equation simplifies to

$$\lambda^2 \frac{d^2 \varphi_x}{dy^2} - \varphi_x = \theta y \tag{3.54}$$

where λ is a *characteristic length*, defined by

$$\lambda^2 = \frac{1-\nu}{2} \frac{D_b}{D_s} \tag{3.55}$$

Accounting for $D_b = Et^3/\{12(1-v^2)\}$ and $D_s = (5/6)Gt$, the expression for λ simplifies to

$$\lambda = t/\sqrt{10} \tag{3.56}$$

Note that the characteristic length λ is of the order of the plate thickness, in fact about one third. The solution of Eq. (3.54) is

$$\varphi_x(y) = -y\theta + C_1 e^{-y/\lambda} + C_2 e^{y/\lambda} \tag{3.57}$$

Without loss of generality, transferring to another constant and taking account of the knowledge that the stress state can be mirrored with respect to the centre of the cross-section, we write this solution alternatively

$$\varphi_x(y) = -y\theta - Ce^{-(a/2+y)/\lambda} + Ce^{-(a/2-y)/\lambda} \tag{3.58}$$

The advantage is that the first exponential term occurs in the edge zone $y = -a/2$ and attenuates from there, and the other in the opposite edge zone $y = a/2$. They start with the value 1 at the edge. Due to the assumption $a/t > 2$, both exponential functions will have completely vanished at a distance a, therefore before they reach the opposite plate edge. This is because the exponent of e is then $-2\sqrt{10}$, which leads to a value less than 0.002. From boundary condition (3.53) it follows, accounting for Eqs. (3.52) and (3.58)

$$\rho_{xy} = 0 \rightarrow \frac{\partial \varphi_x}{\partial y} + \frac{\partial \varphi_y}{\partial x} = 0 \rightarrow \frac{1}{\lambda} C \left[e^{-(a/2+y)/\lambda} + e^{-(a/2-y)/\lambda} \right] = 2\theta \tag{3.59}$$

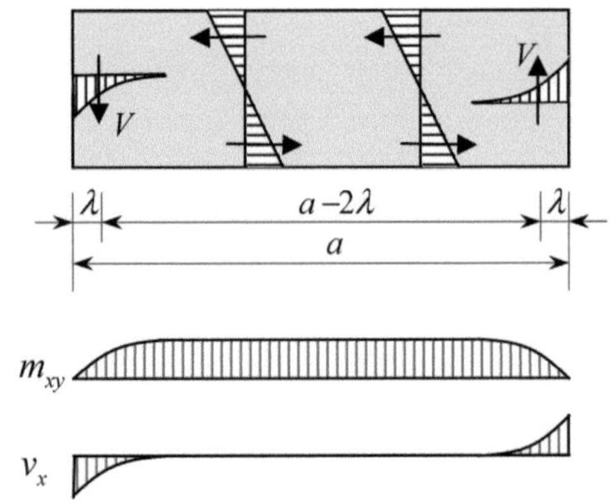

Figure 3.21 Distribution of twisting moment and shear force.

Note that one exponential term is zero for $y = -a/2$ and the other for $y = a/2$. Substitution of either edge $y = -a/2$ or edge $y = a/2$ leads to the solution $C = 2\lambda\theta$. The final shape of Eq. (3.58) becomes

$$\varphi_x(y) = \left(-y - 2\lambda e^{-(a/2+y)/\lambda} + 2\lambda e^{-(a/2-y)/\lambda}\right)\theta \tag{3.60}$$

Now we are able to calculate the twisting moment m_{xy} and shear force v_x in the cross-section. The formulas are, with $\eta = 6/5$, and accounting for Eqs. (3.52) and (3.60)

$$\begin{aligned} m_{xy} &= \frac{Et^3}{24(1+\nu)}\left(\frac{\partial\varphi_x}{\partial y} + \frac{\partial\varphi_y}{\partial x}\right) \\ &= \frac{-Et^3}{12(1+\nu)}\left(1 - e^{-(a/2+y)/\lambda} - e^{-(a/2-y)/\lambda}\right)\theta \end{aligned} \tag{3.61}$$

$$\begin{aligned} v_x &= \frac{Et\lambda}{2(1+\nu)\eta}\left(\varphi_x + \frac{\partial w}{\partial x}\right) \\ &= \frac{5Et\lambda}{6(1+\nu)}\left(-e^{-(a/2+y)/\lambda} + e^{-(a/2-y)/\lambda}\right)\theta \end{aligned} \tag{3.62}$$

The distribution of these functions is shown in Figure 3.21. The twisting moment is practically constant over the full width a of the strip. In the edge zones the moment decreases to zero at the edge. The value is negative, which leads to the direction of the shear stress as shown in the figure. The shear force is negative at the edge $y = -a/2$ (right in the figure) and positive at the edge $y = a/2$ (left). Together a flowing-around stream of shear stresses occurs. The exponential functions reduce over short distance to a negligible value. At the edge their value is 1.00, at distance $t/2$ from the edge 0.21, and at a distance t just 0.04 remains. This is still true when a is large compared to t. Disturbances occur only at the edges over a distance of the order of the plate thickness.

We determine the twisting moment in the strip outside the edge zone and the resultant of the shear forces in the edge zone. The twisting moment follows from Eq. (3.61) accounting for Eq. (3.56)

$$m_{xy} = \frac{-Et^3}{12(1+\nu)}\theta \tag{3.63}$$

The integral over the plate width a of the exponential functions which occur in the expressions for m_{xy} and v_x is λ. Then the resultant V of the shear forces v_x in each edge zone follows from Eq. (3.62)

$$V = \int v_x dx = \left(\frac{5Et\lambda}{6\,(1+\nu)}\theta\right) \times \lambda = \frac{Et^3}{12\,(1+\nu)}\theta \tag{3.64}$$

We find that the twisting moment m_{xy} and the resulting shear force V are equal. In Chapter 4 this will be derived again in another way.

We also can compute the ratio of the maximum vertical shear stress at the edge and the horizontal shear stress due to the twisting moment outside the disturbed edge zone. For the calculation of the vertical shear stress we assume a parabolic distribution over the thickness of the plate. So we must divide the vertical shear force at the edge by $2t/3$. For the horizontal shear stress we divide the twisting moment by the section modulus $t^2/6$. The result is

$$\frac{\sigma_{\text{shear force}}}{\sigma_{\text{twisting moment}}} = \frac{5}{2\sqrt{10}} = 0.79 \tag{3.65}$$

The maximum vertical shear stress is 79% of the horizontal shear stress by the twisting moment. The result is independent of materials properties.

Now we compute the torque M_t in the cross-section from the distributions in Eqs. (3.61) and (3.62). We write $M_t = M_m + M_v$, where M_m is the contribution by the twisting moments m_{xy} and M_v by the shear forces v_x. The integral of the twisting moment in Eq. (3.61) is quickly written

$$M_m = \left| \int_{-a/2}^{a/2} m_{xy}\, dy \right| = \frac{Et^3}{12(1+\nu)}(a - 2\lambda) \tag{3.66}$$

Table 3.1 Comparison of torsional moments of inertia.

	I_t/at^3	
a/t	Thick plate theory	Theory of elasticity
2	0.228	0.229
4	0.281	0.281
10	0.312	0.312
∞	0.333	0.333

The contribution of the shear force v_x is calculated in the following way. Apart from the integral of the exponential functions (value λ) we also need the position of the resultant V. Elementary mathematics shows that the distance from the edge also is λ. Therefore, the lever arm between the two resultants is $a - 2\lambda$, and, accounting for Eq. (3.64), we obtain

$$M_v = \frac{Et^3}{12(1+\nu)}(a - 2\lambda)\theta \tag{3.67}$$

We find that M_m and M_v are equal, as in theory of elasticity. The final result for the torque M_t is the sum of both

$$M_t = M_m + M_v = \frac{Et^3}{6(1+\nu)}(a - 2\lambda)\theta \tag{3.68}$$

We may rewrite Eq. (3.68) as

$$M_t = GI_t\theta, \quad I_t = \frac{1}{3}t^3(a - 2\lambda) \tag{3.69}$$

In this shape a comparison can be made with results of Timoshenko and Goodier [7] on the basis of theory of elasticity. In Table 3.1 the ratio I_t/a_t^3 is given for different values of a/t. The match is striking.

Reissner obtained the same result on the basis of his approach using assumed stress fields. In this case there is no difference between the approaches of Reissner and Mindlin. Differences only occur if vertical stresses σ_{zz} are present, which is not the case in the twisted plate strip.

Lesson of this example

We restricted ourselves in the present example to a plate strip in a state of pure torsion. In other plates subjected to different loads, twisting moments may always occur along free edges. Then we will see the same phenomenon as in the twisted plate.

The twisting moment at free edges must be zero, and there are transverse shear forces in cross-sections normal to the free edge. These shear forces occur in an edge zone with a width of the order of the thickness; the twisting moment has to decrease to zero in this small edge zone. Similar phenomena occur on simply-supported edges.

The resultant of the shear forces in the edge zone is equal to the twisting moment outside the edge zone.

The vertical shear stress at the edge is about 80% of the horizontal shear stress outside the disturbed edge.

3.7 Message of the Chapter

- In thick plates we use three independent degrees of freedom, the vertical displacement and two rotations. An appropriate definition of rotations is chosen, different from what is done in the Finite Element Method. Three independent distributed external loads are associated with the degrees of freedom, one vertical load and two torque loads. The latter are usually zero in applications.

- We distinguish five deformations in thick plates, three plate curvatures and two transverse shear angles. Three plate moments and two shear forces are associated with these deformations.

- The needle hypothesis is adopted, which is a generalization of the supposition in beam theory that plane sections remain plain. In the unloaded state the straight needle is perpendicular to the middle plane of the plate. After application of the load the needle still is straight, but, due to shear deformation it can tilt in both the x-and the y-direction compared to the middle plane. This means that we will work with the straight needle line which is 'the best fit' of the distorted cross-section.

- The plate can be considered as a composition of layers as far as bending and torsion is regarded. Each layer is in a state of plane stress, which leads to a flexural rigidity matrix which is similar to the membrane plate apart from a different stiffness factor.

- We end up with three simultaneous differential equations, each of the second-order. If two degrees of freedom are eliminated a sixth-order differential equation will be found. This means that three boundary conditions must be specified at each edge, which implies, that the vertical displacement, the rotation normal to the edge face and the rotation in the edge face can be prescribed. If one or more of these displacements is unconstrained, the associated distributed edge loads in those directions can be prescribed.

- If the plate has a free edge and a twisting moment is present outside the edge zone, the twisting moment is zero at the free edges and local vertical shear forces occur in the edge zone over a distance of the order of the plate thickness. This is shown on the basis of a twisted plate strip.

- The resulting transverse shear force in the edge zone is equal to the twisting moment outside the edge zone.

- The maximum vertical shear stress at the edge is about 80% of the maximum horizontal shear stress outside the disturbed edge zone.

- FE codes refer to the theory of thick plates as Mindlin theory, though Reissner was the first to publish on the subject.

Chapter 4
Thin Plates in Bending

Classical text books on thin plate bending theory are due to Girkmann [2] and Timoshenko and Gere [9]. Szilard [10] and Reddy [11] are authors of more recent books on the subject. We will not copy their derivation of theory rather we follow a different approach. We make thick plate theory of Chapter 3 our starting point, and derive thin plate theory from it as a limit case. Apart of the different approach, also the aim of this book is different. Plate theory need to be covered only as far as necessary to explain traps and surprises in applications of the Finite Element Method.

4.1 Theory for Thin Plates

As was done earlier for beams, we assume that shear deformation is negligibly small, which allows us to state $\gamma_x = 0$ and $\gamma_y = 0$. Then, the kinematic relationships in Eq. (3.27) imply rotations φ_x and φ_y dependent on the displacement w

$$\varphi_x = -\frac{\partial w}{\partial x}; \quad \varphi_y = -\frac{\partial w}{\partial y} \tag{4.1}$$

which reduce and transform the kinematic relations into three new ones

J. Blaauwendraad, *Plates and FEM: Surprises and Pitfalls*, Solid Mechanics and Its Applications 171, DOI 10.1007/978-90-481-3596-7_4,

$$\left.\begin{aligned} \kappa_{xx} &= -\frac{\partial^2 w}{\partial x^2} \\ \kappa_{yy} &= -\frac{\partial^2 w}{\partial y^2} \\ \rho_{xy} &= -2\frac{\partial^2 w}{\partial x \partial y} \end{aligned}\right\} \quad \textit{Kinematic} \tag{4.2}$$

The shear deformation can be neglected if the span to thickness ratio is larger than 5, which is always the case in normal slabs.

The constitutive relations for the shear forces are meaningless now, and are cancelled out. Therefore, it is no longer necessary to distinguish between D_b and D_s. Only D_b has a meaning, and we can omit the subscript b.

$$\left.\begin{aligned} m_{xx} &= D\left(\kappa_{xx} + \nu\kappa_{yy}\right) \\ m_{yy} &= D\left(\nu\kappa_{xx} + \kappa_{yy}\right) \\ m_{xy} &= \tfrac{1}{2}D\left(1-\nu\right)\rho_{xy} \end{aligned}\right\} \quad \textit{Constitutive} \tag{4.3}$$

We now consider the equilibrium equations (3.38). The rotations φ_x and φ_y are no longer independent degrees of freedom, but depend on the displacement w. In this case the loads q_x and q_y cannot be applied. They must be set to be zero, which transforms the last two equilibrium equations in (3.38), as we have already seen in Eq. (3.41), into

$$v_x = \frac{\partial m_{xx}}{\partial x} + \frac{\partial m_{yx}}{\partial y}, \qquad v_y = \frac{\partial m_{yy}}{\partial y} + \frac{\partial m_{xy}}{\partial x} \tag{4.4}$$

Substitution of this result in the first relationship of Eq. (3.38) leads to

$$-\left(\frac{\partial^2 m_{xx}}{\partial x^2} + 2\frac{\partial^2 m_{xy}}{\partial x \partial y} + \frac{\partial^2 m_{yy}}{\partial y^2}\right) = p \quad \textit{Equilibrium} \tag{4.5}$$

as was found earlier in Eq. (3.42) for thick plates with no torque loads q_x and q_y. The relationships Eqs. (4.2), (4.3) and (4.5) govern the behaviour of thin plates. They are the generalization for thin plates of the three relationships (3.13), (3.14) and (3.16) for beams. So, the general relation scheme for thick plates in Figure 3.9 has been simplified to Figure 4.1. Substitution of Eq. (4.2) into Eq. (4.3) gives

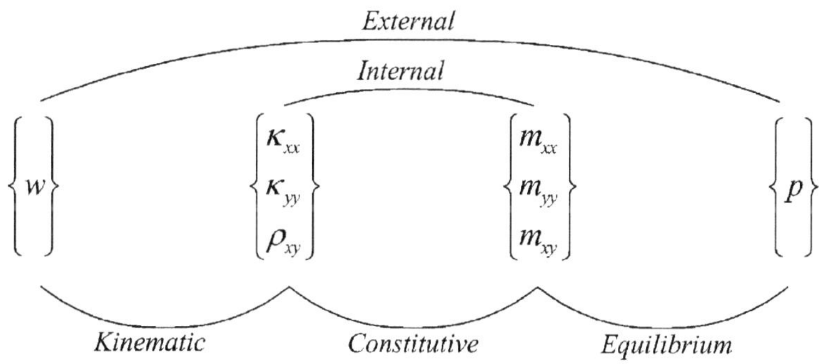

Figure 4.1 Scheme for thin plates (bending only).

$$
\begin{aligned}
m_{xx} &= -D\left(\frac{\partial^2 w}{\partial x^2} + \nu\frac{\partial^2 w}{\partial y^2}\right)\\
m_{yy} &= -D\left(\nu\frac{\partial^2 w}{\partial x^2} + \frac{\partial^2 w}{\partial y^2}\right)\\
m_{xy} &= -\left(1-\nu\right)D\frac{\partial^2 w}{\partial x \partial y}
\end{aligned}
\tag{4.6}
$$

and Eq. (4.7) into Eq. (4.5) delivers a partial differential equation in w only

$$
D\left(\frac{\partial^4}{\partial x^4} + 2\frac{\partial^4}{\partial x^2 \partial y^2} + \frac{\partial^4}{\partial y^4}\right) w - p \tag{4.7}
$$

We have seen the operator in this equation earlier in Chapter 1. With the Laplace operator as defined in Eq. (1.20) we can write

$$
D\,\nabla^2\nabla^2 w = p \tag{4.8}
$$

Again a bi-harmonic equation is obtained. Lagrange (1811) was the first to derive a differential equation for bending of thin plates in this form, be it still with a wrong perception about D.

Correspondence with classical beam theory

The correspondence between the differential equation Eq. (4.7) for plates and the differential equation $EI\,d^4w/dx^4 = p$ for beams is evident. The first term of equation Eq. (4.7) relates to the load bearing capacity in the x-direction, the last term relates to the capacity of the y-direction. The middle term is new, and describes the load bearing capacity due to torsion.

This fourth-order differential equation must be solved subject to the governing boundary conditions (to be discussed later). Once a solution for $w(x, y)$ is found, the bending and twisting moments are calculated with the aid of Eq. (4.7) and finally the shear forces from Eq. (4.4) on the basis of the solution for the moments.

In Eq. (4.7), the moments are expressed in terms of the displacement w. Then this is also possible for the shear forces. If we introduce Eq. (4.7) into Eq. (4.4) we obtain

$$v_x = -D\frac{\partial}{\partial x}\nabla^2 w, \quad v_y = -D\frac{\partial}{\partial y}\nabla^2 w \tag{4.9}$$

The expressions in Eq. (4.9) are plate generalizations of the shear force in a beam as shown in the second relationship of Eq. (3.17).

Remark 1

The expressions in Eq. (4.9) for the shear forces offer us an alternative for the derivation of the bi-harmonic plate equation. Substitution in the first relationship of Eq. (3.38) leads again to the bi-harmonic equation.

$$-\left(\frac{\partial v_x}{\partial x} + \frac{\partial v_y}{\partial y}\right) = p \;\rightarrow\; D\left(\frac{\partial^2}{\partial x^2} + \frac{\partial^2}{\partial y^2}\right)\nabla^2 w = p \;\rightarrow\; D\nabla^2\nabla^2 w = p \tag{4.10}$$

Remark 2

We can show the correspondence to the beam formulas. For beams, we have

$$-\frac{d^2M}{dx^2} = p \tag{4.11}$$

$$M = -EI\frac{d^2w}{dx^2} \tag{4.12}$$

We define again m as the weighted sum of the two bending moments in the plate with respect to Poisson's ratio

$$m = \frac{m_{xx} + m_{yy}}{1+\nu} \tag{4.13}$$

We have seen in Section 2.3 that the sum of the normal stresses in plane stress is invariant under rotation of axes. Likewise, the sum of the bending

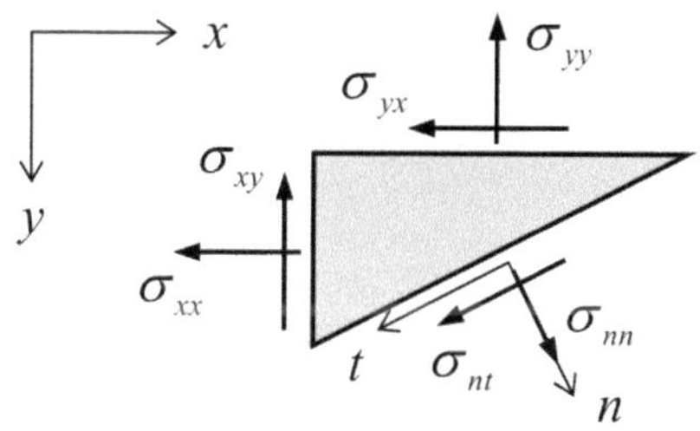

Figure 4.2 Stress transformation.

moments is invariant, and therefore m. We then find that the bi-harmonic equation can be decomposed into

$$-\nabla^2 m = p, \quad m = -D\nabla^2 w \tag{4.14}$$

These two equations show the same structure as Eqs. (4.11) and (4.12) do for beams. The first relationship in (4.14) was obtained earlier for thick plates in Eq. (3.43). The second relationship has a companion in Eq. (3.45) for thick plates. If that equation is divided by D_s, it becomes for infinitely large D_s equal to the second relationship in Eq. (4.14).

4.2 Transformation Rules and Principal Moments

For bending and twisting moments the same transformation rules apply for axes rotation as for the stresses in a plane stress state (see Figure 4.2). For new axes n, t of which the n-axis has an angle α with the x-axis, the following is true in plane stress

$$\begin{aligned}
\sigma_{nn} &= \sigma_{xx}\cos^2\alpha + \sigma_{xy}\sin 2\alpha + \sigma_{yy}\sin^2\alpha \\
\sigma_{tt} &= \sigma_{xx}\sin^2\alpha - \sigma_{xy}\sin 2\alpha + \sigma_{yy}\cos^2\alpha \\
\sigma_{nt} &= -\frac{1}{2}(\sigma_{xx} - \sigma_{yy})\sin 2\alpha + \sigma_{xy}\cos 2\alpha
\end{aligned} \tag{4.15}$$

By integrating the corresponding expressions for the bending and twisting moments over thickness we obtain

$$
\begin{aligned}
m_{nn} &= m_{xx}\cos^2\alpha + m_{xy}\sin 2\alpha + m_{yy}\sin^2\alpha \\
m_{tt} &= m_{xx}\sin^2\alpha - m_{xy}\sin 2\alpha + m_{yy}\cos^2\alpha \\
m_{nt} &= -\frac{1}{2}\left(m_{xx} - m_{yy}\right)\sin 2\alpha + m_{xy}\cos 2\alpha
\end{aligned}
\tag{4.16}
$$

The directions of n and t, for which m_{nt} is zero and m_{nn} and m_{tt} adopt extreme values, are *principal directions*. These moments m_{nn} and m_{tt} are the *principal moments*. The direction of the principle moments is determined by

$$
\tan 2\alpha_o = \frac{2m_{xy}}{m_{xx} - m_{yy}} \tag{4.17}
$$

Instead of formulae (4.17) and (4.17) one may also apply the graphical approach using Mohr's circle. From Eq. (4.17) we find again the invariance of the moment sum under rotation of axes.

4.3 Principal Shear Force

We can write the shear forces of Eq. (4.9) for constant D as

$$
v_x = \frac{\partial}{\partial x}(-D\nabla^2 w), \quad v_y = \frac{\partial}{\partial y}(-D\nabla^2 w) \tag{4.18}
$$

and subsequently with reference to the second part of Eq. (4.14) as

$$
v_x = \frac{\partial m}{\partial x}, \quad v_y = \frac{\partial m}{\partial y} \tag{4.19}
$$

The shear force is the derivative of m. Written in this way, we see even better that the expressions for the shear forces are generalizations of the formula for beams. In order to find v_x, the derivative in x-direction has to be determined and for v_y the derivative in y-direction. In a random direction r the shear force is

$$
v_r = \frac{\partial m}{\partial r} \tag{4.20}
$$

In Figure 4.3 we have sketched iso-lines of the function $m(x,y)$. In the direction t of the iso-lines the value of m does not change, therefore the shear force v_t is equal to zero and in the direction of the slope, perpendicular to the iso-lines in n-direction, the shear force v_n obtains its maximal value. This is called the *maximal shear force*. The trajectories of v_n are orthogonal to the iso-lines of m.

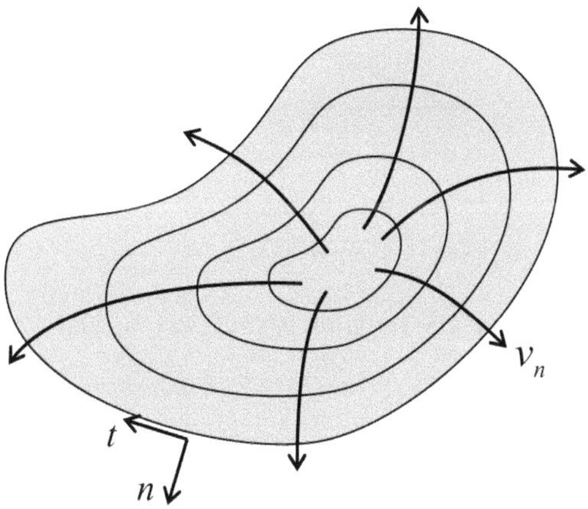

Figure 4.3 Shower analogy. Water flows in the direction of the deepest slope.

W.J. Beranek, former professor of structural mechanics in the architectural department of Delft University of Technology, introduced the *shower analogy* to illustrate the phenomenon of the direction of the load discharge. He considered the function m as 'hill' above the plate surface, and the load p as a rain shower. Then the trajectories of v_n may be compared to the streamlines of the flowing water. All shower water between two trajectories flows parallel to the trajectories and does not pass trajectories (flow lines).

We can compute the value of the maximal shear force and the direction of the trajectories from the values v_x and v_y. Figure 4.4 shows two triangular plate parts with the shear forces acting on their edges. We see two orthogonal sets of axes, the x–y set and the n–t set. The angle between the two

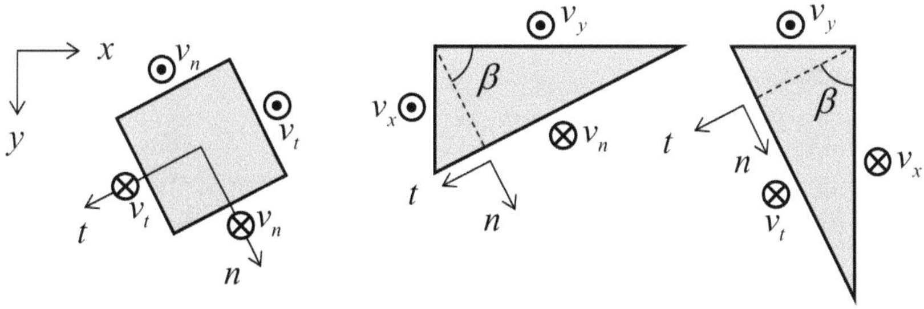

Figure 4.4 Equilibrium of plate parts.

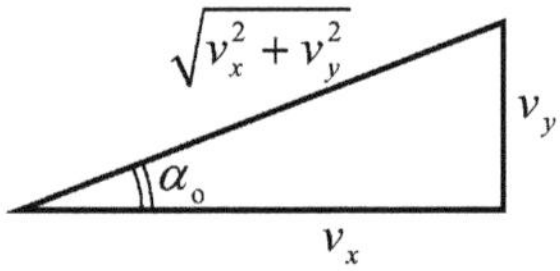

$$\tan\alpha_o = \frac{v_y}{v_x}, \qquad \sin\alpha_o = \frac{v_y}{\sqrt{v_x^2+v_y^2}}, \qquad \cos\alpha_o = \frac{v_x}{\sqrt{v_x^2+v_y^2}}$$

Figure 4.5 Relation between α_0, v_x and v_y.

sets is β. In both sets the z-axis is downwards, so the direction of the shear forces is perpendicular to the plane of drawing. Shear forces are shown as arrows. We can see the top of the arrow (a point) or the back (a cross).An arrow coming towards the reader (the point is seen) indicates shear forces on faces with a negative normal vector; an arrow moving away (the cross is seen) indicates shear forces on faces with a positive normal vector. From the vertical equilibrium of the triangular parts it follows that

$$v_n = v_x \cos\beta + v_y \sin\beta \tag{4.21}$$

$$v_t = -v_x \sin\beta, +v_y \cos\beta \tag{4.22}$$

To determine the value and direction of the maximal shear force v_n we require that $\partial v_n/\partial\beta = 0$. This leads to the condition $-v_x \sin\beta + v_y \cos\beta = 0$. On basis of this condition we conclude two things. First, comparison of this condition with the formula for v_t in Eq. (4.22) shows that v_t is zero when v_n is maximal. Second, we can calculate the direction angle β_o of the trajectory by the formula

$$\tan\beta_o = \frac{v_y}{v_x} \tag{4.23}$$

Next we can compute the value of the maximal shear force v_o from Eq. (4.21). The sine and cosine follow from Figure 4.5. Therefore the maximal shear force becomes

$$v_o = \sqrt{v_x^2 + v_y^2} \tag{4.24}$$

Depending on the boundary conditions, load may flow to the edges at an angle to the edge. This angle need not necessarily be a right angle.

Note the differences

We stress two differences between the principal moments and principal shear forces.

First, the maximum principal moment is in general accompanied by a minimum non-zero principal moment. The maximal shear force is always accompanied by a zero minimum shear force.

Second, the direction β_0 of the maximal shear force is in general different from the direction α_0 of the principal moments. Therefore, trajectory plots for shear forces and moments are different.

4.4 Boundary Conditions for Thin Plates

The bi-harmonic equation contains fourth-order derivatives of the deflection w. For a rectangular plate in both directions eight boundary conditions can be specified, two per edge. Along each edge of a plate we can specify two out of four quantities: w, φ, f and t. The quantity f is a given distributed line load in the w-direction and t a given distributed line torque in the φ-direction. If we prescribe w, we cannot specify f simultaneously. If we specify w, then f will be the computed support reaction. If we specify f on a free edge, then w will be the computed displacement of the edge. Similarly t is a support reaction if φ is specified, and φ is the computed rotation if t is prescribed.

Note on sign convention

Note that the sign convention for f and t differs from the sign convention for shear forces and bending moments. They are loads associated with the displacement w and rotation φ, respectively, and their sign convention is the same as for these degrees of freedom.

4.4.1 Clamped Edge

We consider the plate in Figure 4.6, which is totally restrained along the edge $x = a$. The following kinematic boundary conditions hold for that

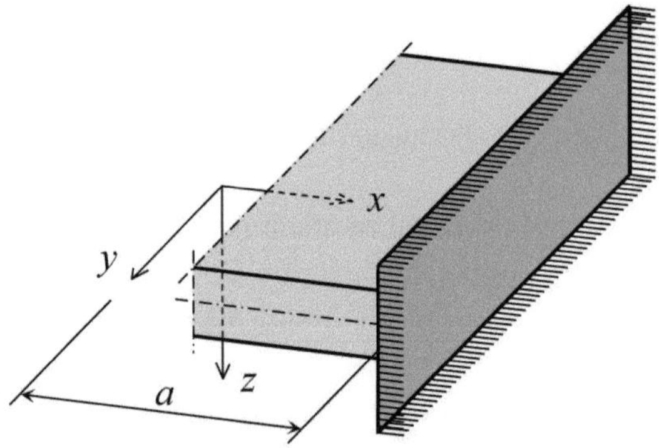

Figure 4.6 Clamped edge.

edge: $w = 0$, $\varphi_x = 0$. Because of $\varphi = -\partial w/\partial x$, we get

$$x = a \ \rightarrow \ \begin{cases} w = 0 \\ \dfrac{\partial w}{\partial x} = 0 \end{cases} \tag{4.25}$$

These are both kinematic conditions. Similarly, at a restrained edge $y = b$,

$$y = b \ \rightarrow \ \begin{cases} w = 0 \\ \dfrac{\partial w}{\partial y} = 0 \end{cases} \tag{4.26}$$

We will also consider the support reactions along the edge $x = a$ that occur at the kinematic boundary conditions. Figure 4.7 shows a small part of the plate and its equilibrium. We generally recommend for formulating boundary or transitional conditions to consider a small plate part, to consider its equilibrium, and to have the width of the plate part reduce to zero.

Zero twisting moment at clamped edge

At the clamped edge, the twisting moment is zero because $\partial w/\partial x$ is zero for every value of y along the edge. Then also the derivative of $\partial w/\partial x$ in y-direction is zero, which is $\partial^2 w/\partial x \partial y$. Therefore the torsion deformation ρ_{xy} is zero, and the moment m_{xy} cannot occur at a clamped edge.

Only a bending moment and a shear force are transmitted to a clamped support. In Figure 4.7 such an edge is depicted parallel to the y-axis, such that

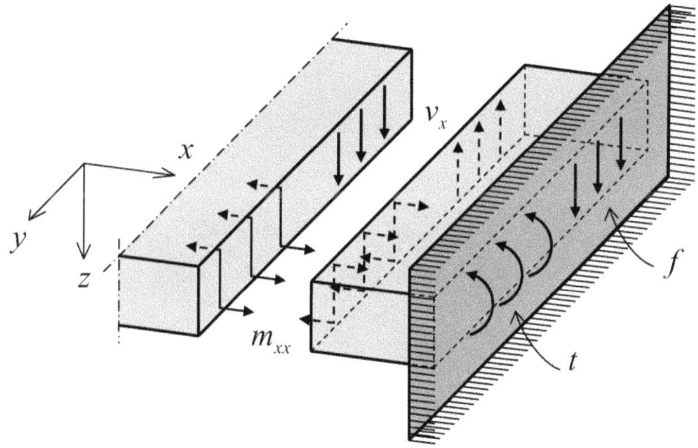

Figure 4.7 Support reactions at clamped edge.

a moment m_{xx} and a shear force v_x are transmitted. The distributed support force f is positive when the force is acting in the w-direction and the distributed fixed-end moment t is positive when it acts in positive φ-direction. The fixed-end moment t is equal to the moment m_{xx}, and the support reaction f is equal to shear force v_x.

Non-zero moment for zero curvature

The bending moment m_{yy} at the edge will not be zero although the curvature is zero. This follows from the relationships between moments and curvatures in Eq. (4.3). If $\kappa_{xx} \neq= 0$ and $\kappa_{yy} = 0$, we find bending moments $m_{xx} = D\kappa_{xx}$ and $m_{yy} = \nu D\kappa_{xx}$. Therefore $m_{yy} = \nu m_{xx}$. This will always occur at a straight clamped edge.

4.4.2 Simply-Supported Edge

If a plate is simply-supported along the edge $x = a$ (see Figure 4.8), we state a condition to the deflection w and the moment t along the edge. In the general case of a non-zero edge moment load t, three plate quantities will be non-zero, the bending moment m_{xx}, the shear force v_x and the twisting

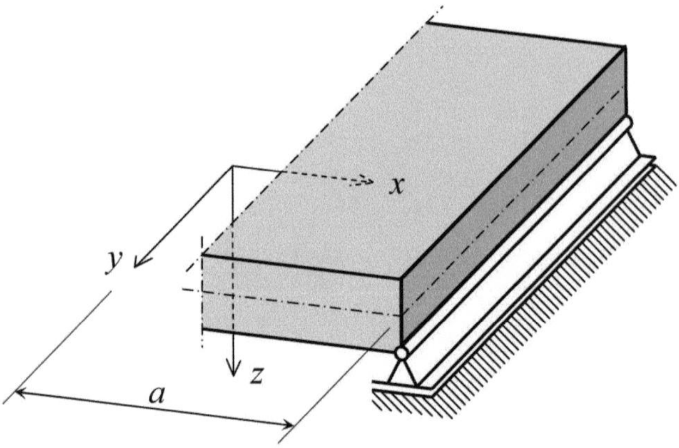

Figure 4.8 Simply-supported edge.

moment m_{xy}. Usually the edge moment t will be zero and therefore m_{xx} also.

Non-zero twisting moment at simply-supported edge

The twisting moment m_{xy} will be non-zero. The slope $\partial w/\partial x$ normal to the edge is free to vary along the edge in the y-direction, therefore its derivative $\partial^2 w/\partial x \partial y$ in the y-direction can have a non-zero value. Then a torsion deformation occurs and a non-zero twisting moment m_{xy}.

So, in practice two plate quantities at the simply supported edge are non-zero, the shear force v_x and the twisting moment m_{xy}. This raises a difficulty, because plate stress resultants must be balanced by one support reaction f.

We have sufficiently explored the problem to be solved. Let us continue and work out the various conditions. With zero value of t we must satisfy the following boundary conditions

$$x = a \ \rightarrow \ \begin{cases} w = 0 \\ t = 0 \end{cases} \tag{4.27}$$

This is a kinematic boundary condition in the displacement direction and a dynamic boundary condition for the rotation direction. The dynamic condition implies $m_{xx} = 0$. With relationship (4.7) we can turn the dynamic

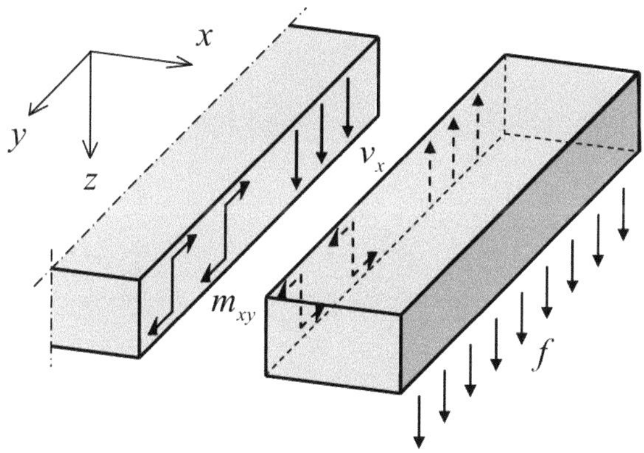

Figure 4.9 Support reaction at simply-supported edge.

condition into a requirement for w

$$\frac{\partial^2 w}{\partial x^2} + \nu \frac{\partial^2 w}{\partial y^2} = 0 \tag{4.28}$$

However, the second derivative $\partial^2 w/\partial y^2$ is zero when w is zero for every value of y on the straight edge $x = a$. Therefore the two conditions in Eq. (4.27) become

$$x = a \rightarrow \begin{cases} w = 0 \\ \dfrac{\partial^2 w}{\partial x^2} = 0 \end{cases} \tag{4.29}$$

Zero moment for zero curvature at simply-supported edge

Thus both second derivatives $\partial^2 w/\partial x^2$ and $\partial^2 w/\partial y^2$ are zero at a simply-supported edge. Therefore, both moments m_{xx} and m_{yy} are zero at a straight simply-supported edge. Here is a difference with the clamped straight edge. There a bending moment m_{yy} occurs which is ν times the clamped moment m_{xx}. Along a simply-supported edge, by contrast, the moment m_{yy} is zero.

The kinematic boundary condition $w = 0$ leads to a support reaction f. Doing the calculation for f we have to take into account both the shear force v_x and the twisting moment m_{xy}. Consider an edge part as in the clamped case for determination of the support reaction f. Figure 4.9 shows such a

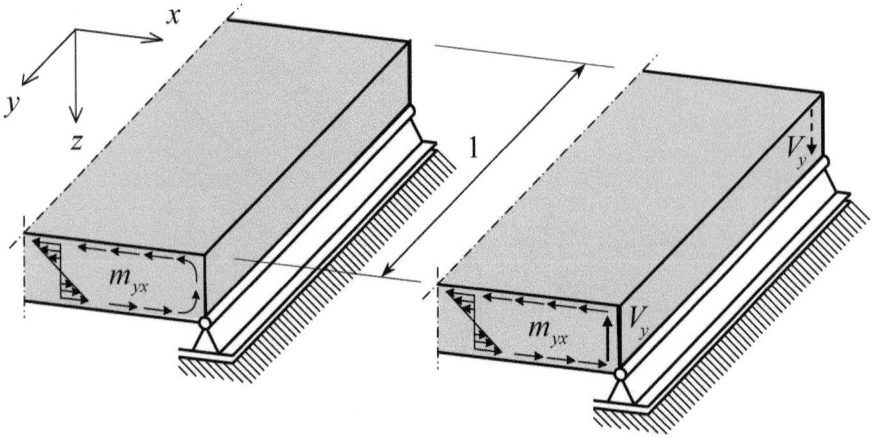

Figure 4.10 Constant twisting moment at the edge.

simply supported edge. The connecting plate transmits a shear force v_x and a twisting moment m_{xy} to this support, but the support can be only a distributed force f on the plate. The question arises what happens to m_{xy} and if f still has the same value as v_x, as is the case for a fully clamped edge. To answer these questions we will have to expand on the phenomenon of torsion.

Consider a plate part in which the twisting moment is constant. To elucidate, in Figure 4.10 the plate parts of Figure 4.9 are now stuck together. In the section perpendicular to the support, a positive twisting moment m_{yx} is present. Figure 4.10 shows a positive twisting moment. The twisting shear stresses, varying linearly over the plate thickness in a section perpendicular to the edge, cannot act on the free edge. Therefore, the twisting shear stresses have to run around at the ends; this happens in an edge zone with a length approximately equal to the thickness of the plate, as explained in Section 3.6. In this small zone, the vertical resultant of the shear stresses is a force V_y, a concentrated shear force in the y-direction. In thin plate theory this force V_y is thought to be concentrated at the very edge. In a section on the side of the positive y-direction (the front side in Figure 4.10) this force faces upward, and in a section on the side of the negative y-direction (the rear side) this force faces downward. In Section 3.6 we have shown that V_y is equal to the twisting moment m_{xy}. There it was the result of a long derivation on the basis of thick plate theory. The magnitude of the concentrated force V_y can also be calculated in a quicker way, using the part of the plate represented in Figure 4.11. In the sections parallel to the x-axis a moment m_{yx} is present, and in the sections parallel to the y-axis a moment m_{xy} is present;

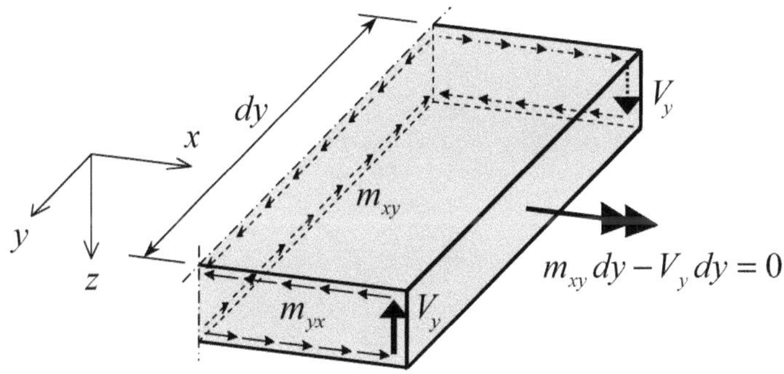

Figure 4.11 Determination of concentrated shear force at edge.

these are equal for isotropic material properties. We now consider the torque equilibrium about the x-axis. In this equilibrium, only m_{xy} and V_y play a part. A possible shear force v_x and support reaction f do not contribute to the moment equilibrium. Shear forces v_y do contribute, but this contribution vanishes when the size of the considered plate part in x-direction is reduced to zero. Therefore, see Figure 4.11, the equilibrium leads to

$$V_y = m_{xy} \tag{4.30}$$

Shear force equals twisting moment

The vertical force V_y that belongs to the shear stresses running round due to m_{yx} has the numerical magnitude of m_{xy}. The force has the dimension N (Newton) and the twisting moment Nm/m, so the dimensions are equal.

After we know what happens when a constant twisting moment occurs, the expansion to a varying twisting moment is no longer difficult. Figure 4.12 shows a part of the edge with all the vertical forces that act on it. Over a distance dy, the concentrated vertical shear force has changed with the amount $(\partial m_{xy}/\partial y)\,dy$. The vertical equilibrium of the considered plate part yields

$$f = v_x + \frac{\partial m_{xy}}{\partial y} \tag{4.31}$$

In the literature on plate theory, f is referred to as the *Kirchhoff shear force*. This is to be regretted, because it is not a shear force, but rather a support

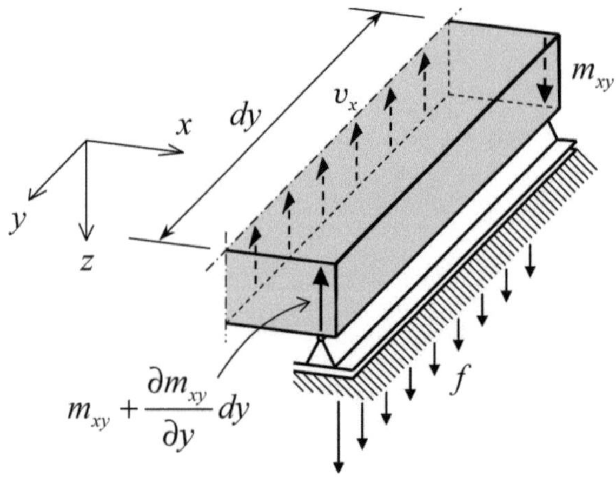

Figure 4.12 Varying twisting moment at edge.

reaction, therefore a line load acting externally on the plate. Analogously, the line load on an edge parallel to the x-axis can be found

$$f = v_y + \frac{\partial m_{yx}}{\partial x} \tag{4.32}$$

The twisting moment is resisted by extra support reactions. The expressions (4.31) and (4.32) are valid for supports in which the outward-pointing normal on the plate edge points in the positive x-direction, or positive y-direction respectively. The definition of f (positive when acting in the positive w-direction) implies that the sign of the right-hand term must change if the normal on the plate edge is pointing in the negative direction

$$f = -\left(v_x + \frac{\partial m_{xy}}{\partial y}\right), \quad f = -\left(v_y + \frac{\partial m_{yx}}{\partial x}\right) \tag{4.33}$$

The formulas for f provide the magnitude of the distributed support reaction per unit length on the edge. Deriving these formulas showed that only the *increase* of the twisting moment plays a part. However, at the end of the edge a special situation occurs. Figure 4.13 shows a corner of a plate in which two simply-supported edges come together. The dimensions Δl of the plate particle are thought to be small. In the vertical equilibrium of this plate particle in the corner of the plate, two concentrated loads $V_y = m_{xy}$ and $V_x = m_{yx}$ play a role, together with the distributed support reactions f and load p on the plate. The reaction f must be multiplied by Δl and p by $(\Delta l)^2$. If we allow Δl to approach zero, the contribution of f and p vanishes. Only the forces $V_y = m_{xy}$ and $V_x = m_{yx}$ persist. Because the vertical equilibrium

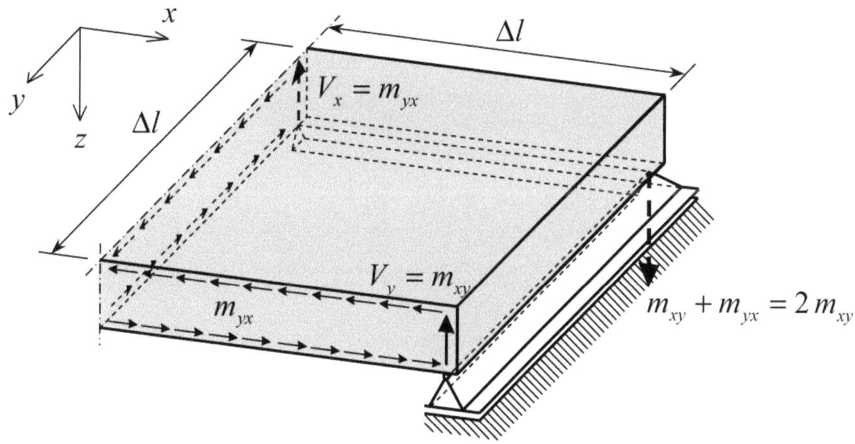

Figure 4.13 Twisting moment at simply-supported corner.

of this plate corner has to be ensured, there has to be a concentrated corner reaction of the magnitude $m_{xy} + m_{yx}$. For isotropic plates $m_{xy} = m_{yx}$, so the corner reaction is $2m_{xy}$. Depending on the sign of the twisting moment, this reaction force can be either compressive or tensile.

Spiritual father or mother?

Lagrange was the first to derive the bi-harmonic differential equation in the correct form. In fact, he was a jury member in a competition organized by the French Academy of Science at the suggestion of emperor Napoleon, and corrected the solution subjected by Sophie Germain. However, in his solution the meaning of constant D is still unclear. Later, Navier derives the differential equation again and obtains for D the plate flexural rigidity of Eq. (3.31). After him Kirchhoff clarifies the boundary problem [12]. Therefore, it is common practice to refer to thin plate theory as Kirchhoff theory.

4.4.3 Free Edge

A free edge is not supported at all. Again, the edge previously drawn simply supported, but now without support, is the starting point. In the stress picture, such a free edge is comparable to a simply supported edge, the only difference being that w is not prescribed, but f is. The structural engineer could also place a line load t on the edge, in which case one will get two equilibrium (dynamic) boundary conditions

$$x = a \rightarrow \begin{cases} m_{xx} = t \\ v_x + \dfrac{\partial m_{xy}}{\partial y} = f \end{cases} \tag{4.34}$$

With Eq. (4.4) we can determine

$$\frac{\partial m_{xx}}{\partial x} + \frac{2\partial m_{xy}}{\partial y} = f \tag{4.35}$$

With Eq. (4.7) this becomes

$$-D\left\{\frac{\partial^3 w}{\partial x^3} + (2-\nu)\frac{\partial^3 w}{\partial x \partial y^2}\right\} = f \tag{4.36}$$

For Eq. (4.34) we can then write

$$x = a \quad \rightarrow \begin{cases} \dfrac{\partial^2 w}{\partial x^2} + \nu \dfrac{\partial^2 w}{\partial y^2} = \dfrac{-t}{D}, \\ \dfrac{\partial^3 w}{\partial x^3} + (2-\nu)\dfrac{\partial^3 w}{\partial x \partial y^2} = \dfrac{-f}{D} \end{cases} \tag{4.37}$$

If the free edge is not being loaded at all, both right-hand members of Eq. (4.37) will be zero.

4.4.4 Discontinuity in Thickness

Let us consider a discontinuity in plate thickness. The boundary between the two parts of different thickness is assumed to be in y-direction. We denote the two plate parts by L (left) and R (right). Then four transition conditions apply

$$w_L = w_R, \quad \varphi_{xL} = \varphi_{xR}, \quad m_{xxL} = m_{xxR}, \quad v_{xkL} = v_{xkR} \tag{4.38}$$

where v_{xK} is the Kirchhoff shear force. That the Kirchhoff shear forces must be equal, can be clarified by showing the physical reality. In a section perpendicular to the boundary the magnitude of the twisting moment abruptly changes, because the mixed second derivative $\partial^2 w/\partial x \partial y$ at the boundary

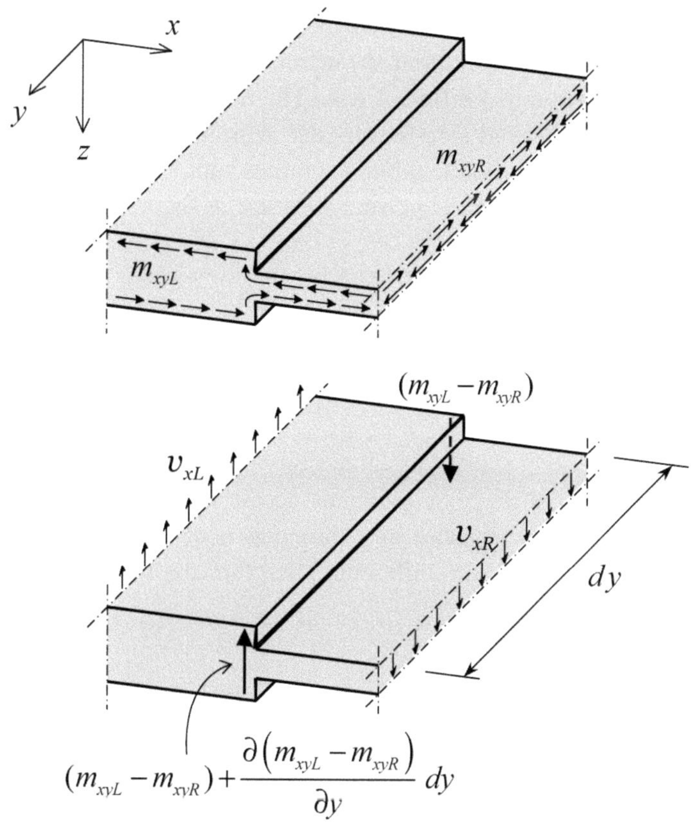

Figure 4.14 Twisting moment at discontinuity in thickness.

is common for both plate parts, but the flexural rigidity D is different. On the boundary there is a vertical force, see Figure 4.14. The size of this force is equal to the difference of the two twisting moments left and right of the boundary. If both plate thicknesses are equal, the difference between the two twisting moments is zero and no force will occur, and if one of the thicknesses is zero (a free edge) the full twisting moment of the present plate part remains. The vertical equilibrium of the shown plate part requires (when no line load is present on the boundary)

$$\begin{gathered} -v_{xL} - \frac{\partial}{\partial y}\left(m_{xyL} - m_{xyR}\right) + v_{xR} = 0 \rightarrow \\ \left(v_{xL} + \frac{\partial m_{xyL}}{\partial y}\right) - \left(v_{xR} + \frac{\partial m_{xyR}}{\partial y}\right) = 0 \end{gathered} \tag{4.39}$$

The final equation clearly states that the Kirchhoff shear force of the left plate part is equal to the Kirchhoff shear force of the right plate part. We can also read the equation in a different way. The first term is exactly f_L for the plate part on the left-hand side for an edge with its normal in a positive x-direction. The second term, including the minus sign, is f_R for the right-hand plate part for an edge with its normal pointing in the negative x-direction. This provides us with

$$f_L + f_R = 0 \tag{4.40}$$

The sum of the two line loads on the boundary is zero. If an external line load f is applied on the boundary, the condition becomes

$$f_L + f_R = f \tag{4.41}$$

The bending moment m_{xx} is continuous across the boundary, but the moment m_{yy} is not. The curvature in y-direction is equal in both plate parts, but the flexural rigidities are different. Therefore the moment m_{yy} will be discontinuous.

Summing up for thickness discontinuity

Summing up, at a boundary between plates of different thickness the bending moment normal to the boundary is continuous, and the bending moment parallel to the boundary and the twisting moment are discontinuous. The Kirchhoff shear force must be continuous. This implies that the shear forces normal to the boundary and parallel to the boundary will be discontinuous.

4.5 Message of the Chapter

- The theory for thin plates is a limiting case of the theory of thick plates in which shear deformation tends to zero. Rotations depend on the vertical displacements. Curvatures become second derivatives of the displacement.

- The bi-harmonic equation, which is obtained for membrane plates, also appears in the plate bending theory.

- FE codes refer to the theory of thin plates as Kirchhoff theory, though Lagrange was the first to derive the correct differential equation.

- A shear force in any direction is the derivative of the bending moment in that direction plus the derivative of the twisting moment in the transverse direction.

- The transformation rules for moments are the same as for membrane forces. Principal moments occur in the direction in which the twisting moment is zero.

- A principal shear force occurs, while the shear force in the transverse direction is zero. The direction of the principal shear force (trajectories) differs from the direction of the principle moments. Therefore, a trajectory plot of the shear forces is different from the trajectory plot of the moments.

- The shower analogy helps us understand the flow of the distributed plate load to the supports. If the moment sum is considered as a hill, the load flows like water in the direction of the steepest slope to the edges.

- At a clamped edge there is a bending moment normal to the edge and a bending moment in the direction of the edge. The value of the latter is Poisson's ratio times the former. The twisting moment is always zero.

- At a (unloaded) simply-supported edge, the bending moments both normal to the edge and in direction of the edge are zero. However, the twisting moment need not be zero.

- At a simply-supported edge, there is a concentrated edge shear force in sections normal to the edge. The size is equal to the size of the twisting moment. (A twisting moment has the same dimensions as a point load, Nm/m and N respectively). This also occurs at free edges.

- The support reaction in a simply-supported edge is equal to the shear force normal to the edge plus the derivative of the twisting moment in the direction of the edge.
- If two simply-supported edges join in an (unloaded) corner, a concentrated vertical corner support reaction occurs with a value twice the twisting moment. This can be upward or downward depending on the sign of the twisting moment.
- In a corner between two free edges, the twisting moment is zero if the corner is not loaded. If a point load is applied in the free corner, the twisting moment will become half the size of the point load.
- At a discontinuity of thickness, the bending moments normal to the line of discontinuity are continuous. However, the bending moment in the direction of the boundary, and the twisting moments are discontinuous.
- At a discontinuity of thickness, the Kirchhoff shear force is continuous. The shear forces normal to and parallel to the boundary are discontinuous.

Chapter 5
Rectangular Plate Examples

We focus on special aspects of the theory of thin plates by discussing a state of constant bending curvature in Section 5.1 and a panel of constant torsion in Section 5.2. In Section 5.3 we show the effectiveness of a square simply-supported plate subject to a distributed load. In Section 5.4 we discuss the special case of a twist-less plate. Finally, we devote Section 5.5 to a viaduct subject to an edge load.

5.1 Basic Bending Cases

5.1.1 Cylindrical Deflection

We consider a cylindrical deflection (see Figure 5.1) with shape

$$w = Cx\,(a - x) \tag{5.1}$$

for a plate with a non-zero Poisson's ratio. Substitution of this expression into the bi-harmonic equation (4.7) gives $p = 0$. This means that the function w is a solution to the differential equation in the absence of a distributed load p. The deflection is zero along the straight edges $x = 0$ and $x = a$. This is where the supports can be thought to be. All lines that run parallel to the supports remain straight. The formulas in (4.7) imply

$$m_{xx} = 2DC, \quad m_{yy} = 2\nu DC, \quad m_{xy} = 0 \tag{5.2}$$

We conclude that there is a bending moment in the y-direction, Poisson's ratio times the moment in the x-direction

J. Blaauwendraad, *Plates and FEM: Surprises and Pitfalls*, Solid Mechanics and Its Applications 171, DOI 10.1007/978-90-481-3596-7_5,

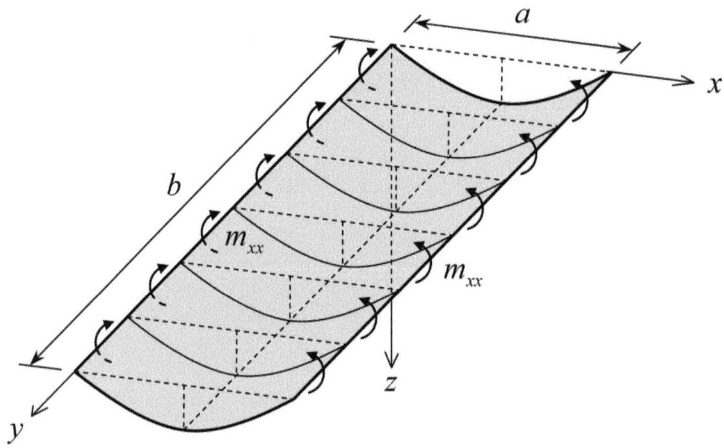

Figure 5.1 Cylindrical deflection plane.

$$m_{yy} = \nu m_{xx} \tag{5.3}$$

Furthermore Eq. (4.4) implies

$$v_x = 0, \quad v_y = 0 \tag{5.4}$$

There is a constant bending moment m_{xx} in the plate, which is caused by an externally applied moment of the same size along the straight edges $x = 0$ and $x = a$. In the direction of the straight generating lines, there is a constant moment m_{yy} of magnitude νm_{xx}. For steel, with a value $\nu = 0.3$, this will lead to $m_y = 0.3 m_{xx}$; for concrete with the value $\nu = 0.2$, $m_y = 0.2 m_{xx}$. Twenty percent of the reinforcement in the span direction is necessary in the lateral direction, even when there is no curvature there!

5.1.2 Cylindrical Deflection of Arbitrary Shape

Now we consider the general shape of the deflection $w = f(x)$ due to a distributed load p. Substitution into the bi-harmonic equation in (4.7) shows that the load is

$$p = D \frac{d^4}{dx^4} f(x) \tag{5.5}$$

For the moments we find, see Eq. (4.7)

$$m_{xx} = -D \frac{d^2}{dx^2} f(x), \quad m_{yy} = \nu m_{xx}, \quad m_{xy} = 0 \tag{5.6}$$

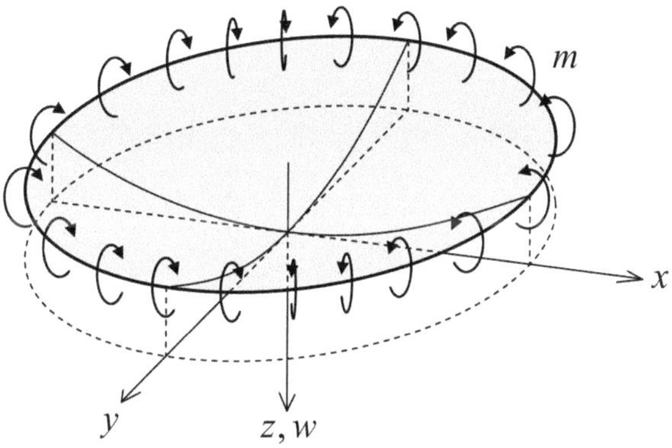

Figure 5.2 Omni-directional bending.

We infer again that if the deflection is constant in the y-direction, a moment m_{yy} is generated even though there is no curvature in the y-direction. This confirms our finding for the clamped edge in Section 4.4.1.

5.1.3 Omni-Directional Bending

We consider the superposition of the solutions $w = -Cx^2$ and $w = -Cy^2$. With $x^2 + y^2 = r^2$ this leads to

$$w = -C(x^2 + y^2) = -Cr^2 \tag{5.7}$$

This is a paraboloid of revolution (see Figure 5.2). The moments are

$$m_{xx} = 2DC(1 + \nu), \quad m_{yy} = 2DC(1 + \nu), \quad m_{xy} = 0 \tag{5.8}$$

Using the transformation formulas (4.17) we find

$$\begin{aligned} m_{nn} &= m_{xx}\cos^2\alpha + m_{yy}\sin^2\alpha = 2DC\,(1 + \nu) \\ m_{tt} &= 2DC\,(1 + \nu) \\ m_{nt} &= 0 \end{aligned} \tag{5.9}$$

The bending moment is equal in all directions. Torsional moments do not appear. Here we have the case of pure bending due to a constant moment m along the perimeter of a circular plate. The constant C follows from

$$m = 2DC\,(1 + \nu) \;\rightarrow\; C = \frac{m}{2D\,(1 + \nu)} \tag{5.10}$$

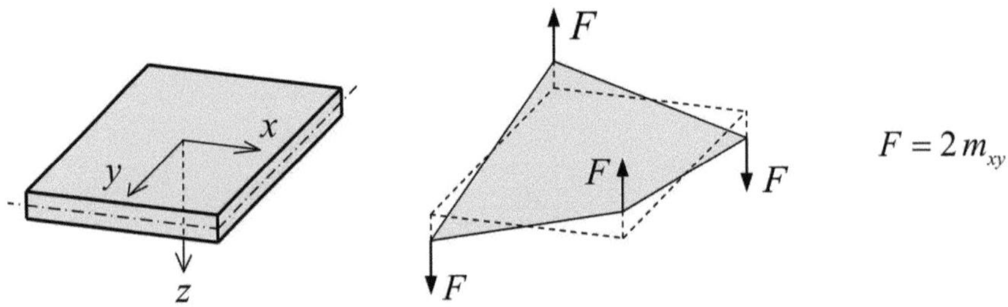

Figure 5.3 Panel with constant torsion.

and the formula for the plane of deflection is

$$w = \frac{-m}{2D\,(1+\nu)}\,r^2 \tag{5.11}$$

5.2 Torsion Panel

We give a rectangular plate a deflection of the shape (see Figure 5.3)

$$w = -Cxy \tag{5.12}$$

This gives a mixed second derivative

$$\frac{\partial^2 w}{\partial x\,\partial y} = -C \tag{5.13}$$

The other two derivatives are zero. The moments are

$$m_{xy} = (1-\nu)\,DC, \quad m_{xx} = 0, \quad m_{yy} = 0 \tag{5.14}$$

Both bending moments are zero and the torsion is constant and positive. According to Eq. (4.4) the derivative of the moments provides the shear forces. The shear forces are zero:

$$v_x = 0, \quad v_y = 0 \tag{5.15}$$

According to Eq. (4.5) we can compute the load from the second derivatives of the moments. As a result the load p is also zero. On the edge x = constant, the following support reaction is obtained:

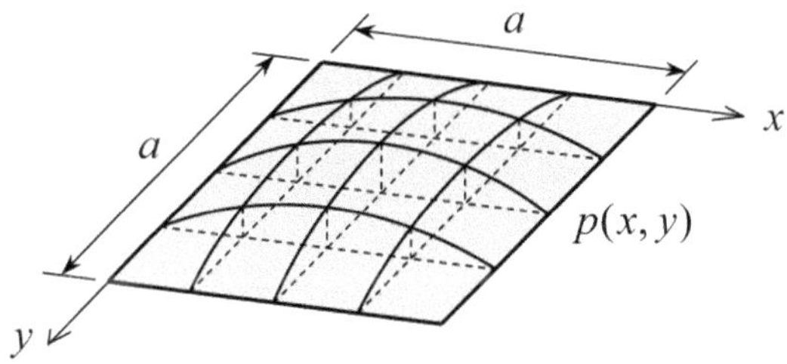

Figure 5.4 Two-way sine load on a simply-supported square plate.

$$f = v_x + \frac{\partial m_{xy}}{\partial y} = 0 \tag{5.16}$$

which is also zero. This is also the case for the other edges. So, no load occurs at the edges and no load over the area of the plate. Yet a twisting moment is present. The twisting moments in the four corners of the plate generate a concentrated support reaction of $2m_{xy}$, as shown in Figure 5.3. The load consists of two couples of point loads in opposite direction. In literature, this plate case is known as the *Nadai's plate*. This stress state may be used to experimentally determine the plate flexural rigidity D. The panel with a constant twisting moment and four corner forces will play a role in Chapter 9 on approximating computational methods in pre-FE days.

5.3 Two-Way Sine Load on Square Plate

A square plate with dimensions a is simply supported along its four edges. The origin of the coordinate system is chosen in a corner, and the axes coincide with the sides of the square, see Figure 5.4. A distributed load is applied of the form

$$p = \hat{p} \sin \frac{\pi x}{a} \sin \frac{\pi y}{a} \tag{5.17}$$

This two-way sine load may be considered to be the first term of a Fourier series of a homogeneously distributed load. The amplitude $\hat{p}$ is the value of the load at the plate centre ($x = y = a/2$).

5.3.1 Displacement

We assume

$$w = \hat{w} \sin \frac{\pi x}{a} \sin \frac{\pi y}{a} \tag{5.18}$$

where $\hat{w}$ is the centre value. This choice satisfies the boundary conditions along the simple supports as discussed in Section 4.4.2.

$$x = 0 \quad \text{and} \quad x = a \rightarrow \begin{cases} w = 0 \\ \dfrac{\partial^2 w}{\partial x^2} = 0 \end{cases} \tag{5.19}$$

$$y = 0 \quad \text{and} \quad y = a \rightarrow \begin{cases} w = 0 \\ \dfrac{\partial^2 w}{\partial y^2} = 0 \end{cases} \tag{5.20}$$

Substitution of this trial solution in the differential equation (4.7) yields

$$\left[\frac{\pi^4}{a^4} + 2\frac{\pi^4}{a^4} + \frac{\pi^4}{a^4}\right]\hat{w} = \frac{\hat{p}}{D} \rightarrow \hat{w} = \frac{\hat{p}a^4}{4\pi^4 D} \tag{5.21}$$

The solution is therefore

$$w = \frac{\hat{p}a^4}{4\pi^4 D} \sin\frac{\pi x}{a} \sin\frac{\pi y}{b} \tag{5.22}$$

This is a particular solution that satisfies the boundary conditions of the simply-supported plate. Then the particular solution is the complete solution and we do not need find a homogenous solution of the differential equation. The plane of deflection is similar in shape to the load distribution. This is visualized in Figure 5.5. For the maximum deflection of the square plate we find

$$\hat{w} = \frac{\hat{p}a^4}{4\pi^4 D} \tag{5.23}$$

For the maximum deflection of a beam with unit width, flexural rigidity D and the same span and subjected to a one-way sine line load with maximum $\hat{p}$, we find

$$\hat{w} = \frac{\hat{p}a^4}{\pi^4 D} \tag{5.24}$$

The deflection of the plate is a quarter of the deflection of the beam. The beam solution applies for a very wide plate that spans in one direction; that plate is only a quarter as stiff as the square plate.

One might expect a square plate to be twice as stiff as a beam, at a first look, noticing that a plate can transfer loads in two directions, so beam-action in the x-direction and beam-action in the y-direction may cooperate. However, the square plate receives additional stiffness by two other ‘beams’, which act in the diagonal direction. The length of these ‘beams’ is longer, but

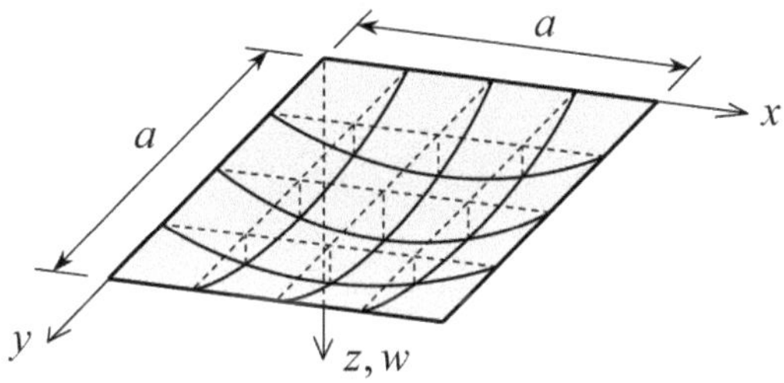

Figure 5.5 Deflection of the square plate.

the ends act as clamped ends in the corners. Because of the straight edges, both $\partial w/\partial x$ and $\partial w/\partial y$ are zero at the corners. Therefore, the derivative of w must be zero in all directions at that position, which explains the apparent clamped ends of the diagonal 'beams' and their important contribution to the stiffness.

Effectiveness of plate

The middle term in the bi-harmonic differential equation due to torsion contributes to the same extent as the first and last term due to bending. This shows that a square or nearly square plate is a very effective load-carrying structure. A factor of about four in effectiveness is also to be expected for a homogeneously distributed load.

5.3.2 Moments and Shear Forces

The formulas in (4.7) lead to the moments

$$\begin{aligned} m_{xx} &= \frac{1+\nu}{4\pi^2}\hat{p}a^2 \sin\frac{\pi x}{a}\sin\frac{\pi y}{a} \\ m_{yy} &= \frac{1+\nu}{4\pi^2}\hat{p}a^2 \sin\frac{\pi x}{a}\sin\frac{\pi y}{a} \\ m_{xy} &= -\frac{1-\nu}{4\pi^2}\hat{p}a^2 \cos\frac{\pi x}{a}\cos\frac{\pi y}{a} \end{aligned} \tag{5.25}$$

The distributions of these moments are drawn in Figure 5.6. The solution for the moments confirms that the boundary conditions are satisfied. The distributions of the bending moments have the same shape as the deflection and

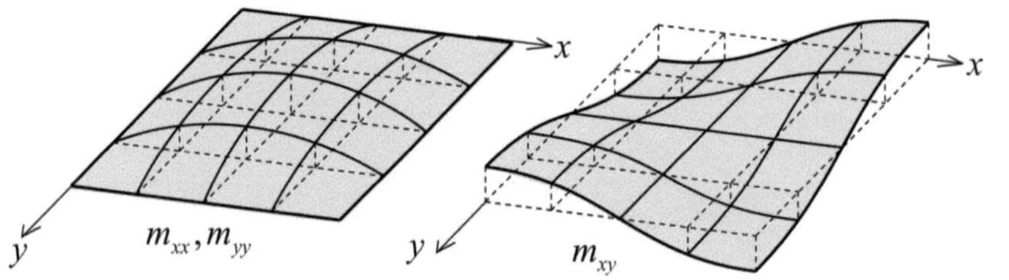

Figure 5.6 Distribution of moments under two-way sine loading.

the load. The shape of the twisting moments is different. Where the bending moment is at a maximum, the twisting moment is zero, and where the bending moment is zero the twisting moment is at a maximum. The maximum bending moment in the plate is

$$\hat{m}_{xx} = \hat{m}_{yy} = \frac{1+\nu}{4\pi^2}\,\hat{p}a^2 \tag{5.26}$$

For a very wide plate that is supported only in one direction (x-direction) we find

$$\hat{m}_{xx} = \frac{1}{\pi^2}\hat{p}a^2 \tag{5.27}$$

This is also the moment in a beam with a unit width under a comparable load. The maximum moment in the plate is a factor $(1+v)/4$ smaller than in a beam. Again, the force action in a square plate is very effective. The largest twisting moments arise in the four corners. In the corner $x = 0$, $y = 0$ of the plate the twisting moment is

$$\hat{m}_{xy} = -\frac{(1-\nu)}{4\pi^2}\,\hat{p}a^2 \tag{5.28}$$

This moment is of the same order of magnitude as the maximum bending moment in the centre of the plate. For a zero Poisson's ratio it is even equal. The shear forces can be derived from the moments by applying Eq. (4.4)

$$\begin{aligned} v_x &= \frac{1}{2\pi}\hat{p}a\cos\frac{\pi x}{a}\sin\frac{\pi y}{a} \\ v_y &= \frac{1}{2\pi}\hat{p}a\sin\frac{\pi x}{a}\cos\frac{\pi y}{a} \end{aligned} \tag{5.29}$$

Their distribution over the plate area is depicted in Figure 5.7. The correctness of the shear forces can be checked as follows. We can compute the total shear force that flows to the edges. Along edge $x = 0$ the shear force is

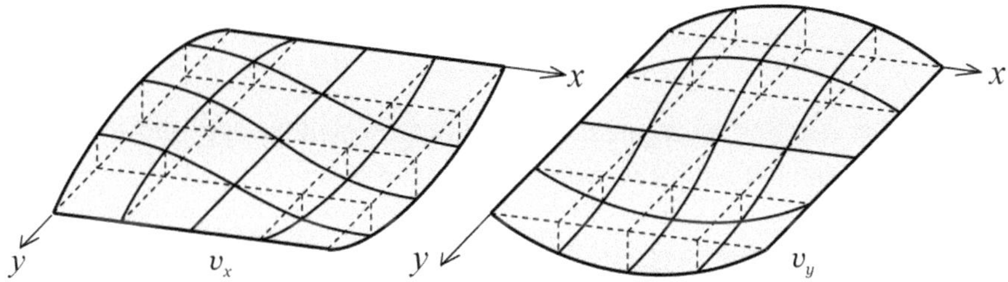

Figure 5.7 Distribution of shear forces under two-way sine load.

$$v_x = \frac{1}{2\pi}\hat{p}a \sin\frac{\pi y}{a} \tag{5.30}$$

and the total shear force S along this edge is

$$S = \frac{1}{2\pi}\hat{p}a \int_0^a \sin\frac{\pi y}{a}\,dy = \frac{1}{2\pi}\hat{p}a \cdot \frac{2a}{\pi} = \frac{1}{\pi^2}\hat{p}a^2 \tag{5.31}$$

For reasons of symmetry the total shear force which flows to the four edges is four times S

$$4S = \frac{4}{\pi^2}\hat{p}a^2 \tag{5.32}$$

This total shear force should be equal to the total load P, which is applied to the plate

$$P = \hat{p}\int_0^a \sin\frac{\pi x}{a}\,dx \int_0^a \sin\frac{\pi y}{a}\,dy = \hat{p}\frac{2a}{\pi}\cdot\frac{2a}{\pi} = \frac{4}{\pi^2}\hat{p}a^2 \tag{5.33}$$

Indeed $4S$ equals P correctly.

5.3.3 Support Reactions

We continue the analysis of the square plate by computing the distributed support reactions. Along the edge $x = 0$ the formula is, see Eq. (4.33)

$$f = -\left(v_x + \frac{\partial m_{xy}}{\partial y}\right)_{x=0} \tag{5.34}$$

The earlier results for v_x and m_{xy} lead to

$$f = -\left(\frac{3-\nu}{4\pi}\hat{p}a\cos\frac{\pi x}{a}\sin\frac{\pi y}{a}\right)_{x=0} = -\frac{3-\nu}{4\pi}\hat{p}a\sin\frac{\pi y}{a} \tag{5.35}$$

The support reaction is negative, so its direction will be opposite to the direction of w and the load p (compressive reactions).

Surprising support reaction

The support reaction f is larger than the shear force v_x. For zero Poisson's ratio the difference is a factor of 1.5.

The sum of the total support reaction along the four edges is

$$4R = 4\left(-\frac{3-\nu}{4\pi}\hat{p}a\int_0^a \sin\frac{\pi y}{a}dy\right) = -\left(\frac{6-2\nu}{\pi^2}\right)\hat{p}a^2 \tag{5.36}$$

The absolute value of this is evidently much larger than the total load given in Eq. (5.33); again a factor of 1.5 exists between load and support reactions for zero Poisson's ratio. This difference is fully explained by the existence of balancing concentrated reactions in the four plate corners. In the left-upper corner ($x = 0$, $y = 0$) the value of the twisting moment is

$$m_{xy} = -\frac{1-\nu}{4\pi^2}\hat{p}a^2 \tag{5.37}$$

This is a negative value, so the direction of the shear stresses in sections perpendicular to the edges is as shown in Figure 5.8. Therefore, the two concentrated edge shear forces V_x and V_x are directed upward. For vertical equilibrium, a downward lumped corner reaction F is needed

$$F = |m_{xy} + m_{yx}| = 2\left|-\frac{1-\nu}{4\pi^2}\hat{p}a^2\cos\frac{\pi x}{a}\cos\frac{\pi y}{a}\right|_{\substack{x=0\\y=0}} = \frac{1-\nu}{2\pi^2}\hat{p}a^2 \tag{5.38}$$

Apparently a local downward force is needed to keep the square plate on the simple support. If the plate is not fixed to the support, it will lift up in the corner. To fix the corner, a tensile reaction force is needed. The same force occurs in all corners. Now we should compare $4R + 4F$ to the applied load P

$$4R + 4F = -\frac{6-2\nu}{\pi^2}\hat{p}a^2 + \frac{2-2\nu}{\pi^2}\hat{p}a^2 = \frac{-4}{\pi^2}\hat{p}a^2 = -P \tag{5.39}$$

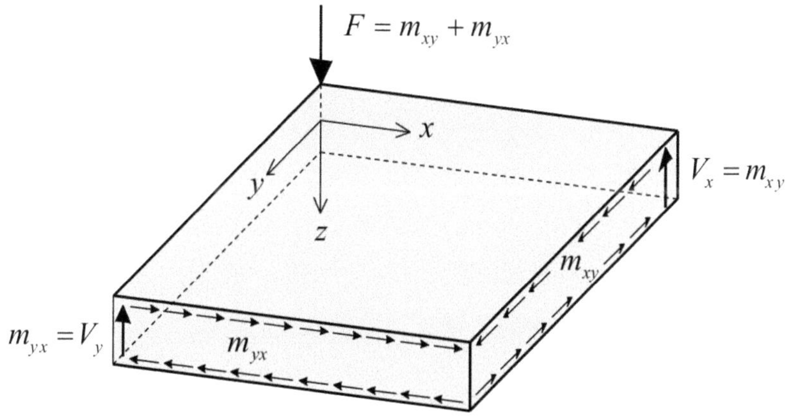

Figure 5.8 Direction of shear stresses for negative m_{xy} in left-upper corner

The sum of all reactions is equal to the load. The sign has become negative because of the sign convention for support reactions.

It is interesting to examine how the shower analogy must be interpreted in this case. Figure 5.9 is a picture of the trajectories. Let us consider the 'hill' as a roof. The diagonals and the horizontal and vertical lines through the middle of the roof are lines of symmetry and therefore trajectories. For this combination of load and boundary conditions the trajectories between these lines of symmetry end perpendicular to the edges. We may consider the edges as open gutters that are perforated at their lower side over their full length

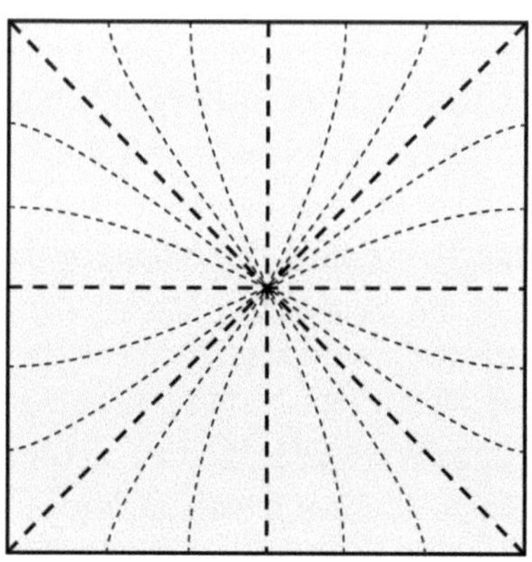

Figure 5.9 Trajectories for shear forces.

in order to let the rain that flows from the roof through immediately. At the same time a lumped well in each corner is bringing up water. This additional water flows through the gutters off the corner and also disappears through the perforations. In this way the water that flows through the perforations is more than the water that falls upon the roof in a rain shower. The additional part comes from the wells in the corners.

Remark

If the simply-supported plate is not rigidly connected to the support, and tensile forces cannot be carried, corners will lift and tilt, which makes the plate less stiff and leads to higher bending moments. A known example of the tilting of a corner of the plate occurs at lock gates produced as double mitre gate. There is leakage at the lower corners of each single gate, because no tensile reaction force can occur.

5.3.4 Stiff Edge Beams

We have seen that the support reaction f is a factor of 1.5 larger than the shear force v_x near the edge for zero Poisson's ratio. We could imagine that the simple support is realized by edge beams of infinite flexural and shear rigidity and zero torsion rigidity, supported by columns at the corners. These beams are subjected to higher loads than might be expected at first glance, and this needs to be kept in mind when detailing such beams. We will now elaborate on the maximum moment and shear force in the beams.

Figure 5.10c shows the plan of a square plate on edge beams. The edge beams are supported by ball supports at the four corners. An edge beam is supposed to be an I-section. A side view is made in section A-A, shown in Figure 5.10b. At both ends of the section, other edge beams are crossed. The plate (thickness t) fits nicely between the flanges of the I-section and is perfectly glued to the web of the edge beam. The connection is able to transfer the shear force v_x and twisting moment m_{xy} from plate to web. These plate actions are the loading of the edge beam. No concentrated vertical shear force need occur in the plate edge zone, as is the case at a simply-supported edge. At the end of the section A-A, there are shear forces V_b in the edge beams which are crossed by the section. We show such a force in Figure 5.10b in the web of the I-section.

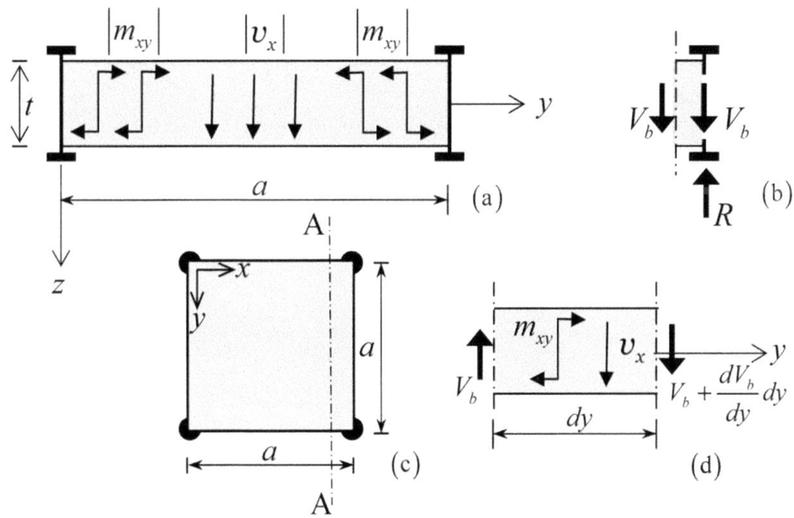

Figure 5.10 Loading of rigid edge beam due to plate.

The maximum bending moment M_b in the edge beam occurs at mid-span, and the maximum shear force V_b at the beam ends. The beam moment is due to both the shear force v_x and twisting moment m_{xy}. The shear force in the beam is due to the shear force v_x only. Accounting for the cosine shape of v_x and the sine shape of m_{xy} in Eqs. (5.28) and (5.29) we obtain

$$\begin{aligned} M &= \frac{\hat{v}_x a^2}{\pi^2} + \frac{1}{\pi}\hat{m}_{xy} a = \frac{\hat{p}a^3}{2\pi^3} + \frac{(1-\nu)\,\hat{p}a^3}{4\pi^3} = \frac{(3-\nu)\,\hat{p}a^3}{4\pi^3} \\ V &= \frac{1}{\pi}\hat{v}_x a = \frac{\hat{p}a^2}{2\pi^2} \end{aligned} \tag{5.40}$$

Surprising large beam moment

In the expression for the beam moment M we notice the factor $(3-\upsilon)/4$ again as seen earlier in Eq. (5.35) for the support reaction! For zero Poisson's ratio the moment is 50% larger than expected on the basis of the shear force that acts on the beam. For $\nu = 0.2$ it is 40%.

The column reaction R is computed as follows:

$$R = -2V = -\frac{\hat{p}a^2}{\pi^2} \tag{5.41}$$

which indeed is one quarter of the total load P on the plate, and it is a compressive force. No corner tensile force occurs if the simple support is realized through a flexure-rigid edge beam.

The introduction of the edge beam provides an alternative way to derive the boundary condition at a free edge. For that purpose we have to consider an elementary beam part of length dy as depicted in Figure 5.10d. For force equilibrium in the z-direction and moment equilibrium about the x-axis, respectively, we obtain

$$\frac{dV_b}{dy} - v_x = 0, \qquad V_b + m_{xy} = 0 \tag{5.42}$$

From the second equation we learn $V_b = -m_{xy}$. Substitution in the first equation and sign change leads to

$$\frac{\partial m_{xy}}{\partial y} + v_x = 0 \tag{5.43}$$

which, for zero load f, is identical to the condition we derived earlier in Eq. (4.34).

5.4 Twist-Less Plate

In the preceding section we considered a square plate subjected to distributed load and supported by flexure-stiff and torsion-weak edge beams. For the two-way sine load we found maximum bending moments in the plate centre and maximum twisting moments at the corners. The values of these moments are of about the same size and for zero Poisson's ratio exactly equal. If the flexural rigidity of the edge beams decreases, the deflections will increase and the distribution of moments will change. The bending moments will become larger and the twisting moments smaller. For a sufficiently small flexural rigidity the twisting moments become zero. This can best be shown for a homogeneously distributed load p. No twisting moments in the plate means that the middle term in the bi-harmonic differential equation can be skipped. Then the plate behaves as a grid of orthogonal strips in which only bending occurs. The displacement field in this case becomes

$$w(x, y) = \hat{w}[f(x) + f(y)] \tag{5.44}$$

Here $\hat{w}$ is the maximum deflection of the edge beams; the shape function $f(x)$, with maximum 1, is the deflected shape of a simply-supported beam

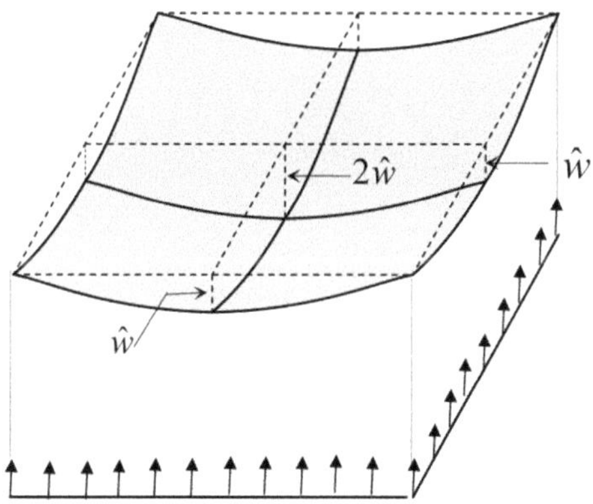

Figure 5.11 Deflection and support reactions in twist-less plate.

subjected to a homogeneously distributed load. The same applies for $f(y)$. The displacement field is shown in Figure 5.11. The consequence of this choice is a zero torsion deformation ρ_{xy} and zero twisting moments m_{xy}. Furthermore the curvature κ_{xx} depends only on x and the curvature κ_{yy} only on y. This leads to moments m_{xx} and shear forces v_x which are constant in the y-direction, and moments m_{yy} and shear forces v_y which are constant in the x-direction.

So, each edge beam is loaded by a homogeneously distributed load $pa/4$ and must have the same deflected shape as the adjacent plate. This can be the case only when the flexural rigidity EI of the edge beam equals the bending stiffness of a plate strip of width $a/2$. Therefore $EI = aD/2$. If we choose this beam rigidity, no twisting moments will occur in the plate.

Twistless slab

For a proper choice of the edge beam stiffness no twisting moments will occur. Then the distributed load p is half transferred in the x-direction and half in the y-direction.

5.5 Edge Load on Viaduct

Consider a bridge slab with span a, and width b. The bridge is simply supported at $x = a/2$, and $x = -a/2$, and has free edges at $y = b/2$ and

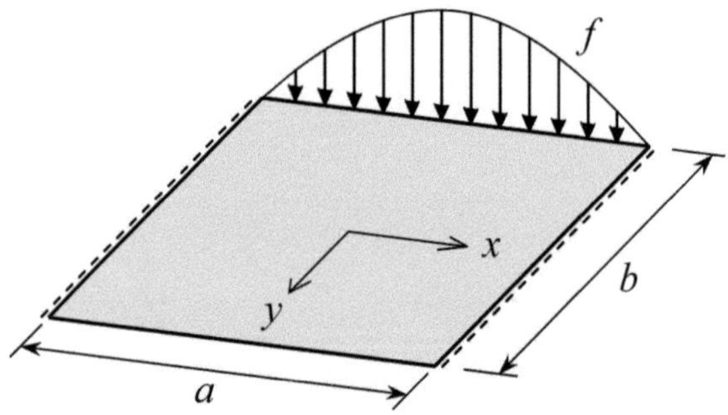

Figure 5.12 Bridge with load on edge.

$y = -b/2$. A cosine-shaped load is applied on the edge $y = -b/2$, see Figure 5.12. This load can be considered to be an approximation to a distributed line load with some heavy vehicles in the middle part of the span. In this case, a distributed load p is not taken into account. The distributed edge load is

$$f(x) = \hat{f} \cos \alpha x, \quad \alpha = \pi/a \tag{5.45}$$

The boundary conditions require

$$x = \pm\frac{1}{2}a \rightarrow \begin{cases} w = 0 \\ m_{xx} = 0 \end{cases} \tag{5.46}$$

$$y = -\frac{1}{2}b \rightarrow \begin{cases} m_{yy} = 0 \\ -\left(v_x + \dfrac{\partial m_{xy}}{\partial y}\right) = f \end{cases} \tag{5.47}$$

$$y = +\frac{1}{2}b \rightarrow \begin{cases} m_{yy} = 0 \\ v_x + \dfrac{\partial m_{xy}}{\partial y} = 0 \end{cases} \tag{5.48}$$

Applying the method of separation of variables, we can describe w as a product of two functions, $w = w(y) \cos \alpha x$. The function $w(y)$ is the distribution of the deflection along the vertical line at mid-span. This choice for w satisfies the boundary conditions at the supports. Substitution into the bi-harmonic equation (4.7) delivers an ordinary differential equation for $w(y)$ of the fourth order.

$$D\left(\alpha^4 w - 2\alpha^2 \frac{d^2 w}{dy^2} + \frac{d^4 w}{dy^4}\right) = 0 \tag{5.49}$$

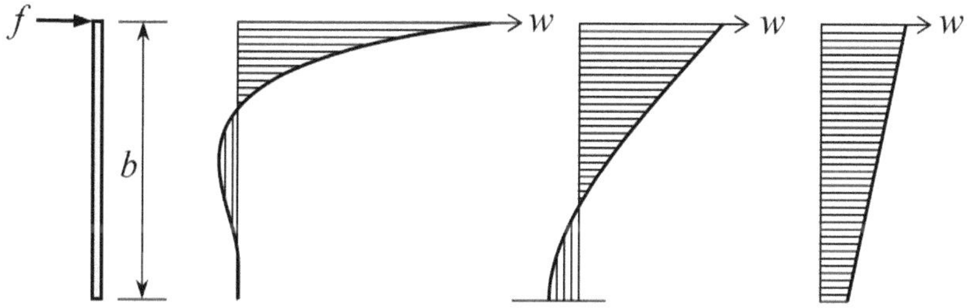

Figure 5.13 Deflection curve at mid-span.

In Section 2.2.2 we have already determined the solution of this homogenous equation. It follows that $w(x, y)$ becomes

$$w = \left(A_1 e^{\alpha y} + A_2 \alpha y e^{\alpha y} + A_3 e^{-\alpha y} + A_4 \alpha y e^{-\alpha y}\right) \cos \alpha x \tag{5.50}$$

The four constants A_1 to and included A_4 follow from two boundary conditions in the edge $y = -b/2$ and two in the edge $y = b/2$. At $y = -b/2$ the moment in y-direction is zero and the Kirchhoff shear force is $-f$. At $y = b/2$ both the bending moment and the Kirchhoff shear force are zero. We will now outline the solution for w on the line $x = 0$ (see Figure 5.13) for three special cases. This solution is closely related to the distribution of the bending moment over the width of the bridge at the middle of the span.

Case 1

The plate is supposed infinitely long in the y-direction, so the viaduct is very wide. Then A_1 and A_2 must be zero, for fading away of the first two terms in Eq. (5.50) to take place. As stated, the picture for the displacement is also a measure for the bending moment in the span direction. At sufficient distance from the loaded edge there is no deflection and bending moment.

Case 2

The plate is a square. This is a practical shape, as could appear in a viaduct. All four constants now are involved, and therefore all four terms $e^{\alpha y}$, $\alpha y e^{\alpha y}$,

$e^{-\alpha y}$ and $\alpha y e^{-\alpha y}$ are present in the solution. The plate sags at the loaded edge and lifts at the opposite edge.

Case 3

The plate (viaduct) is so narrow that it turns into a strip-shaped beam. The plate contributes over all its width in carrying the load, though the part close to the load carries most. In Section 2.2.2 we showed the solution in another form (Taylor expansion).

$$w = \left\{B_1 + B_2\alpha y + B_3(\alpha y)^2 + B_4(\alpha y)^3 + \cdots\right\} \cos\alpha x \tag{5.51}$$

We shall limit ourselves to the case that $\nu = 0$. The curvature is the second derivative, so the moment m_{yy} is linear in y; only the coefficients B_1, B_2, B_3 and B_4 need be considered. Because the moment m_{yy} has to be zero on both edges, it is zero everywhere. This means that B_3 and B_4 are zero. As a result, the deflection becomes linear in y. The two constants B_1 and B_2 follow from Kirchhoff shear force at the two edges, $-f$ and 0 respectively. The same solution would follow from beam theory. The beam is subjected to bending in x-direction by a line load f and to torsion about the x-axis by a distributed torque $bf/2$. The line load causes the constant deflection B_1 and the torque load the rotated part $B_2\alpha y$.

Effective Width

For the convenience of structural design the concept of *effective width* is introduced in codes of practice, because designers normally prefer to do a beam analysis. Suppose that Figure 5.14 is the distribution of the bending

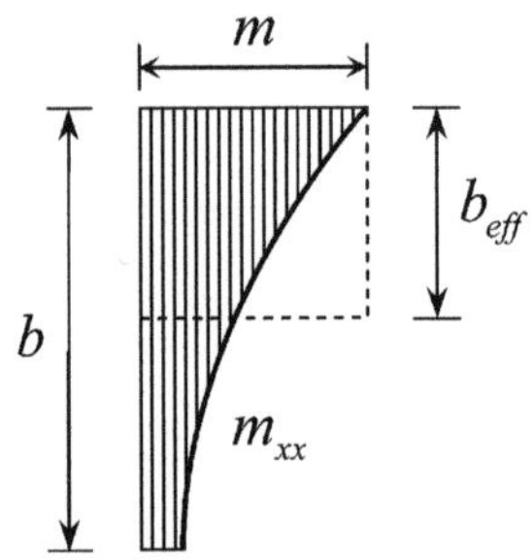

Figure 5.14 Definition of the effective width b_{eff}.

moment m_{xx} over the width b and the maximum value at the edge is m. If the sine-shaped load were applied to a beam with the same span a and a rectangular cross-section with the same depth as the plate, the maximum moment at mid-span would be

$$M = \frac{\hat{f}\, a^2}{\pi^2} \tag{5.52}$$

The ratio of M and m is the effective width b_{eff} of the plate. If engineers can make a good guess for this width, it suffices to do a beam analysis and to spread the total moment over the effective width in order to calculate the edge moment m. Codes of practice offer practical rules for the determination of the effective width.

5.6 Message of the Chapter

- A bending moment can occur for zero curvature. This is due to the effect of a non-zero Poisson's ration and a curvature in the transverse direction.
- A rectangular plate can be brought in a state of constant twisting moment by a set of four equilibrating corner forces.
- A simply-supported square plate under distributed load is about four times more effective than a one-way plate for the same load. Torsion in the corner zones takes care of half the load.
- The support reaction in a simply-supported plate under distributed load is about 50% higher than we expect on the basis of the shear force. The too-large compressive support reactions are balanced by concentrated tensile forces at the four corners. A large concentrated shear force occurs along the edges.
- If the simple support is materialized by stiff edge beams, the moment in the edge beams is about 50% larger than expected on the basis of the load that flows to the edge.

- The flexural stiffness of edge beams can be chosen such that zero twisting moments occur. In a twist-less square plate the bending moments in the edge beams must be calculated on the basis of one quarter of the load on the plate.

- Plate theory helps us to make estimates of the effective width in case of line edge loadings.

Chapter 6
Circular Membrane Plates

In this chapter circular plates in a membrane state require our attention. In Section 6.1 we study plates for axisymmetric load. Section 6.2 is devoted to non-axisymmetric load. Plate bending will be the subject of Chapter 7.

6.1 Axisymmetric Circular Membrane Problems

Figure 6.1 shows a homogeneous circular plate of constant thickness t and axisymmetric loading. For this type of problems it is convenient to change to polar coordinates. The position in the plate is specified by means of the radius r and the angle θ. The state of stress and strain is independent of θ, and there is just one displacement u in radial direction. Ordinary derivatives can be used since u depends only on the coordinate r, as does the load p per unit area. Only two membrane forces are present, n_{rr} and $n_{\theta\theta}$. The shear stress $n_{r\theta}$ cannot occur. Therefore, only two strains exist, ε_{rr} and $\varepsilon_{\theta\theta}$. The scheme for the essential quantities is displayed in Figure 6.2.

The strain ε_{rr} and the displacement u both act in the r-direction; they are related by $\varepsilon_{rr} = du/dr$. For the derivation of the tangential strain, $\varepsilon_{\theta\theta}$, a circle is considered with radius r. The circumference of this circle is $2\pi r$. After application of the axisymmetric load, each point of the circle displaces over a radial distance u. The new radius of the circle is $r + u$ and the circumference $2\pi(r + u)$. The increase of the circumference is $2\pi u$. Division of this increase by the original length $2\pi r$ provides the required strain $\varepsilon_{\theta\theta} = u/r$. So, the constitutive relations for plane stress are

$$\varepsilon_{rr} = \frac{du}{dr}, \quad \varepsilon_{\theta\theta} = \frac{u}{r} \quad \textit{Kinematic} \tag{6.1}$$

J. Blaauwendraad, *Plates and FEM: Surprises and Pitfalls*, Solid Mechanics and Its Applications 171, DOI 10.1007/978-90-481-3596-7_6,

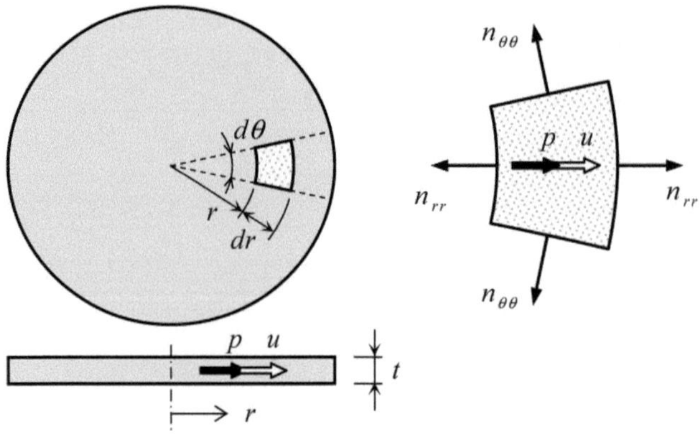

Figure 6.1 Displacement, load and membrane forces in axisymmetric plate.

The constitutive relations of Eq. (1.13) reduce to

$$\left.\begin{aligned} n_{rr} &= \frac{Et}{1-\nu^2}(\varepsilon_{rr} + \nu\varepsilon_{\theta\theta}) \\ n_{\theta\theta} &= \frac{Et}{1-\nu^2}(\varepsilon_{\theta\theta} + \nu\varepsilon_{rr}) \end{aligned}\right\} \quad \textit{Constitutive} \tag{6.2}$$

For the equilibrium equations we consider an elementary plate part of length dr and aperture angle $d\theta$ as shown in Figure 6.1. The length of the edge at the inside of the element is $r\,d\theta$. The total force on this edge is $n_{rr} r\,d\theta$ and points to the left. At the outside of the element, at a distance dr further, the force has increased $d(n_{rr} r\,d\theta)\,dr$, pointing to the right. The angle $d\theta$ is independent of r, which means that the increment can be written as $d(n_{rr} r)\,dr\,d\theta$. A force $n_{\theta\theta}\,dr$ is acting perpendicular to each straight edge of the element. Since the angle between the two forces is $d\theta$, there is a force $-n_{\theta\theta}\,dr\,d\theta$, where the minus sign indicates the direction of the force (negative r-direction).The distributed load p provides an outward-pointing force. For that purpose, p

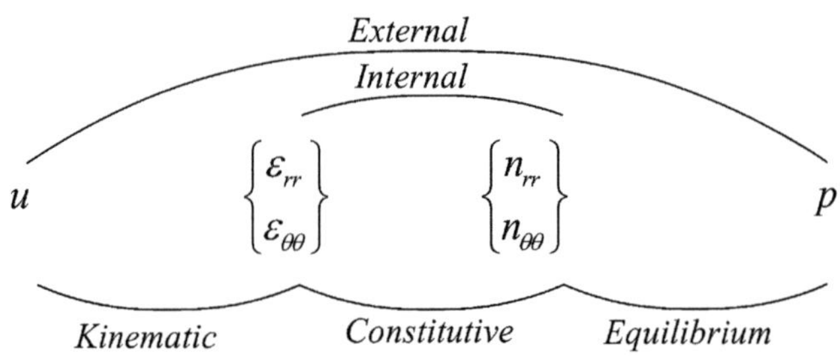

Figure 6.2 Scheme of relationships.

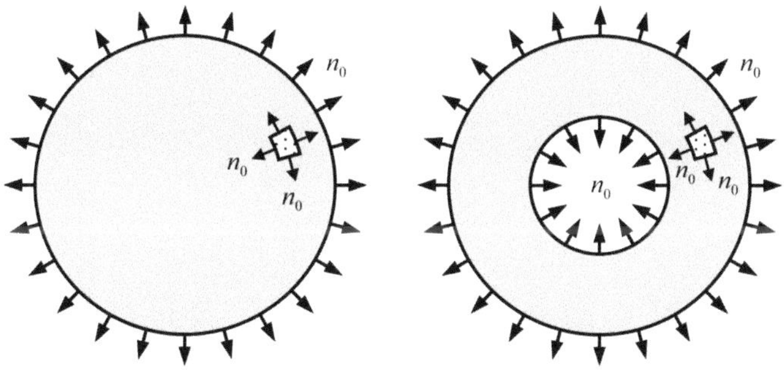

Figure 6.3 Only one constant stress-state exists with equal n_{rr} and $n_{\theta\theta}$.

has to be multiplied by the area $r\, rmd\theta\, dr$ of the plate element. The force equals $pr\, d\theta\, dr$. For equilibrium, the sum of the three forces has to be zero. After division by $d\theta\, dr$ the following equilibrium equation is obtained:

$$-\frac{d}{dr}(r\; n_{rr}) + n_{\theta\theta} = rp \quad \textit{Equilibrium} \tag{6.3}$$

In this stress problem, there are no rigid body displacements. For any displacement u there is a strain field. Further, only one combination of constant strains is possible. For the strain $\varepsilon_{\theta\theta}$ to have a constant value ε_0 a displacement is required of $u = \varepsilon_0 r$. The strain ε_{rr} then also equals ε_0. The only possible constant strains are identical strains $\epsilon_{\theta\theta}$ and ϵ_{rr}. Then, from the constitutive relations in Eq. (6.2) it follows that the membrane forces n_{rr} and $n_{\theta\theta}$ are equal and constant too. When the constant values $n_{rr} = n_0$ and $n_{\theta\theta} = n_0$ are substituted in the equilibrium equation (6.3) it appears that the distributed load p across the plate area has to be zero. The plate can be loaded only along the edge. Figure 6.3 shows two situations, a circular plate with and without a hole. In both plates, in each point, a membrane force n_0 is present and Mohr's circle is reduced to a point.

An alternative to deriving the three basic sets of equations is the consideration of work. Slightly different quantities are used, which we will show here. The equilibrium equation (6.3) comprises the terms rn_{rr} and rp. It makes sense to introduce new variables for these combinations. This will be done for $rn_{\theta\theta}$ too. We define

$$N_{rr} = r\, n_{rr}; \quad N_{\theta\theta} = r\, n_{\theta\theta}; \quad f = rp \tag{6.4}$$

The two quantities N_{rr} and $N_{\theta\theta}$ are normal forces with the dimension of force; f is a line load with the dimension of force per unit of length. Application of the transformations in Eq. (6.4) keeps the kinematic equations

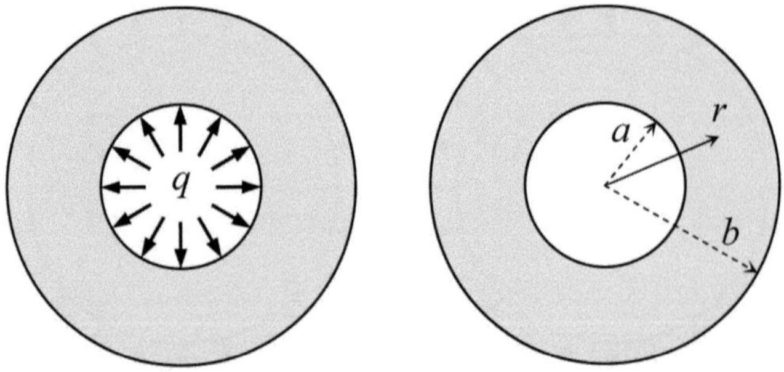

Figure 6.4 Thick-walled pipe with load at inner face.

(6.1), and changes the constitutive equations (6.2) and equilibrium equation (6.3) into

$$\varepsilon_{rr} = \frac{du}{dr}, \quad \varepsilon_{\theta\theta} = \frac{u}{r} \qquad \textit{Kinematic} \tag{6.5}$$

$$\left.\begin{aligned} N_{rr} &= \frac{Etr}{1-\nu^2}(\varepsilon_{rr} + \nu\varepsilon_{\theta\theta}) \\ N_{\theta\theta} &= \frac{Etr}{1-\nu^2}(\varepsilon_{\theta\theta} + \nu\varepsilon_{rr}) \end{aligned}\right\} \qquad \textit{Constitutive} \tag{6.6}$$

$$-\frac{dN_r}{dr} + \frac{N_{\theta\theta}}{r} = f \qquad \textit{Equilibrium} \tag{6.7}$$

We will continue with these three equations for the derivation of the differential equation. Substitution of the kinematic equations (6.5) into the constitutive equations (6.6) leads to

$$N_{rr} = \frac{Etr}{1-\nu^2}\left(\frac{du}{dr} + \nu\frac{u}{r}\right); \quad N_{\theta\theta} = \frac{Etr}{1-\nu^2}\left(\frac{u}{r} + \nu\frac{du}{dr}\right) \tag{6.8}$$

Substitution of this result into the equilibrium equation (6.7) leads to the differential equation

$$\frac{Et}{1-\nu^2}Lu = f \tag{6.9}$$

where the operator L is

$$L = r\frac{d}{dr}\frac{1}{r}\frac{d}{dr}r \tag{6.10}$$

This differential equation is of the second order.

6.1.1 Thick-Walled Tube

The differential equation can be used to determine the stress state in a thick-walled tube subjected to an internal gas pressure q. Figure 6.4 defines the tube. Flat sections remain flat after deformation, but the strain ε_{zz} in the axial direction will not be zero. On average, σ_{zz} will be equal to zero. Therefore, the problem will be treated as a plane stress state. A slice of unit thickness of the tube is considered, which is cut perpendicularly to the axial direction. This means that n_{rr} and $n_{\theta\theta}$ are equal to σ_{rr} and $\sigma_{\theta\theta}$, respectively. For this load case it holds that the distributed load f is zero and therefore

$$Lu = 0 \tag{6.11}$$

In the general solution of this second-order differential equation there are two coefficients which must be determined from the boundary conditions. Choosing the trial solution Cr^m, we obtain two roots $m = -1$ and $m = 1$. Therefore

$$u = C_1 \frac{1}{r} + C_2 r \tag{6.12}$$

The coefficients C_1 and C_2 follow from the two boundary conditions

$$\begin{aligned} r = a \;\rightarrow\; n_{rr} &= -q; \quad N_{rr} = -qr \\ r = b \;\rightarrow\; n_{rr} &= 0; \quad N_{\theta\theta} = 0 \end{aligned} \tag{6.13}$$

The results for the displacement u and the stress quantities n_{rr} and $n_{\theta\theta}$ become

$$u(r) = \frac{a^2}{b^2 - a^2} \frac{q}{Et} \left\{ (1+\nu) \frac{b^2}{r} + (1-\nu) r \right\} \tag{6.14}$$

$$n_{rr} = \frac{a^2}{b^2 - a^2} \left(-\frac{b^2}{r^2} + 1 \right) q; \quad n_{\theta\theta} = \frac{a^2}{b^2 - a^2} \left(\frac{b^2}{r^2} + 1 \right) q \tag{6.15}$$

The results are presented in Figure 6.5. Both stresses σ_{rr} and $\sigma_{\theta\theta}$ are nonlinear over the thickness of the tube.

6.1.2 Circular Hole in a Homogeneous Stress State

We want to compute the stress concentration factor on the edge of a circular hole in a large plate with a homogeneous stress state of equal principal membrane forces n. The radius of the hole is a. The homogeneous membrane forces without the hole are

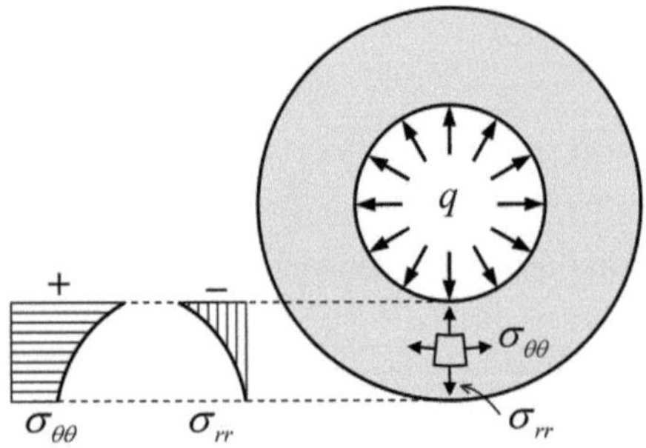

Figure 6.5 Stresses in a thick-walled pipe due to gas pressure.

$$n_{rr} = n; \quad n_{\theta\theta} = n \tag{6.16}$$

In order to make the edge of the proposed hole to be stress-free a loading case must be superimposed in which the edge is loaded with an opposite load on the boundary $r = a$. The boundary condition is $n_{rr} = -n$. Further it is known that the stresses will vanish for large radius r, which requires C_2 to be zero. The result is

$$n_{rr} = n\left(-\frac{a^2}{r^2}\right); \quad n_{\theta\theta} = n\left(\frac{a^2}{r^2}\right) \tag{6.17}$$

Still these membrane forces must be superimposed on the constant equal stresses for the case without a hole. The final result for the large plate with hole is

$$n_{rr} = n\left(1 - \frac{a^2}{r^2}\right); \quad n_{\theta\theta} = n\left(1 + \frac{a^2}{r^2}\right) \tag{6.18}$$

Due to the hole, the maximum value of the membrane force $n_{\theta\theta}$ is twice the value n of the homogeneous stress state. The stress concentration factor is 2, see Figure 6.6.

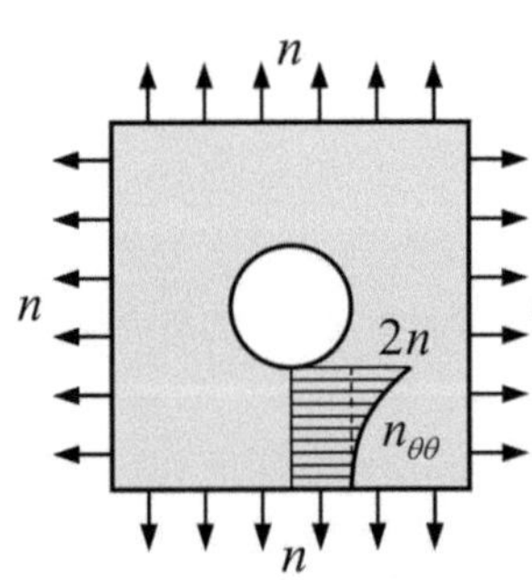

Figure 6.6 Concentration factor 2 for equal principal stresses.

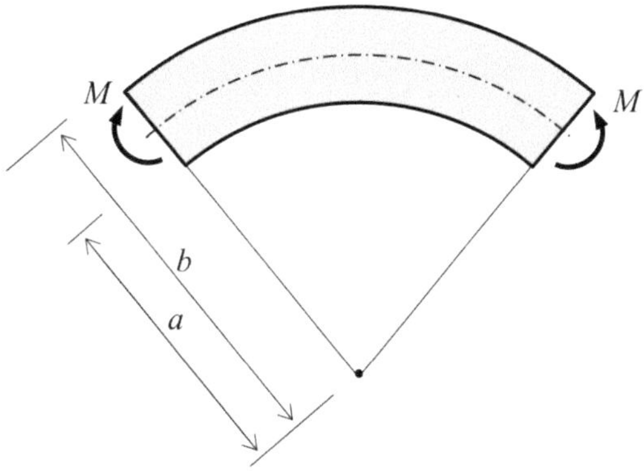

Figure 6.7 Curved beam subjected to constant moment.

6.1.3 Curved Beam Subjected to Constant Moment

Consider a curved beam with a constant radius of curvature as shown in Figure 6.7. The inner and outer radii are a and b, respectively, and the beam is subjected to a constant moment M. The stress state in this axisymmetric structure will also be axisymmetric. The beam has a rectangular cross-section of small width t. This curved bar is modelled as a thin plate with thickness t. This problem can be solved with the findings of Section 6.1. An alternative solution procedure is given in [7]. The stresses are

$$\sigma_{rr} = f_{rr}\frac{M}{C}; \quad \sigma_{\theta\theta} = f_{\theta\theta}\frac{M}{C} \tag{6.19}$$

where

$$C = t\left(\frac{1}{4}\left(b^2 - a^2\right) - \frac{a^2b^2}{b^2 - a^2}\left(\ln\frac{b}{a}\right)^2\right) \tag{6.20}$$

$$\begin{aligned} f_{rr} &= -\frac{ab}{b^2 - a^2}\ln\left(\frac{b}{a}\right)\frac{ab}{r^2} + \frac{a^2}{b^2 - a^2}\ln\left(\frac{r}{a}\right) - \frac{b^2}{b^2 - a^2}\ln\left(\frac{r}{b}\right), \\ f_{\theta\theta} &= -1 + \frac{a\,b}{b^2 - a^2}\ln\left(\frac{b}{a}\right)\frac{a\,b}{r^2} + \frac{a^2}{b^2 - a^2}\ln\left(\frac{r}{a}\right) - \frac{b^2}{b^2 - a^2}\ln\left(\frac{r}{b}\right) \end{aligned} \tag{6.21}$$

In these expressions, C depends only on a, b and the thickness t, the geometry data. The functions $f_{rr}(r)$ and $f_{\theta\theta}(r)$ are dimensionless and provide the distribution of stresses over the height of the cross-section. In Figure 6.8 this distribution is displayed for two different values of the ratio a/b, a value

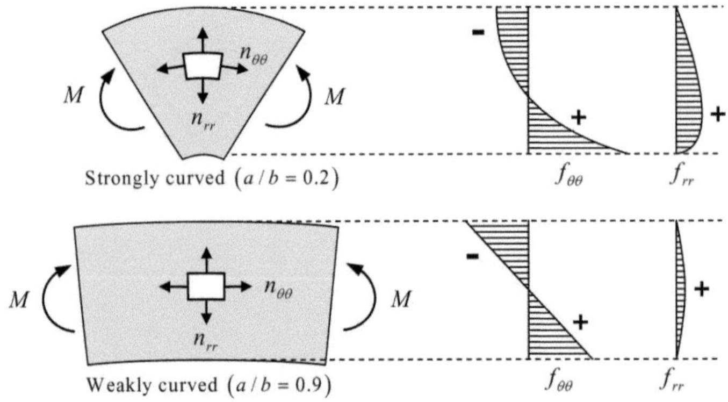

Figure 6.8 Stress distribution in curved beam for different curvatures.

that is small compared to unity (strong curvature) and a value close to unity (weak curvature). For a strong curvature, the bending stress distribution deviates severely from a linear distribution, irrespective of the fact that flat cross-sections remain flat.

We note that for pure bending, there are also stresses σ_{rr} in the height direction. This can be made clear if we consider the equilibrium of the part of the beam below the neutral line. Integration of the tensile stresses $\sigma_{\theta\theta}$ over the height of the beam part leads to a tensile force. The two tensile forces acting on both ends of the beam part have different directions and work line. Equilibrium is possible only if there is a radial outward-pointing stress σ_{rr} in the neutral line, acting over the whole length of the beam part. Therefore, it can be concluded that σ_{rr} is a tensile stress. The same conclusion is obtained if we consider the part of the beam outside the neutral line, where compressive stresses $\sigma_{\theta\theta}$ are present. If we translate this finding to a curved reinforced concrete beam, we conclude that stirrups are needed in a curved beam even when there is no shear force. Figure 6.9 demonstrates this by drawing struts and ties in the curved beam. Red lines are in tension and green lines in compression.

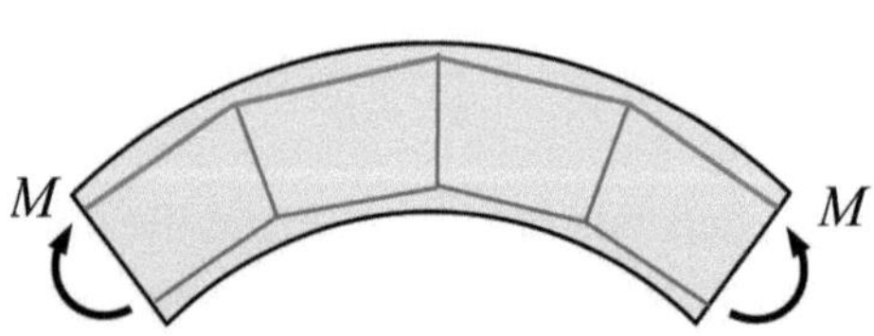

Figure 6.9 A constant moment in concrete curved beam may ask for stirrups.

6.2 Non-Axisymmetric Circular Membrane Problems

If the stress state is not axisymmetric, we have to account for three stresses σ_{rr}, $\sigma_{\theta\theta}$ and $\sigma_{r\theta}$. It appears advantageous to switch to the force method for this type of problem. We will discuss problems where load is applied only on the edges. We start the discussion in the orthogonal set of axes x, y. In Chapter 1 we derived the kinematic equations (1.9), the constitutive equations (1.13) and the equilibrium equations (1.14), and we substituted them in each other, starting from the kinematic equations and ending with the equilibrium equations. In the force method we make use of these three set of equations in the opposite order. We first construct a stress solution that satisfies equilibrium, after that we use the constitutive relations, to end up with expressions for strains and finally we construct a compatibility condition for the strains from the kinematic relations. In Eq. (1.14) we have three unknown stresses in two equilibrium equations. Therefore the stress state is statically indeterminate and we introduce a stress function $\varphi(x, y)$, as initially proposed by Airy [13]. The following set of stresses satisfies the two equilibrium equations in (1.6) for zero distributed area forces p_x and p_y

$$n_{xx} = \frac{\partial^2 \varphi}{\partial y^2}; \quad n_{yy} = \frac{\partial^2 \varphi}{\partial x^2}; \quad n_{xy} = -\frac{\partial^2 \varphi}{\partial x \partial y} \quad \textit{Equilibrium} \tag{6.22}$$

The constitutive relations are now used in the shape of Eq. (1.12)

$$\begin{Bmatrix} \varepsilon_{xx} \\ \varepsilon_{yy} \\ \gamma_{xy} \end{Bmatrix} = \frac{1}{Et} \begin{bmatrix} 1 & -\nu & 0 \\ -\nu & 1 & 0 \\ 0 & 0 & 2(1+\nu) \end{bmatrix} \begin{Bmatrix} n_{xx} \\ n_{yy} \\ n_{xy} \end{Bmatrix} \quad \textit{Constitutive} \tag{6.23}$$

The required compatibility relation for the strains is derived from the kinematic relations in (1.9). There three strain relations are expressed in terms of in two degrees of freedom u_x and u_y. Elimination of these two displacements leads to one relation between the three strains

$$\frac{\partial^2 \varepsilon_{xx}}{\partial y^2} + \frac{\partial^2 \varepsilon_{yy}}{\partial x^2} - \frac{\partial^2 \gamma_{xy}}{\partial x \partial y} = 0 \quad \textit{Compatibility} \tag{6.24}$$

Substitution of the three relationships (6.22) and (6.23) into Eq. (6.24) leads to a differential equation for the stress function φ. Again we find the biharmonic equation which was obtained in Chapter 1.

$$\left(\frac{\partial^4}{\partial x^4} + 2\frac{\partial^4}{\partial x^2 \partial x^2} + \frac{\partial^4}{\partial y^4} \right) \varphi = 0, \quad \nabla^2 \nabla^2 \phi = 0 \tag{6.25}$$

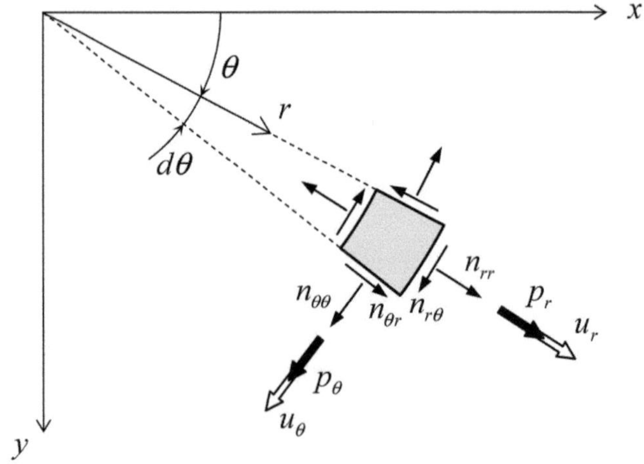

Figure 6.10 Displacements and stresses in polar coordinates.

Due to Eq. (6.22), the quantity $\nabla^2\varphi$ is the sum of the two normal forces. Figure 6.10 shows which displacements and stresses are involved for the description in polar coordinates. The three second derivatives in (6.22) need be transformed to derivatives with respect to r and θ. More precisely, we have to change from the orthogonal set of axis (x, y) to the orthogonal set (r, t) of directions, where t is the direction of the tangent line to a circle of radius r, see Figure 6.10. Formally the transformation is done with aid of the chain rule for derivatives; This leads to

$$\begin{aligned} \frac{\partial^2\varphi}{\partial x^2} &\to \frac{\partial^2\varphi}{\partial r^2} \\ \frac{\partial^2\varphi}{\partial y^2} &\to \frac{1}{r}\frac{\partial\varphi}{\partial r} + \frac{1}{r^2}\frac{\partial^2\varphi}{\partial\theta^2} \\ \frac{\partial^2\varphi}{\partial x\partial y} &\to \frac{\partial}{\partial r}\left(\frac{1}{r}\frac{\partial\varphi}{\partial\theta}\right) \end{aligned} \tag{6.26}$$

The transfer from the second derivative with respect to x to the second derivative with respect to r is simple. We just replace x by r. The mixed second derivative with respect to x and y is also simple, if we notice that dy is equal to $r\,d\theta$. However, the transformation of the second derivative with respect to y needs more explication. The result consists of two contributions. The last one, which has $r^2\,d\theta^2$ in the numerator, is expected; but the first one may be a surprise. This term is independent of θ. Figure 6.11 helps to explain this term. To understand the second derivative in t-direction we consider the

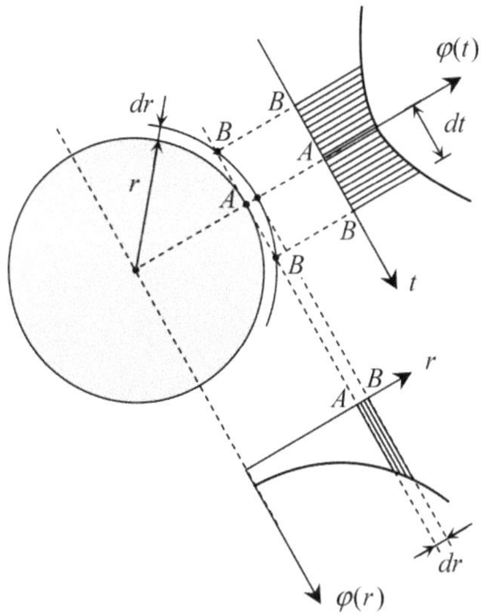

Figure 6.11 Second derivative in tangent line in axisymmetric state.

value of φ at point A on the circle with radius r and the two points B at the circle with radius $r + dr$. If $d\varphi/dr$ is not zero the value of φ in the points B will be different from the value in point A, and therefore a non-zero second derivative exists in the tangent line at point A. This second derivative always occurs, in axisymmetric as well as non-axisymmetric cases.

Now it is clear how we must replace Eq. (6.22). For polar coordinates we use

$$n_{rr} = \frac{1}{r}\frac{\partial \varphi}{\partial r} + \frac{1}{r^2}\frac{\partial^2 \varphi}{\partial \theta^2}; \quad n_{\theta\theta} = \frac{\partial^2 \varphi}{\partial r^2}; \quad n_{r\theta} = -\frac{\partial}{\partial r}\left(\frac{1}{r}\frac{\partial \varphi}{\partial \theta}\right) \tag{6.27}$$

The Laplace operator ∇^2 can be determined from the sum of n_{rr} and $n_{\theta\theta}$. The equation of Airy (6.25) then becomes:

$$\left(\frac{\partial^2}{\partial r^2} + \frac{1}{r}\frac{\partial}{\partial r} + \frac{1}{r^2}\frac{\partial^2}{\partial \theta^2}\right)\left(\frac{\partial^2}{\partial r^2} + \frac{1}{r}\frac{\partial}{\partial r} + \frac{1}{r^2}\frac{\partial^2}{\partial \theta^2}\right)\varphi = 0 \tag{6.28}$$

6.2.1 Point Load on a Half Plane

The obtained differential equation (6.28) can be used to find the stress distribution in a half plane due to a point load F on the edge, as shown in

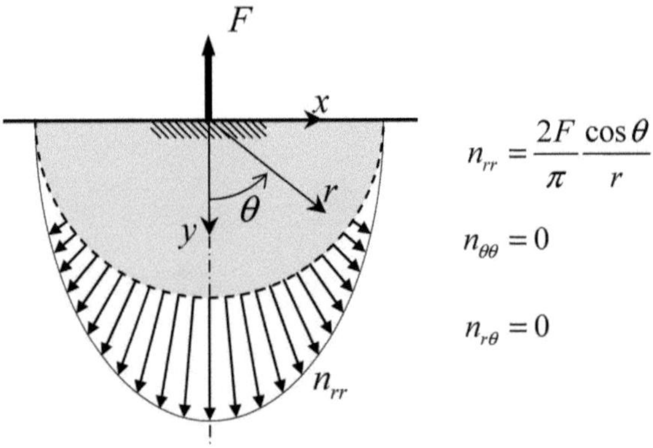

Figure 6.12 Stresses in half-plane with vertical point load.

Figure 6.12. The boundary conditions are zero stresses $\sigma_{\theta\theta}$ and $\sigma_{r\theta}$ at the free edge and zero stresses for infinitely large radius r. It can be shown that the trial function

$$\varphi = Cr\theta \sin\theta \tag{6.29}$$

satisfies Airy's biharmonic equation and the boundary conditions. By application of Eq. (6.27) the membrane forces become

$$n_{rr} = 2C\frac{1}{r}\cos\theta; \quad n_{\theta\theta} = 0; \quad n_{r\theta} = 0 \tag{6.30}$$

This result is very special. The stresses $\sigma_{\theta\theta}$ and $\sigma_{r\theta}$ are not zero just at the edge ($\theta = \pi/2$), but for any value of the angle θ. At a half circle in the half plane the shear stress $\sigma_{r\theta}$ and tangential stress $\sigma_{\theta\theta}$ are zero. Just a membrane force n_{rr} is present. The value of C can be calculated from the condition that vertical equilibrium must exist between the point load F and the membrane forces n_{rr}. This condition leads to

$$n_{rr} = \frac{2F}{\pi} \frac{\cos\theta}{r} \tag{6.31}$$

Boussinesq [14] even found such a type of solution for a compressive point-load F on an infinite 3D half-space, from which Flamant [15] obtained the stated solution. Therefore, the solution for a point-load on a half-plane is also called Boussinesq's solution. In each point (r, θ) a transformation can be made from the stresses σ_{rr}, $\sigma_{\theta\theta}$ and $\sigma_{r\theta}$ to the stresses σ_{xx}, σ_{yy} and σ_{xy}. These three stresses are all different from zero. Figure 6.13 shows the distribution of the vertical membrane force n_{yy}. The deeper the section, the more n_{yy} is spread.

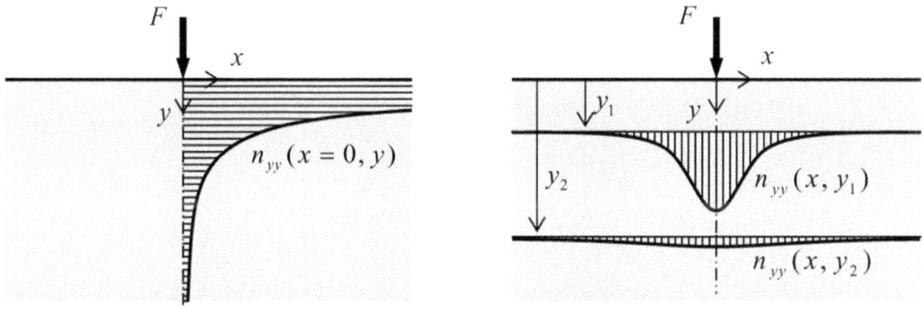

Figure 6.13 Vertical stress under point load on half-space.

6.2.2 Brazilian Splitting Test

A well-known method for determining the splitting strength of brittle materials like concrete is the so called Brazilian splitting test. In this test a circular cylinder is loaded by two opposite line loads as shown in Figure 6.14. Direct tensile tests on concrete are difficult to perform, because a special tensile test set-up needs to be available. Fortunately, there is a simple relation between the vertical line load and the tensile stress in a Brazilian splitting test. It is assumed that the stresses do not vary along the axial direction of the cylinder so that a slice of unit thickness can be considered. The solution of the point load on an half space of Section 6.2.1 can be used to determine the stress state in the cylinder. The solution for a compressive force F on a half-plane becomes very simple when it is presented by eccentric circles as done in Figure 6.15. For all points on a circle $r = d \cos\theta$. Then in each circle the stress σ_{rr} is constant while the other stress components $\sigma_{\theta\theta}$ and $\sigma_{r\theta}$ are zero. The constant value of σ_{rr} is $-\sigma_o$ in which $\sigma_o = 2F/\pi d$ is positive. In the vertical line of symmetry, the horizontal stress, $\sigma_{\theta\theta}$, is zero. The material outside

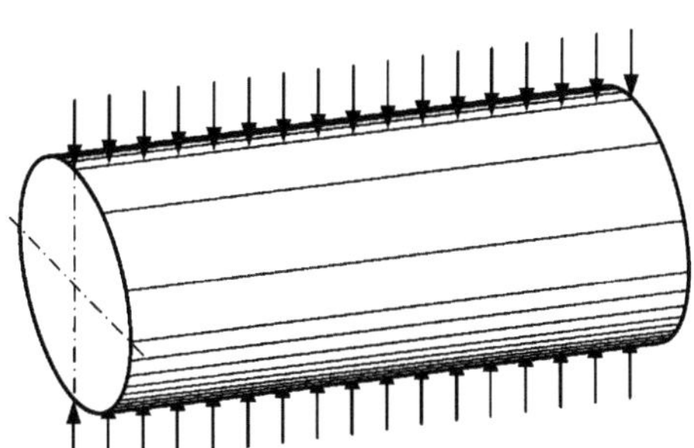

Figure 6.14 Loading scheme on solid cylinder in Brazilian splitting test.

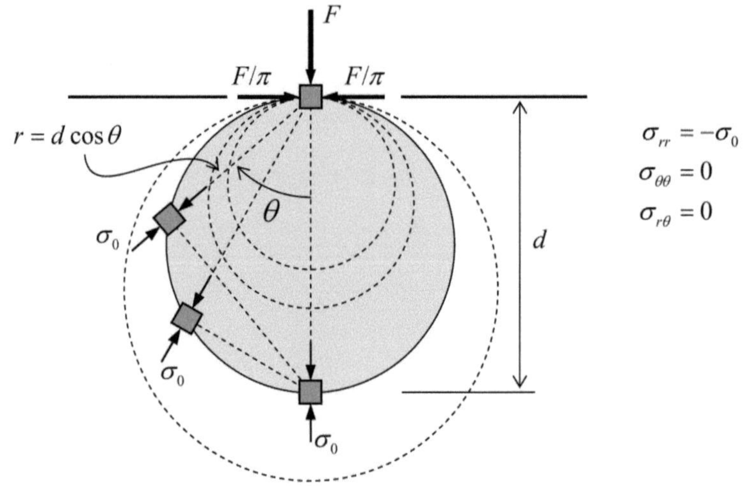

Figure 6.15 Alternative presentation of stresses in half-space.

the circle has been removed and replaced by the edge loading σ_o. In Figure 6.16 the same figure is presented together with the mirror image of the solution. When both solutions are superimposed a circular disk is obtained that is loaded by two concentrated forces F and a radial edge stress $-\sigma_o$. Note that no horizontal stresses are present in the vertical line of symmetry.

The final step in the derivation is to remove the edge stress by adding the axisymmetric solution of a disk with a constant tensile stress σ_o on the edge depicted in Figure 6.3. In this load case there is a hydrostatic stress state with a tensile stress σ_o acting in all directions. Therefore, the horizontal stress, $\sigma_{\theta\theta}$, in the vertical line of symmetry is σ_o. The result of the superposition is

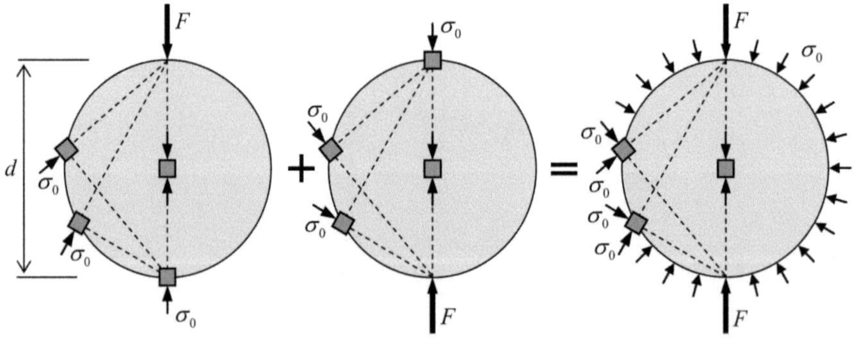

Figure 6.16 Sum of half-space solution and its mirror image.

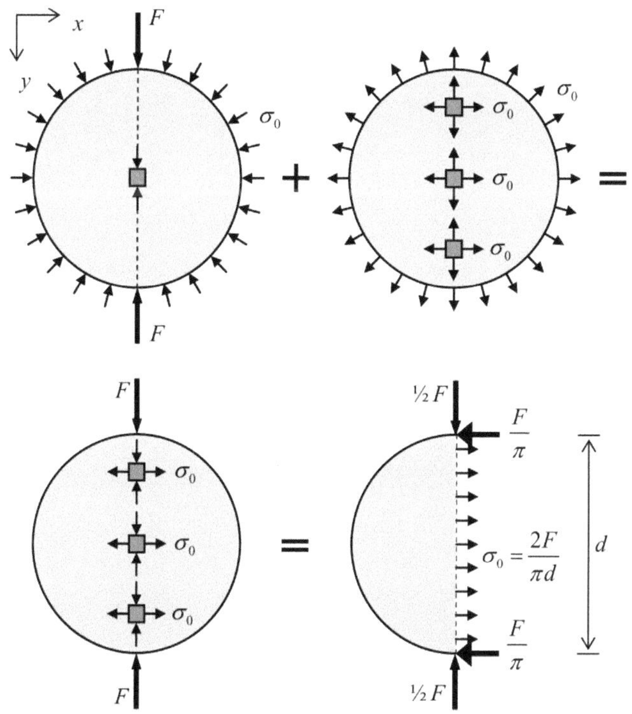

Figure 6.17 Superposition of two stress states yields the final solution.

a circular disk subjected to two diametrically opposite point-loads F with a free unloaded circular edge, see Figure 6.17. On the vertical line of symmetry there is a constant tensile stress σ_0. Therefore, the ultimate result is a homogeneous tensile stress along the vertical line of symmetry, $\sigma_{xx} = 2F/\pi d$. The total horizontal force on the line of symmetry has to be zero. Therefore, there must be local horizontal compressive forces at the point of action of the forces F, equal to $\frac{1}{2}\sigma_0 d = F/\pi$.

In this linear-elastic solution the homogeneous tensile stress in the vertical line of symmetry is balanced by a concentrated horizontal force F/π at each end of the line of symmetry. This implies infinitely large compressive stresses at those positions. In reality the elasticity limits will be surpassed and nonlinear material effects will enter in the zones where the loads are applied.

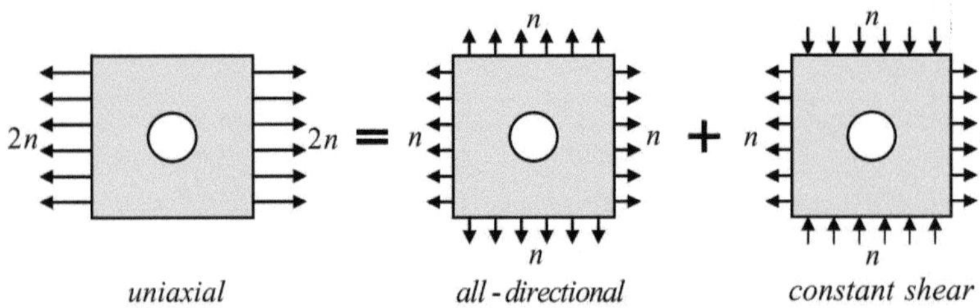

Figure 6.18 Uniaxial stress state as combination of two basic cases.

6.2.3 Hole in Plates with Shear and Uniaxial Stress

In Section 6.1 for axisymmetric problems we were able to determine the stress peak near a hole in a plate in a radially homogeneous (is hydrostatic) stress state. Now we will investigate the case of a hole in a plate with a constant shear stress. We do it for a plate which has a tensile stress in the x-direction and a compressive stress in the y-direction. After we have solved that problem we also can determine the stress concentration in an uniaxially stressed plate. That case is a superposition of the hydrostatic case of Section 6.1 and the present constant shear case. Figure 6.18 shows this superposition.

Shear Stress

We consider a large plate with a hole, in which equal principal stresses of opposite sign occur, see Figure 6.19. We choose the origin of the axes r and θ at the centre of the hole. The value of the principal membrane forces is n. It can be verified from Mohr's circle or the transformation rules that the homogeneous membrane forces at each position in the plate in absence of the hole would be

$$\begin{aligned} n_{rr} &= \quad n \cos 2\theta \\ n_{\theta\theta} &= -n \cos 2\theta \\ n_{r\theta} &= -n \sin 2\theta \end{aligned} \tag{6.32}$$

If a hole is created, the membrane forces n_{rr} and $n_{r\theta}$ have to be made zero on the edge of the hole. This means that an edge loading has to be superimposed, which causes the same membrane forces but with an opposite sign:

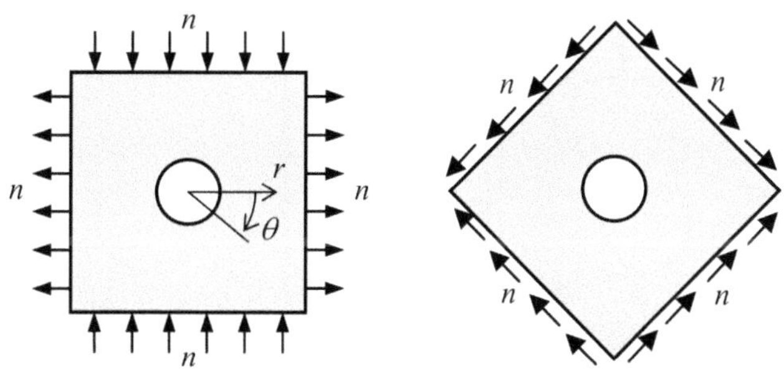

Figure 6.19 Circular hole in constant shear field.

$$\begin{aligned} n_{rr} &= -n \cos 2\theta \\ n_{r\theta} &= \quad n \sin 2\theta \end{aligned} \tag{6.33}$$

The bi-harmonic differential equation (6.28) has to be solved. This can be done by choosing a solution for φ of the form

$$\varphi = \varphi(r) \cos 2\theta \tag{6.34}$$

This means that the variables r and θ are separated. Substitution in the differential equation yields an ordinary fourth-order differential equation for $\varphi(r)$:

$$\left(\frac{d^2}{dr^2} + \frac{1}{r}\frac{d}{dr} - \frac{4}{r^2}\right)\left(\frac{d^2}{dr^2} + \frac{1}{r}\frac{d}{dr} - \frac{4}{r^2}\right)\varphi = 0 \tag{6.35}$$

The general solution of this fourth-order differential equation will have four constants, to be determined from the boundary conditions. Substitution of the trial function Cr^m leads to four roots $m = -2$, $m = 0$, $m = 2$ and $m = 4$, therefore the solution is

$$\varphi = \left(C_1 r^4 + C_2 r^2 + C_3 + C_4 \frac{1}{r^2}\right) \cos 2\theta \tag{6.36}$$

From Eq. (6.27) we derive the expressions for the membrane forces. We must determine the four coefficients from the boundary conditions, two on the edge of the hole, see Eq. (6.33), and two from the condition that all membrane forces vanish for large r. The result for the membrane forces is

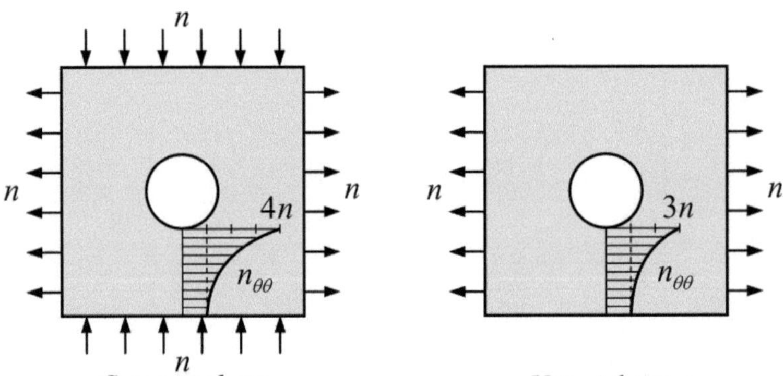

Figure 6.20 Stress concentration factors for constant shear and uniaxial stress.

$$\begin{aligned} n_{rr} &= n\left(-4\frac{a^2}{r^2}+3\frac{a^4}{r^4}\right)\cos 2\theta \\ n_{\theta\theta} &= n\left(-3\frac{a^4}{r^4}\right)\cos 2\theta \\ n_{r\theta} &= n\left(-2\frac{a^2}{r^2}+3\frac{a^4}{r^4}\right)\sin 2\theta \end{aligned} \tag{6.37}$$

This solution still has to be superimposed on the homogenous stresses of Eq. (6.33) for the situation without hole. The final result is

$$\begin{aligned} n_{rr} &= n\left(1-4\frac{a^2}{r^2}+3\frac{a^4}{r^4}\right)\cos 2\theta \\ n_{\theta\theta} &= n\left(-1-3\frac{a^4}{r^4}\right)\cos 2\theta \\ n_{r\theta} &= n\left(-1-2\frac{a^2}{r^2}+3\frac{a^4}{r^4}\right)\sin 2\theta \end{aligned} \tag{6.38}$$

The maximum tensile stress $n_{\theta\theta}$ at the hole edge in peripheral direction appears for $r = a$, $\theta = \pm\pi$ and is equal to $4n$. This value is four times the applied principal membrane stresses; the stress concentration factor is 4, see Figure 6.20.

Uni-Axial Stress

The uni-axial stress state is found from the superposition of solution (6.39) and the solution for the axisymmetric case in Section 6.1.2, divided by 2 in order to relate it to an applied stress of the magnitude n

$$\begin{aligned} n_{rr} &= \frac{n}{2}\left\{\left(1-\frac{a^2}{r^2}\right)+\left(1-4\frac{a^2}{r^2}+3\frac{a^4}{r^4}\right)\cos 2\theta\right\} \\ n_{\theta\theta} &= \frac{n}{2}\left\{\left(1+\frac{a^2}{r^2}\right)-\left(1+3\frac{a^4}{r^4}\right)\cos 2\theta\right\} \\ n_{r\theta} &= \frac{n}{2}\left\{-1-2\frac{a^2}{r^2}+3\frac{a^4}{r^4}\right\}\sin 2\theta \end{aligned} \tag{6.39}$$

For this stress state the maximum tensile membrane force $n_{\theta\theta}$ is three times the value of the uniaxial membrane force. The stress concentration factor is 3. The distribution of the stresses is shown in the right-hand part of Figure 6.20.

6.3 Message of the Chapter

- In thick-walled tubes under internal pressure the stress in tangential direction is not constant over the thickness. There is a nonlinear distribution.
- In thick-walled tubes under internal pressure we cannot neglect the stresses in thickness direction.
- A constant moment in a curved beam evokes tensile stresses in the depth direction of the beam. For a reinforced beam, stirrups may be needed in absence of a shear force.
- At a round hole in a homogeneous (hydrostatic) stress field the stress concentration factor is 2.

- At a round hole in a uniaxial stress field the stress concentration factor is 3.

- For a constant shear stress field the stress concentration factor even gets the value 4.

- Stresses in a half plane due to a point load normal to the free edge are very special. In polar coordinates just normal stresses in radial direction occur. The tangential stress and shear stress are zero.

- In the Brazilian splitting test there is a homogeneous tensile stress over the vertical plane of symmetry of the cylinder. These stresses are accompanied by balancing compressive lumped forces, close to the applied loads.

Chapter 7
Circular Thin Plates in Bending

In this chapter we elaborate on the theory of thin plates for circular plates resisting an axisymmetric load. We can do that in a compact way by using the derivation of the bi-harmonic equation for rectangular coordinates in Chapter 4 and its transformation to polar coordinates in Chapter 6.

7.1 Derivation of the Differential Equation

The vertical deflection w will solely depend on the radius r. The moments and shear forces are defined in the direction of the polar coordinates r and θ. The moments to be distinguished are the radial moments mrr and the tangential moments $m_{\theta\theta}$. Due to the axisymmetry no twisting moments $m_{r\theta}$ occur. Only one shear force component is present, the radial shear force v_r acting on areas of constant radius r. Because of axisymmetry, there is no shear force v_θ in sections of constant θ. We conclude that the bending moments m_{rr} and $m_{\theta\theta}$ are principal moments. The moment trajectories coincide with the radii and meridians of the plate. The shear force v_r is the principal shear force, and the shear trajectories are in the directions of the radii.

The second derivatives for polar coordinates in Eq. (6.27) can be used to find the curvatures in the circular plate. Because of axisymmetry the derivative with respect to θ is zero. The two curvatures become

$$\kappa_{rr} = -\frac{d^2 w}{dr^2}; \qquad \kappa_{\theta\theta} = -\frac{1}{r}\frac{dw}{dr} \tag{7.1}$$

J. Blaauwendraad, *Plates and FEM: Surprises and Pitfalls*, Solid Mechanics and Its Applications 171, DOI 10.1007/978-90-481-3596-7_7,

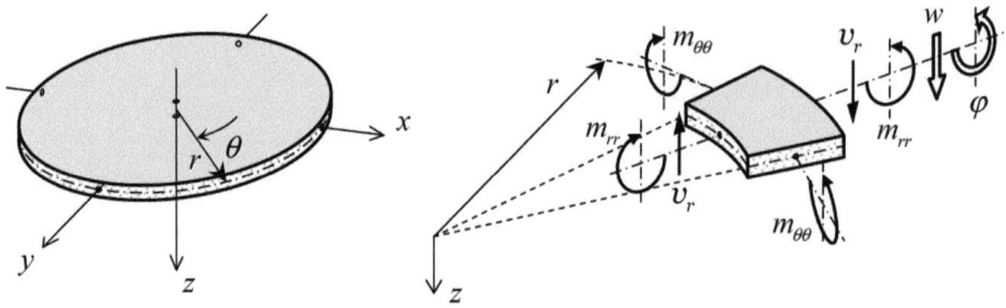

Figure 7.1 Displacements, moments and shear force in axisymmetric plate.

The constitutive relationships of (4.3) are replaced by and reduced to

$$m_{rr} = D(\kappa_{rr} + \nu\kappa_{\theta\theta}); \quad m_{\theta\theta} = D(\nu\kappa_{rr} + \kappa_{\theta\theta}) \tag{7.2}$$

On the basis of Eq. (6.28) we can transform the bi-harmonic equation (4.8) for a rectangular plate under a homogeneous distributed load p into an equation for polar coordinates, omitting terms which are dependent on θ

$$\left(\frac{\partial^2}{\partial r^2} + \frac{1}{r}\frac{\partial}{\partial r}\right)\left(\frac{\partial^2}{\partial r^2} + \frac{1}{r}\frac{\partial}{\partial r}\right) w = p. \tag{7.3}$$

The general solution of this fourth-order differential equation is

$$w = C_1 + C_2 r^2 + C_3 \ln r + C_4 r^2 \ln r + \frac{pr^4}{64D}. \tag{7.4}$$

The last term is the particular solution. The four integration constants have to be determined from the boundary conditions. After that, we obtain the moments and the shear force from the formulae

$$\begin{aligned} m_{rr} &= -D\left(\frac{d^2w}{dr^2} + \nu\frac{1}{r}\frac{dw}{dr}\right), \\ m_{\theta\theta} &= -D\left(\frac{1}{r}\frac{dw}{dr} + \nu\frac{d^2w}{dr^2}\right) \\ v_r &= -D\left(\frac{d^3w}{dr^3} + \frac{1}{r}\frac{d^2w}{dr^2} - \frac{1}{r^2}\frac{dw}{dr}\right) \end{aligned} \tag{7.5}$$

7.2 Simply-Supported Circular Plate with Edge Moment

Figure 7.2 shows a simply-supported circular plate subjected to a moment m_o at the outer edge $r = a$. The inner edge is the centre point $r = 0$. The

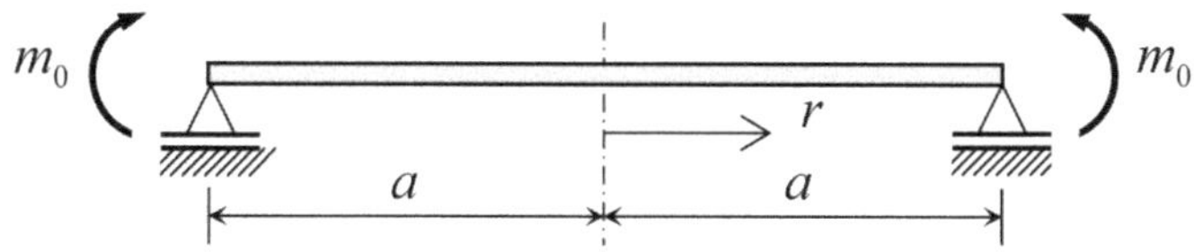

Figure 7.2 Simply supported circular plate with edge moment.

boundary conditions at the outer edge are $w = 0$ and $m_{rr} = m_o$. At the inner edge the conditions are zero rotation and zero shear force, so $dw/dr = 0$ and $v_r = 0$. On this basis we find $C_3 = 0$ and $C_4 = 0$ and obtain a parabolic deflection shape

$$w = \frac{m_0}{2(1+\nu)D}(a^2 - r^2) \tag{7.6}$$

Further, we find two equal curvatures, two equal moments, and a zero shear force

$$m_{rr} = m_{\theta\theta} = m_o, \quad v_r = 0 \tag{7.7}$$

Apparently, there is a homogeneous moment field, equal to the applied edge moment. In hindsight, the solution for the deflection w is to be expected. Over a diagonal of the plate there is a constant moment, and therefore a constant curvature, which corresponds to a parabolic displacement diagram.

7.3 Clamped Circular Plate with Distributed Load

Next we consider a plate which is clamped at the edges and subjected to a uniformly distributed load p, see Figure 7.3. Again the outer edge of the plate is given by $r = a$ and the centre by $r = 0$. Now the boundary conditions at the supported edge are a zero displacement and rotation, so $w = 0$ and $dw/dr = 0$. At the centre point the same conditions hold as in the previous case, a zero rotation and zero shear force, so $dw/dr = 0$ and $v_r = 0$. The four conditions lead to next function for the deflection (again C_3 and C_4 are zero)

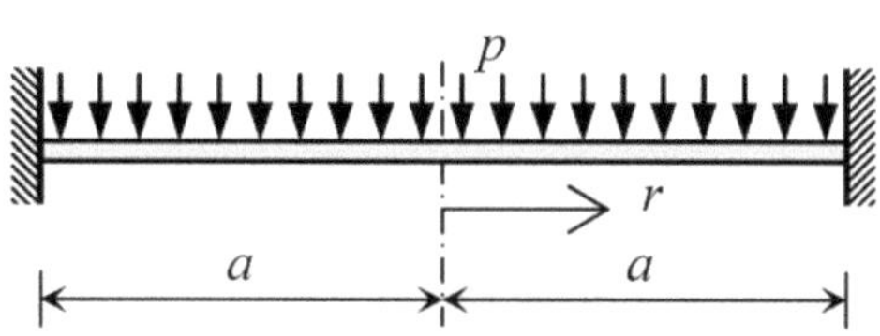

Figure 7.3 Clamped circular plate with uniformly distributed load.

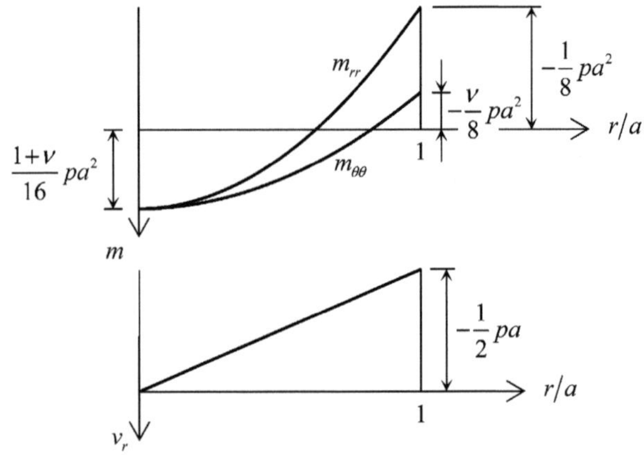

Figure 7.4 Results for clamped plate due to uniformly distributed load.

$$w = \frac{p}{64D}(a^2 - r^2)^2. \tag{7.8}$$

The displacement function is of the fourth degree in r and, of course, the maximum of the vertical deflection occurs at the centre of the plate. This is $w_{\max} = pa^4/(64D)$. The bending moments and shear force in this case are

$$\begin{aligned} m_{rr} &= \frac{1}{16}pa^2\left((1+\nu) - (3+\nu)\frac{r^2}{a^2}\right) \\ m_{\theta\theta} &= \frac{1}{16}pa^2\left((1+\nu) - (1+3\nu)\frac{r^2}{a^2}\right) \\ v_r &= -\frac{1}{2}pr \end{aligned} \tag{7.9}$$

Figure 7.4 is the graphical representation of these results. Similar to what we know from beams under a comparable distributed load, the moment distribution is quadratic in r. As we may expect, the moments m_{rr} and $m_{\theta\theta}$ are equal at the centre of the plate. Of course, there m_{rr} for an angle θ becomes $m_{\theta\theta}$ for an angle $\theta + \pi/2$. Here the distinction between the moments disappears. At the clamped edge the tangential moment is just v times the radial moment m_{rr}. This result is in agreement with the moments in a rectangular plate with clamped edges. The tangent line to the clamped circular edge remains a straight horizontal line, and yet a non-zero bending moment $m_{\theta\theta}$ occurs, due to Poisson's ratio of the material. For zero Poisson's ratio the moment at the centre is half the moment at the edge, as is the case for a clamped beam. In a clamped beam of unit width and length $2a$ the sum

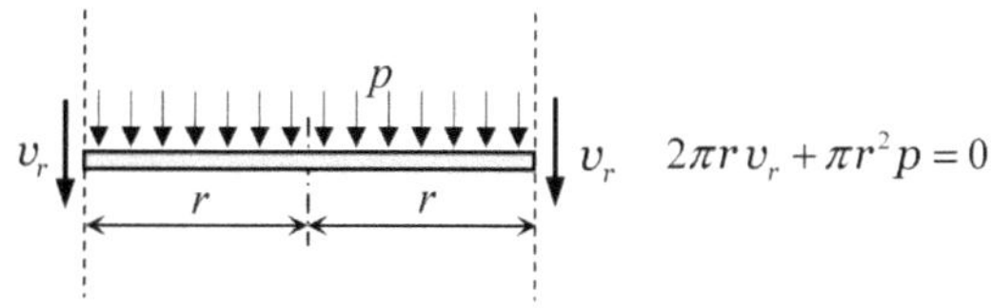

Figure 7.5 Equilibrium between load and shear force.

of the absolute values of the mid-span and end moment would be $pa^2/2$. In the clamped circular plate it is $pa^2/5$ for Poisson's ratio $\upsilon = 0.2$, so the clamped circular plate is a factor 2.5 more moment-efficient than a beam.

The shear force has a linear distribution. We can find this result without solving the differential equation. Figure 7.5 shows the cross-section of a part of the plate with radius r. Consider the free-body equilibrium. The distributed load p over the area πr^2 must be carried by the shear force vr over the circumference with length $2\pi r$. So, accounting for the sign conventions, we must solve the simple equation $\pi r^2 p + 2\pi r v_r = 0$, which leads indeed to $v_r = -pr/2$. Because we have considered an inward plate part around the centre, the result is independent of the boundary conditions at the edge. The same result will be found for a simply-supported edge. In a beam of unit width with the same span $2a$ and the same distributed load, the shear force at the support would be pa, twice as much. The shear-efficiency of the circular plate is doubled.

7.4 Simply-Supported Circular Plate with Distributed Load

The solution for the simply-supported plate drawn in Figure 7.6 can be obtained from the solution of the clamped plate. Along the outer edge of the clamped plate there is a support moment equal to $m_{rr} = pa^2/8$. For the simply-supported plate this moment has to be made zero. Therefore, the solution for the simply-supported plate can be found by superposition of the

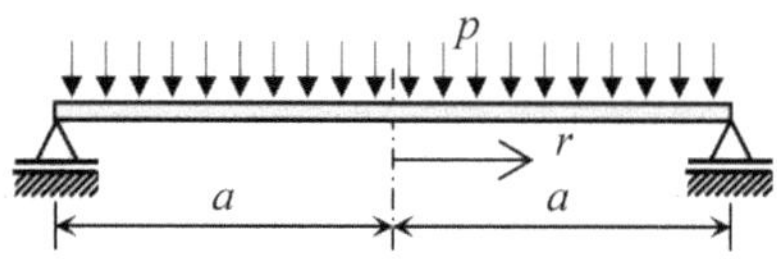

Figure 7.6 Simply-supported circular plate under uniformly distributed load.

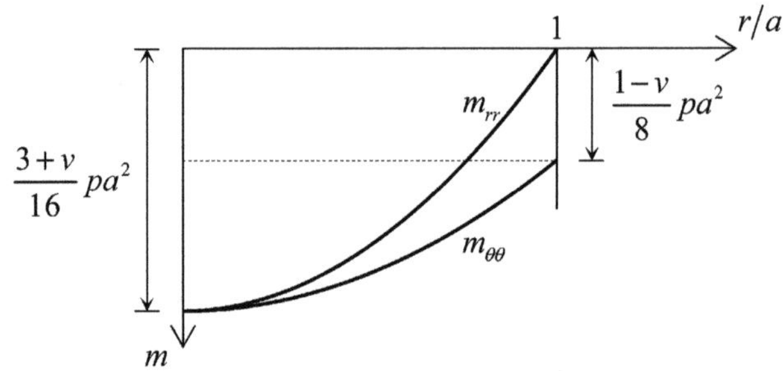

Figure 7.7 Moments in simply-supported circular plate.

solution for the clamped plate and the solution for a circular plate with just having an edge moment $m_{rr} = pa^2/8$ (see Section 7.2):

$$\begin{aligned} m_{rr} &= \frac{3+\nu}{16} pa^2 \left(1 - \frac{r^2}{a^2}\right) \\ m_{\theta\theta} &= \frac{1}{16} pa^2 \left((3+\nu) - (1+3\nu)\frac{r^2}{a^2}\right) \\ v_r &= -\frac{1}{2} pr \end{aligned} \tag{7.10}$$

This means that the horizontal axis in Figure 7.4 just has to shift upward over a distance of $pa^2/8$, leading to Figure 7.7. The shear force diagram need not be re-drawn, for it is the same as for the clamped plate. At the edge, there is a positive moment $m_{\theta\theta}$, both for zero and non-zero Poisson's ratio. This is different from what occurs at the edge of a simply-supported rectangular plate. There, a zero moment m_{xx} normal to the edge will always lead to a zero moment myy in the direction parallel to the edge, as explained in Section 4.4.2.

We can make an engineer's check on the moments $m_{\theta\theta}$. For that purpose we consider a half plate as shown in Figure 7.8 and introduce the symbol P

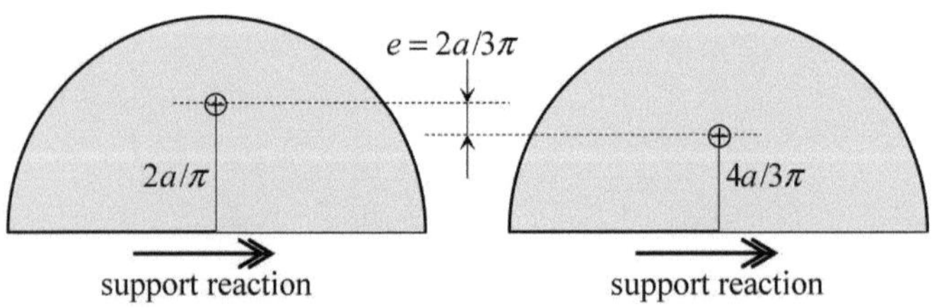

Figure 7.8 Position of centres of gravity.

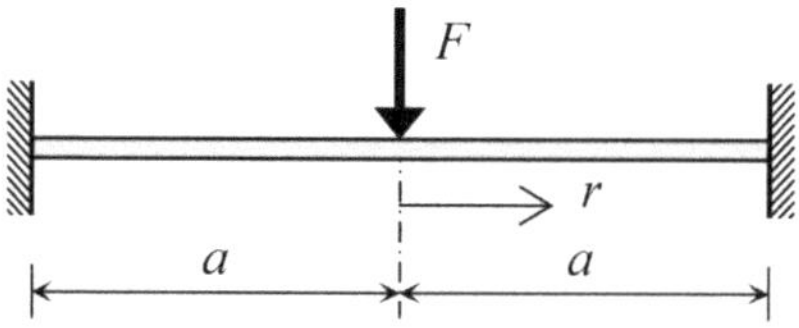

Figure 7.9 Clamped circular plate with point load.

for the total load $\pi a^2 p$ on the plate. The sum $M_{\theta\theta}$ of all the moments $m_{\theta\theta}$ along the straight edge can be calculated by integration of the formula in Eq. (7.10). We then find $M_{\theta\theta} = a^3 p/3$. This moment must be balanced by the moments due to the support reactions and load. The support reactions on the half plate sum to $P/2$ and so does the load. An elementary calculation shows that the centre of gravity of the support reactions is at a distance $2a/\pi$ of the straight edge and at a distance $4a/3\pi$ for the distributed load. This is depicted in Figure 7.8. The distance e between the two centres of gravity is $2a/3\pi$. The moment due to the support reactions and the load is $Pe/2 = a^3 p/3$. Indeed, this is just the value of $M_{\theta\theta}$. The average value of $m_{\theta\theta}$ along the diagonal is $a^2 p/6$.

7.5 Clamped Circular Plate with Point Load

In this case the plate is loaded only at the centre by a point load F, as shown in Figure 7.9. No distributed load is present, so $p = 0$. The boundary conditions at the outer edge are the same as in Section 7.3 for a distributed load. The displacement and the rotation are zero, so $w = 0$ and $dw/dr = 0$. At the plate centre, the two conditions involve the rotation and the shear force. The first one is the same as in Section 7.3 for a distributed load, and is $dw/dr = 0$. The second one is different and is derived from the free-body diagram in Figure 7.10, and is $v_r = F/(2\pi r)$. On the basis of these boundary conditions, the solution becomes

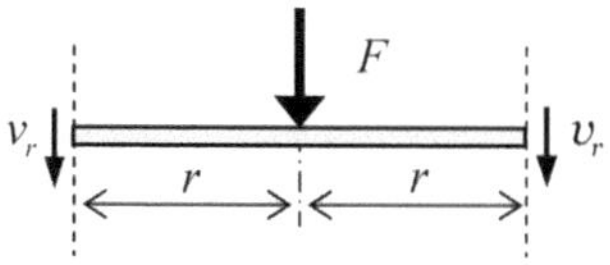

Figure 7.10 Free-body diagram of central plate part.

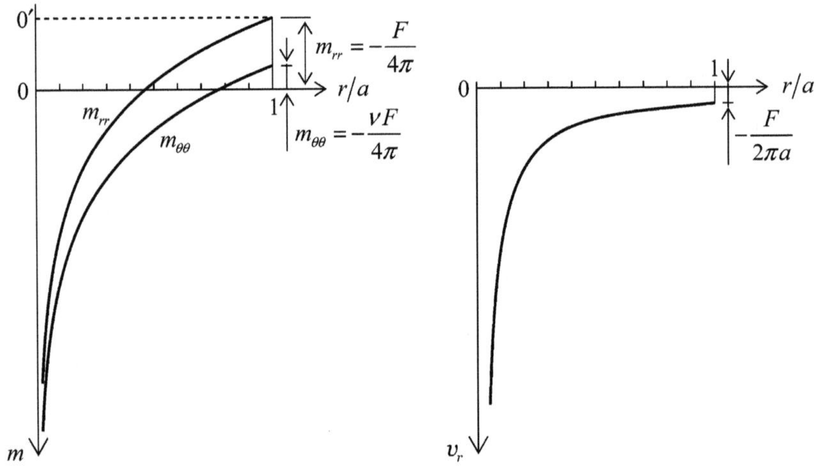

Figure 7.11 Moments and shear force due to point load.

$$w = \frac{Fa^2}{16\pi D}\left(1 - \frac{r^2}{a^2}\right) + \frac{Fr^2}{8\pi D}\ln\left(\frac{r}{a}\right). \tag{7.11}$$

The maximum displacement at the centre of the plate is $w_{\max} = Fa^2/(16\pi D)$. Comparison of this result with the maximum vertical deflection for a uniformly distributed load in Section 7.3 shows that the deflection increases by a factor of 4 when the full load ($P = \pi a^2 p$) is concentrated at the centre of the plate. For a clamped beam this factor is 2. For the moments and the shear force we find

$$\begin{aligned} m_{rr} &= \frac{F}{4\pi}\left\{-1 + (1+\nu)\ln\left(\frac{a}{r}\right)\right\} \\ m_{\theta\theta} &= \frac{F}{4\pi}\left\{-\nu + (1+\nu)\ln\left(\frac{a}{r}\right)\right\} \\ v_r &= -\frac{F}{2\pi r} \end{aligned} \tag{7.12}$$

These results are displayed in Figure 7.11. Again we find moments $m_{\theta\theta}$ along the clamped edge which are ν times the clamped moment m_{rr}. At the position of the point load F the moments are infinitely large. In reality, point loads will be spread over some area which makes the moments finite. We will return to this in Section 7.7. The shear force at the plate centre also becomes infinitely large. For point loads spread over some area, it will in reality be finite.

7.6 Simply-Supported Circular Plate with Point Load

The solution for the simply-supported plate with point load F can be found by shifting the horizontal axis in Figure 7.11 for the moments in the vertical direction, such that mrr becomes zero at the edge $r = a$ as in Section 7.4 for a distributed load. The origin then shifts from O to O'. The graph for the shear force remains unchanged for it is independent of the boundary condition at the outer edge. The formulas in (7.12) for the moments and shear force now change into

$$\begin{aligned} m_{rr} &= \frac{F}{4\pi}(1+\nu)\ln\left(\frac{a}{r}\right) \\ m_{\theta\theta} &= \frac{F}{4\pi}\left\{1-\nu+(1+\nu)\ln\left(\frac{a}{r}\right)\right\} \\ v_r &= -\frac{F}{2\pi r} \end{aligned} \tag{7.13}$$

It is striking again that the moments and the shear force at the centre of the plate are infinite. The bending and shear stresses become infinite too. The same holds for the vertical stress σ_{zz} just beneath the point load. In reality infinite stresses never develop because theoretical point loads do not exist. In the neighbourhood of a point load, the assumptions of plate-theory are no longer satisfied. At a distance of approximately the plate thickness from the point load, the plate theory becomes valid.

Point loads on plates in general

We have derived the singular character of the moments and shear force for a circular plate. For other plate shapes and boundary conditions the moments and shear force will also be very large in the neighbourhood of concentrated loads. It can be stated that the behaviour in the neighbourhood of any concentrated load is of the same character. For example, the formulas (7.12) and (7.13) can be used for the calculation of the moments in the neighbourhood of a concentrated load on a square plate. Close to the point load the difference between these two sets of formula is negligible, because the term $(1+\nu)\ln(a/r)$ by far exceeds the constant term in the formulas. Then the difference between m_{rr} and $m_{\theta\theta}$ vanishes and so does the influence of the boundary condition.

We can do the same equilibrium check for the sum of bending moments $m_{\theta\theta}$ in the plate diagonal as we did for a distributed load. Now only the support reactions lead to bending moments $m_{\theta\theta}$ in this line. In this case the total moment is $M_{\theta\theta} = Fe/2$ where $e = 2a/\pi$. We obtain therefore $M_{\theta\theta} = 2aF/\pi$. Even when the moment $m_{\theta\theta}$ becomes infinitely large at the centre, its integral is finite, with an average value F/π. In Chapter 14 we will continue on this subject when we discuss the design of reinforcements on top of columns. For a distributed load we obtained the result $M_{\theta\theta} = aP/3$, so the value for the point load, $M_{\theta\theta} = 2aF/\pi$, is about twice as large for the same total load ($P = F$). In a beam the difference is exactly a factor two. In this respect, the circular plate is apparently not particularly more effective for point loads.

7.7 Circular Plate Part on Top of Column

Floors in buildings often consist of reinforced slabs supported by a regular pattern of columns. It appears hard for structural engineers to decide which plate moment must be assumed on the tops of the columns. Anticipating Chapter 14 where we will address this subject in more detail, here we make a preliminary assessment. We consider a column grid which is equidistant in two directions, and in this grid we consider a column which is sufficiently remote from the edge of the grid. The distance between rows of columns is $2a$. Figure 7.12 shows a square plate part around such a column. The boundaries of the plate part are lines of symmetry at which a positive moment occurs, requiring reinforcement in the bottom layer of the slab. Negative moments occur on top of the column, requiring reinforcement in the top layer of the slab. The column has a square cross-section. We replace the slab part by a circular plate of radius a which is supported by a circular column of radius b. This radius is chosen such that the cross-section area of the replacing column is equal to the area of the replaced square cross-section. We do not precisely know how the support reaction between the slab and the column is spread; therefore we make two different assumptions. First we assume that the column load is a homogeneously distributed load p over a circular area of radius b. Secondly, we assume that the column is transferred to the slab as a circumferential line load f with radius b. In the analysis we must connect two different plate parts, a circular plate with radius $0 \leq r < b$ and an annular plate $b \leq r \leq a$. At the edge $r = 0$ the displacement and rotation must be zero, at the boundary $r = b$ between the two plates, four conditions hold, equal displacements, equal rotations and equal moments in the r-direction;

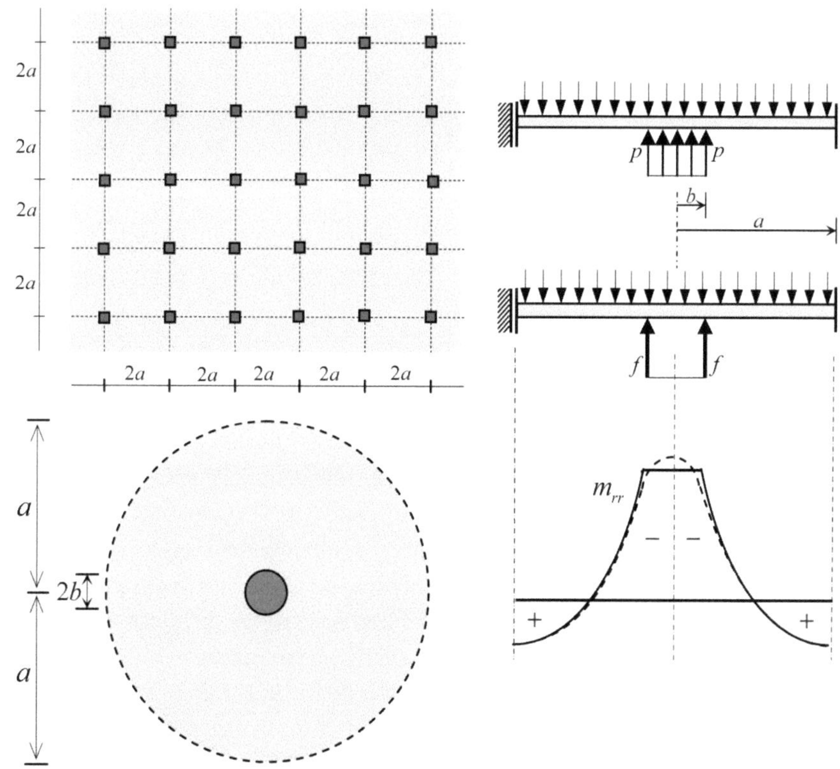

Figure 7.12 Investigation of moments above column in a grid.

the fourth is related to the shear forces. At the boundary of the two late parts, shear forces are equal in the first analysis, and balance the circumferential line load in the second analysis. At the outer edge $r = a$ the rotation and the shear force are zero.

We have sketched in Figure 7.12 what type of results are to be expected. On top of the column some differences occur between one load type and the other, but at some distance from the column there are no noticeable differences. It appears that we can make a fair guess of the maximum moments on top of the column by the following formulas

$$\begin{aligned} m_{\max} &\approx \frac{F}{4\pi}\left(-\frac{1}{4} + (1+\nu)\ln\frac{a}{b}\right) \quad \textit{p-load} \\ m_{\max} &\approx \frac{F}{4\pi}\left(-\frac{3}{4} + (1+\nu)\ln\frac{a}{b}\right) \quad \textit{f-load} \end{aligned} \tag{7.14}$$

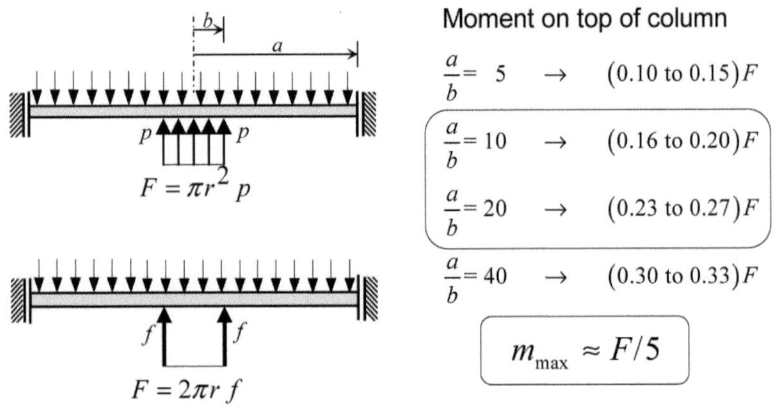

Figure 7.13 Expected moments above column.

The difference between the two formulas is independent of the ratio a/b. For large a/b the term $\ln(a/b)$ is dominant, so the relative difference decreases. For Poisson's ratio $\nu = 0.2$ we find values for different ratios a/b as shown in Figure 7.13. On the basis of this exploration we expect plate bending moments on top of a column with reaction force F in the band with between $F/6$ and $F/4$, dependent on the ratio a/b. A practical rule of thumb for m_{max} might be $F/5$. A safe rule is $F/4$.

7.8 Message of the Chapter

- In a circular plate with axisymmetric load, principal bending moments occur in the direction of the radius and the meridian. The twisting moment is zero. Only one shear force occurs in the direction of the radius.
- Moment trajectories coincide with the radii and the meridians. Shear trajectories coincide with the radii.
- For the same axisymmetric load, moment diagrams for a clamped plate and a simply-supported plate are the same, apart from a shift of the base line of the moment diagram.

- At a circular clamped edge the moment response is the same as at a straight clamped edge. There is a moment in the direction of the edge equal to the product of Poisson's ratio and the clamped moment.

- Moments at a circular simply-supported edge are different from moments at a straight simply-supported edge. At a straight edge, both moments (normal to and in direction of the edge) are zero. At a circular edge, the moment normal to the edge is zero, but the moment in the direction of the edge is not.

- The shear force due to an axisymmetric load is independent of the boundary conditions.

- A point load leads to infinitely large bending moments at the plate centre. However, if the moments are integrated over the diagonal, we find a finite result.

- For a point load, the total moment about a diagonal is about twice that for a distributed load of the same total size.

- The moment on top of a column can be expressed in terms of the size of the support reaction R. Dependent on the area of the column cross-section, the maximum moment lies between $R/6$ and $R/4$. A practical rule of thumb is $R/5$; a safe rule is $R/4$.

Part 2
Didactical Discrete Models

Chapter 8
Discrete Model for Membrane Analysis

This chapter and the next chapter (Chapter 9) are intended as an intermediate step between the classical approach with differential equations and the current computational Finite Element analysis. In pre-FE days, differential equations were solved approximately by Finite Difference (FD) analysis. In that method a grid is chosen over the area of the plate, and the differential equation at each grid point (node) is displaced by an algebraic equation. Solving the set of linear equations leads to an approximate solution of the problem, a solution which becomes more accurate as the mesh is chosen finer. The Finite Element Method (FEM) is the successor of the Finite Difference Method (FDM), in a way which makes it much easier to model plates of any shape and to satisfy boundary conditions. The model for membrane states to be discussed in the present chapter serves two educational goals. First, the discussion is a simple preparation to the stiffness method. The concept of stiffness matrix is introduced, boundary effects are easily accounted for, and often the same solution is obtained as in a classic FD-analysis. Second, the structural engineer sees that different transfer mechanisms are present in a plane stress state. Some members carry horizontal axial forces in the horizontal direction, other members vertical axial forces and special members shear forces. In Chapter 9 comparable members will occur for bending in two directions, and a special member for torsion.

J. Blaauwendraad, *Plates and FEM: Surprises and Pitfalls*, Solid Mechanics and Its Applications 171, DOI 10.1007/978-90-481-3596-7_8,

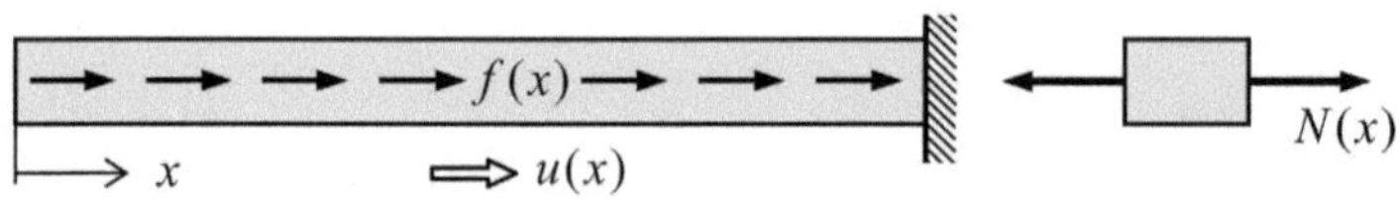

Figure 8.1 Bar loaded in axial direction by a distributed load.

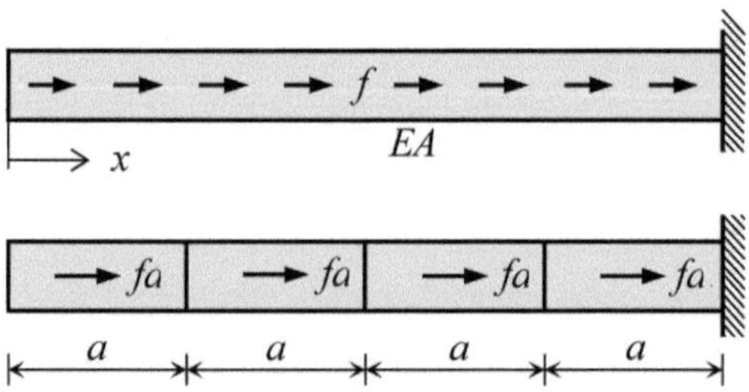

Figure 8.2 Structure divided into four discrete elements.

8.1 Truss Model

To achieve a good approximation to membrane plate behaviour, we first examine an approximation of a bar. We can restrict ourselves to a truss element that is loaded axially, so only normal forces N occur (a truss element). Consider the structure shown in Figure 8.1. The cross-sectional area A, the distributed line load f and the elastic modulus E are constant over the length of the structure.

In order to approximate this continuous system, we select nodes. We keep the load distributed, but lump the elastic deformation into discrete springs. This means that we consider the structure as an assemblage of mass-less rigid parts and springs. We divide the total length l into elements. In the example we choose four equal parts of length a, see Figure 8.2. Next we replace each element by the structure shown in Figure 8.3, a rigid bar with a spring at each end and load f_a. The spring stiffness is chosen such that the new structure experiences the same axial deformation as the original element for a constant normal force. This leads to the spring rigidity $D = 2EA/a$. If we link the four elements, we obtain the structure of Figure 8.4, with spring rigidities $D = EA/a$ and $D = 2EA/a$. Figure 8.5 shows the result for this approximation. The normal force $N(x)$ is continuous and the displacements $u(x)$ discontinuous.

This is a very simple example because the structure is statically determinate. In more general statically indeterminate cases we apply the *stiffness method.* We look upon the structure as an assembly of four elements as

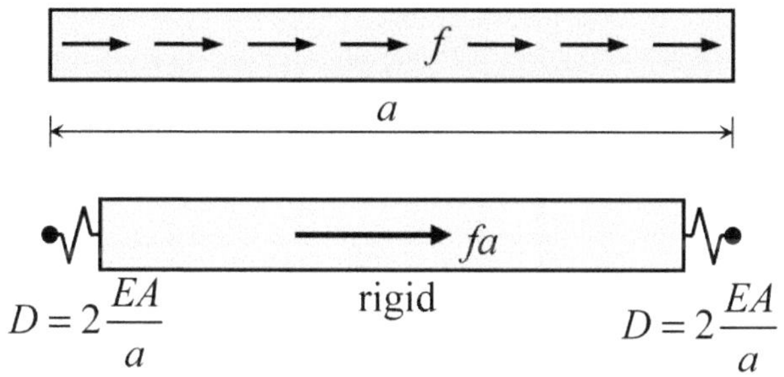

Figure 8.3 Element discretization.

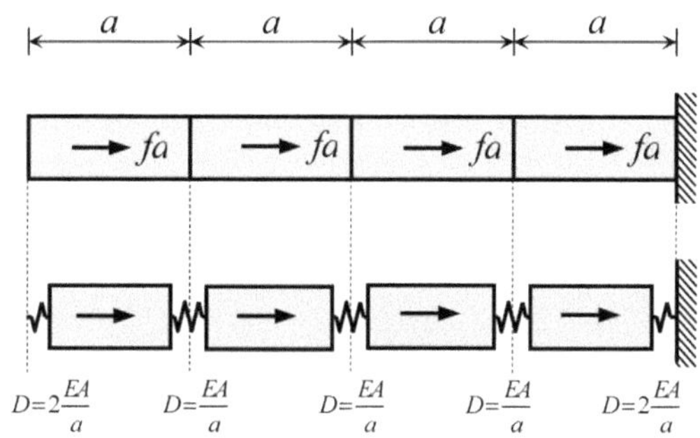

Figure 8.4 Structure divided into four discrete elements.

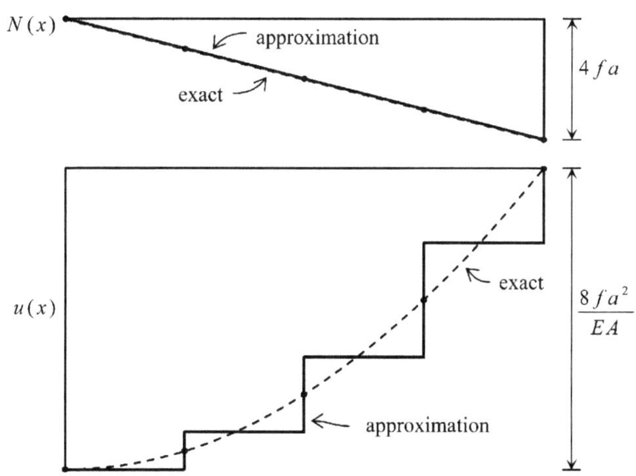

Figure 8.5 Displacement and normal force for the exact solution and approximation.

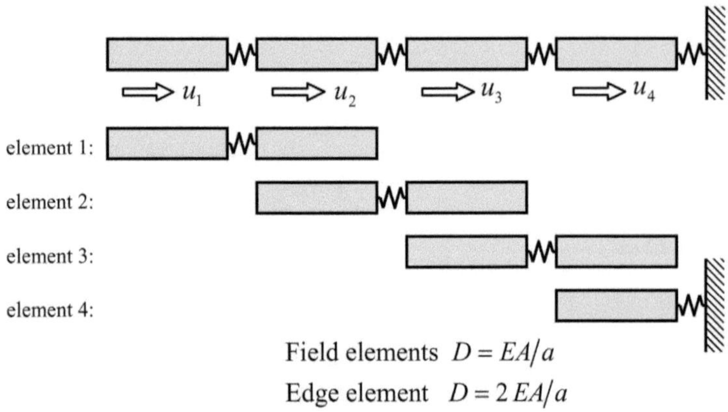

Figure 8.6 Application of the stiffness method.

shown in Figure 8.6. The discrete *field elements* 1, 2 and 3 have a *stiffness matrix* consisting of two rows and two columns. For an *edge element* 4 the stiffness matrix consists of one number

$$\underbrace{\frac{EA}{a}\begin{bmatrix} 1 & -1 \\ -1 & 1 \end{bmatrix}}_{\text{element 1}};\quad \underbrace{\frac{EA}{a}\begin{bmatrix} 1 & -1 \\ -1 & 1 \end{bmatrix}}_{\text{element 2}};\quad \underbrace{\frac{EA}{a}\begin{bmatrix} 1 & -1 \\ -1 & 1 \end{bmatrix}}_{\text{element 3}};\quad \underbrace{\frac{EA}{a}[2]}_{\text{element 4}} \tag{8.1}$$

The assemblage of the *global stiffness matrix* of the total structure leads to

$$\frac{EA}{a}\begin{bmatrix} 1 & -1 & 0 & 0 \\ -1 & 2 & -1 & 0 \\ 0 & -1 & 2 & -1 \\ 0 & 0 & -1 & 3 \end{bmatrix}\begin{Bmatrix} u_1 \\ u_2 \\ u_3 \\ u_4 \end{Bmatrix} = \begin{Bmatrix} fa \\ fa \\ fa \\ fa \end{Bmatrix} \tag{8.2}$$

The contribution of all four elements is shown by the four squares of dashed lines from the top left-hand corner to the bottom right-hand corner of the global stiffness matrix. The terms in the right-hand load vector consist of the load on the rigid parts of the structure. The solution of this set of equations is

$$\{u_1\ u_2\ u_3\ u_4\} = \frac{fa^2}{EA}\{8\ \ 7\ \ 5\ \ 2\} \tag{8.3}$$

From these displacements, one can easily calculate the stress resultants (normal forces) in the springs of the four elements. To distinguish the element

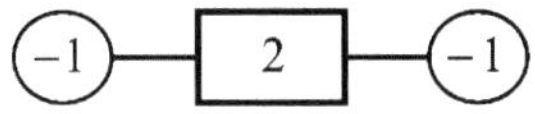

Figure 8.7 Discretization scheme. Also FDM-scheme.

number from the number of a degree of freedom (d.o.f.), we use element numbers as superscript.

$$\{N^1 \;\; N^2 \;\; N^3 \;\; N^4\} = fa\{1 \;\; 2 \;\; 3 \;\; 4\} \tag{8.4}$$

Pattern of coefficients as in Finite Difference Method

In the rows of the stiffness matrix which correspond to u_2 and u_3 we notice the pattern as shown in Figure 8.7. If more elements were used, more such rows would appear in the global stiffness matrix. In this particular case the same scheme is found when a discretization of the governing differential equation $-EA\, d^2u/dx^2 = f$ of Eq. (1.14) is made on basis of finite differences.

The linking of elements is not restricted to elements of the same length and the same EA. The spring rigidity D may vary from one element to another. The serial linking of springs of different rigidities is a well-known procedure.

8.2 Membrane Plate Model

A proper approximation must accurately model the membrane forces n_{xx}, n_{yy} and n_{xy}. We start from the results of the previous section for truss elements and make the extension to plates. In this section our primary goals are insight and understanding, therefore, in order to keep it simple, we exclude lateral contraction. We take into consideration a rectangular plate element with sizes a and b. To approximate the behaviour of the element we replace it by four *truss elements*, one along each edge of the element, and one *shear panel* between the trusses. Figure 8.8 shows this composition. The shear panel is drawn by dotted lines and the size is slightly reduced.

First, we focus on the normal forces n_{xx} and n_{yy}. Membrane forces n_{xx} on the edge with length b are replaced by two lumped forces $n_{xx}b/2$. The spring properties are chosen such that strains ε_{xx} for a homogeneous field

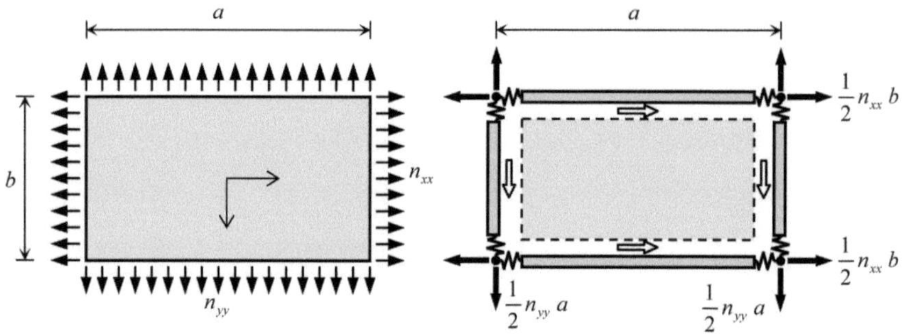

Figure 8.8 Approximation of plate element by trusses and shear panels.

of n_{xx} equal in the actual plate element and in the spring model. The same is done for the y-direction. This leads to the following stiffnesses for the springs in the x-direction and the y-direction respectively

$$D_x = \alpha Et, \quad D_y = \beta Et, \quad \text{where} \quad \alpha = \frac{b}{a}, \quad \beta = \frac{a}{b} \tag{8.5}$$

In an assemblage of a large number of elements, several elements may surround a specific node. In the x-direction at each node one or two couples of two springs may occur, from which we obtain one composed spring. The same holds true for the y-direction.

Next we model the so far neglected n_{xy}-action of the plate element by the introduced shear panel. We choose a constant shear stress in the panel, then the shear strain γ_{xy} will be constant, and therefore the deformed panel will have straight edges, which do not elongate. This enables us to express the state of stress and strain by the use of four degrees of freedom as shown in Figure 8.9. The shear panel is a discrete element that fits perfectly in an orthogonal assemblage of axial discrete spring elements. The stiffness matrix of the panel is

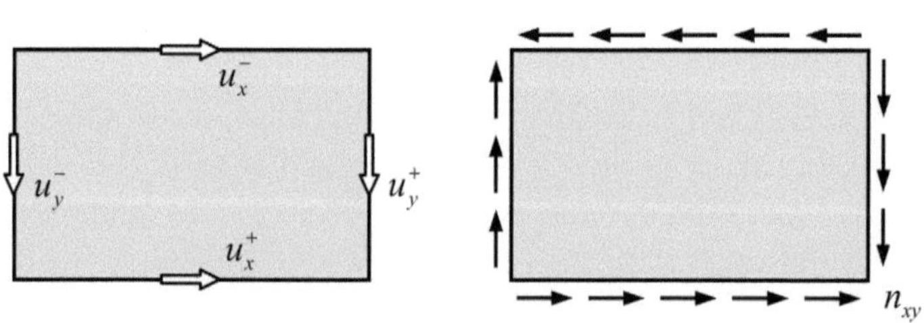

Figure 8.9 Shear panel in the discrete model.

$$\begin{Bmatrix} F_x^- \\ F_x^+ \\ F_y^- \\ F_y^+ \end{Bmatrix} = \frac{1}{2}Et \begin{bmatrix} \beta & -\beta & 1 & -1 \\ -\beta & \beta & -1 & 1 \\ 1 & -1 & \alpha & -\alpha \\ -1 & 1 & -\alpha & \alpha \end{bmatrix} \begin{Bmatrix} u_x^- \\ u_x^+ \\ u_y^- \\ u_y^+ \end{Bmatrix} \tag{8.6}$$

8.2.1 Example. Deep Beam Subjected to Own Weight

The simply-supported deep beam in Figure 8.10 is a symmetric plate in a membrane state subjected to its own weight w per unit area. Because of symmetry we need consider only half the deep beam. We have drawn the degrees of freedom in this half structure and the different discrete element types which we need for this example. The weight is lumped to forces wab in the vertical degrees of freedom; the forces at the edge and line of symmetry are $wab/2$.

Some results for this example are displayed in Figure 8.11. They are for $a = 2$ and $b = 1$. The membrane shear force n_{xy} is constant over the height of an element. The normal membrane force n_{xx} is discontinuous over the height of an element. The analysis leads to lumped normal forces in the horizontal springs; we have to spread these forces over an appropriate width, $b/2$ for the springs at the top and bottom edge, and b for the inward springs. The stress distribution according to the theory for slender beams is included by dotted lines. If judged judiciously, the result of the current analysis comes rather close to the outcome of classical beam theory.

Rows in the global stiffness matrix which are not influenced by boundary conditions have a scheme of coefficients as shown in Figure 8.12. The same scheme holds for the Finite Difference Method.

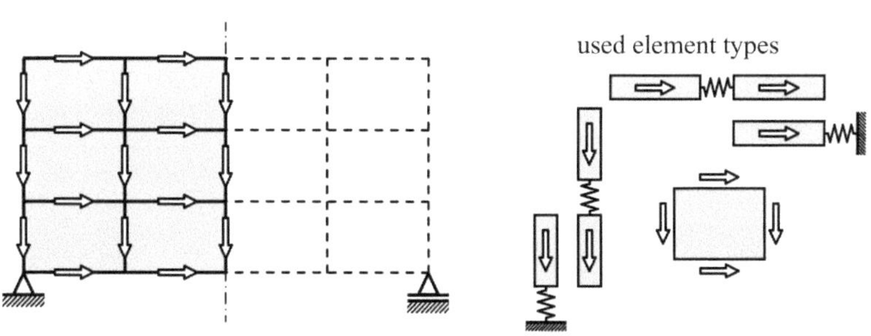

Figure 8.10 Deep beam example with used discrete elements.

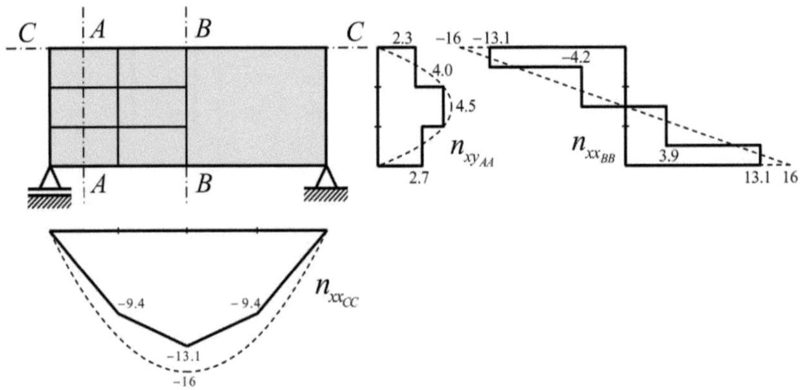

Figure 8.11 Numerical result for the deep beam discrete model.

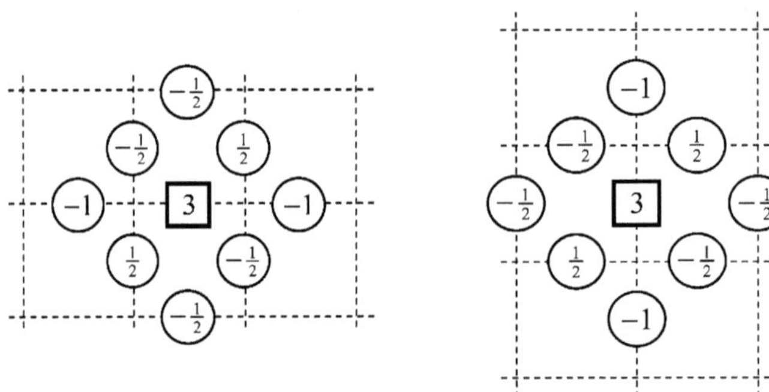

$$-Et\left(\frac{\partial^2 u_x}{\partial x^2}+\frac{1}{2}\frac{\partial^2 u_x}{\partial y^2}+\frac{1}{2}\frac{\partial^2 u_y}{\partial x\partial y}\right)=p_x \qquad -Et\left(\frac{\partial^2 u_y}{\partial y^2}+\frac{1}{2}\frac{\partial^2 u_y}{\partial x^2}+\frac{1}{2}\frac{\partial^2 u_x}{\partial x\partial y}\right)=p_y$$

Figure 8.12 Discretization scheme; also FDM-scheme.

Advantage of discrete model

The conventional Finite Difference Method for plates is not easy to apply to nodes close to or on the boundary. Also, abrupt changes in thickness are difficult to deal with. The concept of springs and shear panels provides an elegant way to overcome such difficulties.

Remark

In the discrete model for membrane plates we have used two different element types, truss elements and shear panels, respectively. So, the axial force action and the shear force action are handled with separate elements. In the Finite Element Method the two actions are integrated into one and the same element. For an orthogonal grid, each element coincides with one rectangle in the grid; this has an advantage for programmers and users. We will return to this in Chapter 10.

8.3 Message of the Chapter

- A structure in a membrane state (plane stress) can be modeled by trusses (spring elements) and shear panels. Each element type has its own stiffness matrix. The global stiffness matrix of a structure is assembled from the stiffness matrices of the individual elements. A load vector is composed from nodal loads.
- Equations in the global stiffness matrix which are not influenced by boundary conditions, are similar to those found in the classical Finite Difference Method.
- Stress distributions over sections of the structure can be discontinuous in a numerical analysis.
- The spring-panel model nicely demonstrates the separate action of normal forces and shear forces.

Chapter 9
Discrete Model for Plate Bending

9.1 Beam Model

In Chapter 8 we discussed an approximation for trusses. The elastic deformations were lumped in springs. We may apply a similar model to beams in bending as a first step to a discrete model for plate bending. A rigid element that has rotational springs at its ends replaces a beam element of length a. This model is depicted in Figure 9.1. The rotational spring is considered to be composed of two parallel springs, for the compression and tension zones, respectively. The rotational rigidity at each end is D. It is required that the beam-ends in both the model and the actual beam have the same rotation e for a constant moment M. This requirement is met if $D = 2EI/a$. When two beam elements are linked together, the two rotational end springs are connected in series. The rigidity of the resulting rotational spring is $D = EI/a$.

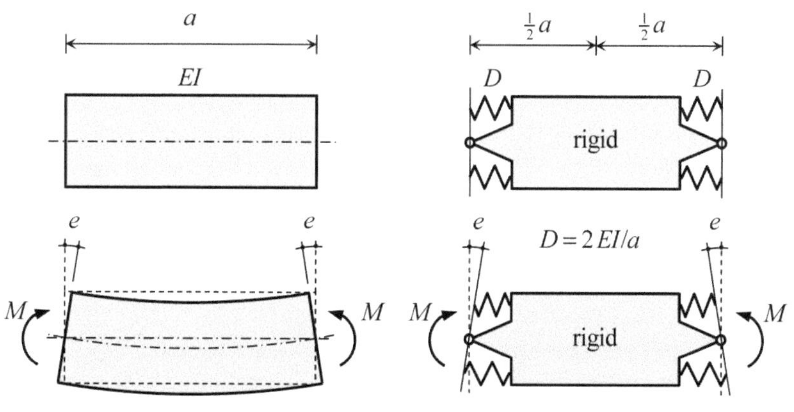

Figure 9.1 Discrete bending model for beam.

J. Blaauwendraad, *Plates and FEM: Surprises and Pitfalls*, Solid Mechanics and Its Applications 171, DOI 10.1007/978-90-481-3596-7_9,

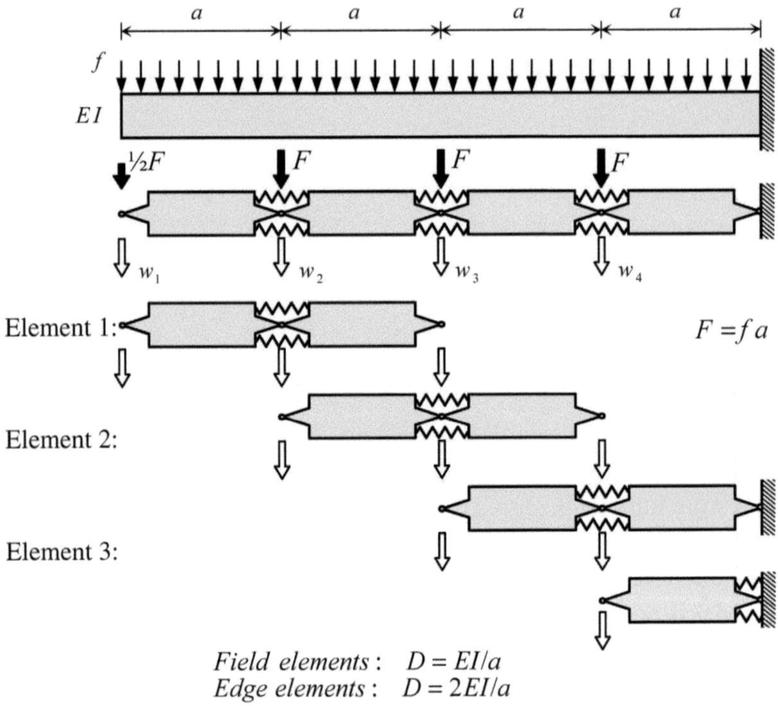

Figure 9.2 Modelling of a cantilever beam.

Figure 9.2 shows a cantilever beam modelled by four beam parts with length a. The homogeneous load is lumped at the hinges between the four sections. The deflections of the hinges are the unknowns. The structure is considered to be an assembly of four discrete elements and to have four degrees of freedom. The elements 1, 2, and 3 each have three degrees of freedom, element 4 only has one. This latter element is used at the clamped end. It need also to be used in a line of symmetry. Then the element has two degrees of freedom. Naturally, this element may occur with the rotational spring at the other end as well. First, the different stiffness matrices of the two element types are derived. The element types will be named *field element* and *edge element* respectively. The stiffness matrices for a field element with nodes i, j, k and an edge element with nodes i, j are respectively

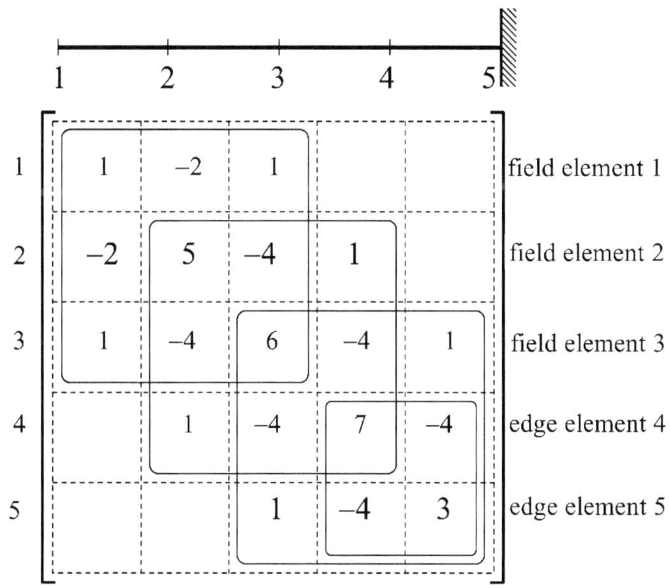

Figure 9.3 Composition of the global stiffness matrix.

$$\left\{ \begin{array}{c} F_i \\ F_j \\ F_k \end{array} \right\} = \frac{D}{a^2} \left[\begin{array}{rrr} 1 & -2 & 1 \\ -2 & 4 & -2 \\ 1 & -2 & 1 \end{array} \right] \left\{ \begin{array}{c} w_i \\ w_j \\ w_k \end{array} \right\}$$

$$\left\{ \begin{array}{c} F_i \\ F_j \end{array} \right\} = \frac{D}{a^2} \left[\begin{array}{rr} 1 & -1 \\ -1 & 1 \end{array} \right] \left\{ \begin{array}{c} w_i \\ w_j \end{array} \right\} \qquad (9.1)$$

The stiffness matrix of the cantilever beam structure is an assembly of three field elements and one edge element. The result is shown in Figure 9.3. The matrix needs to be multiplied by EI/a^3. As is seen in the figure, the displacements are not yet constrained. If displacements are prescribed, corresponding rows and columns are omitted. Figure 9.4 shows a number of possibilities. The third row of the stiffness matrix is complete; that is to say, there is no influence of the boundaries on the coefficients of this row. If we use a finer mesh in the model, then the global stiffness matrix would contain more of these complete rows. Apart from the multiplication factor EI/a^3, the scheme shown in Figure 9.5 applies for nodes not affected by any edge conditions.

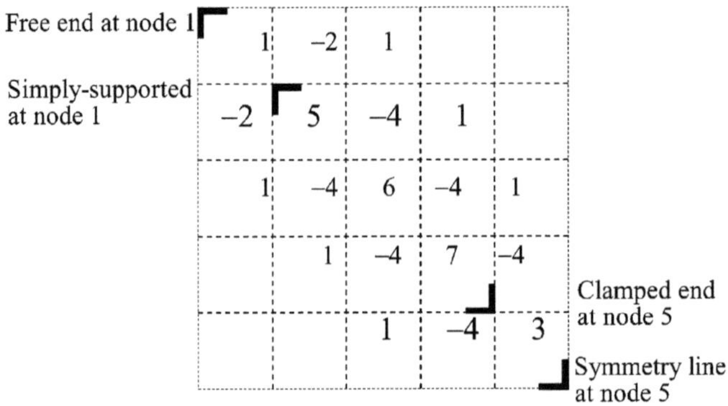

Figure 9.4 Effect of various boundary conditions.

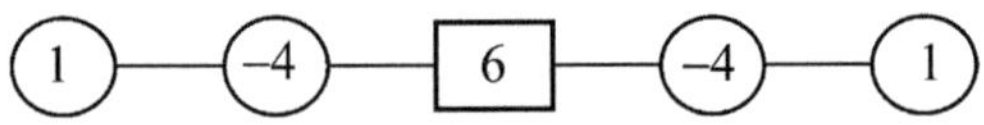

Figure 9.5 Scheme for a complete field row.

9.1.1 Example. Cantilever Beam

The set of equations has been solved for the cantilever beam of Figure 9.6. The nodal force is $F = fa$. The matrix equation reads:

$$\frac{EI}{a^3}\begin{bmatrix} 1 & -2 & 1 & & \\ -2 & 5 & -4 & 1 & \\ 1 & -4 & 6 & -4 & 1 \\ & 1 & -4 & 7 & -4 \\ & & 1 & -4 & 3 \end{bmatrix}\begin{Bmatrix} w_1 \\ w_2 \\ w_3 \\ w_4 \\ w_5 \end{Bmatrix} = \begin{Bmatrix} \frac{1}{2}fa \\ fa \\ fa \\ fa \\ \frac{1}{2}fa + R \end{Bmatrix} \tag{9.2}$$

The dotted lines hold for the matrix equation in which boundary conditions have not yet been introduced, the full lines after accounting for the boundary conditions. The solution of the set of equations is

$$\{w_1 \;\; w_2 \;\; w_3 \;\; w_4\} \;\; w_5\} = \frac{a^4 f}{EI}\{34 \;\; 23 \;\; 12\tfrac{1}{2} \;\; 4\} \;\; 0\} \tag{9.3}$$

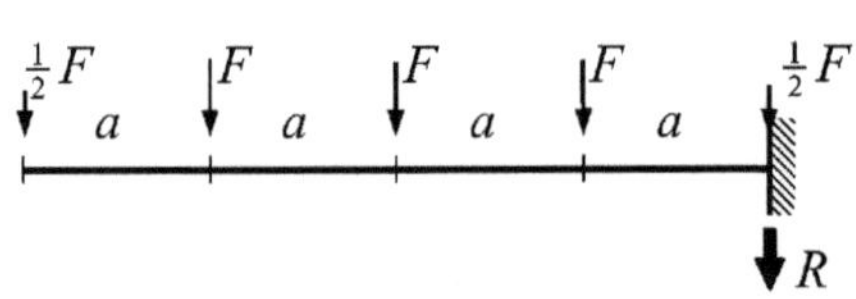

Figure 9.6 Cantilever beam with four elements.

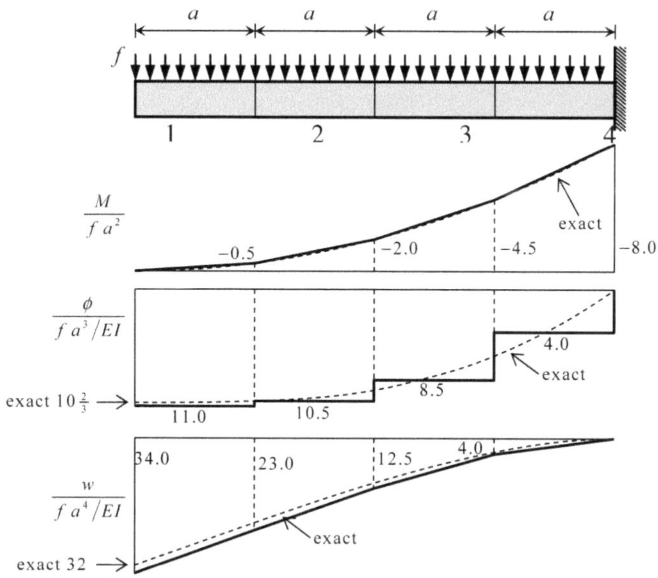

Figure 9.7 Cantilever results for exact solution and approximation

The support reaction, calculated from the fifth row in Eq. (9.2), is $R = -4fa$. The moments are determined from the spring equations

$$\begin{aligned} M &= \frac{EI}{a^2}(-w_i + 2w_j - w_k) \quad \textit{Field element} \\ M &= 2\frac{EI}{a^2}(w_j - w_i) \qquad\quad\;\; \textit{Edge element} \end{aligned} \tag{9.4}$$

The results are depicted in Figure 9.7. Even this course mesh and simple elements give a good approximate result.

Same pattern of coefficients in Finite Difference Method

When the classical Finite Difference Method (FDM) is applied, the differential equation $EI\, d^4w/dx^4 = f$ is replaced by a set of algebraic equations, one for each node. The FDM-equation related to node 3, which is not influenced by the boundary conditions, is exactly the same as the third row in the above matrix equation. So, the scheme of Figure 9.5 is both the scheme in FDM and the present discrete model. The rotational spring method is not restricted to equal element dimensions and constant bending rigidity EI for all elements. Serial linking of different spring stiffness is allowed.

9.2 Plate Bending Model

The spring model for bending of the previous section is a building block for the discrete plate bending model, just as the truss model was for the discrete membrane model. We consider a rectangular plate element of length a and width b, subjected to a homogeneously distributed load p. The plate thickness is t, Young's modulus is E, and Poisson's ratio is zero. The relation between the moments and curvatures is

$$\begin{Bmatrix} m_{xx} \\ m_{yy} \\ m_{xy} \end{Bmatrix} = D \begin{bmatrix} 1 & 0 & 0 \\ 0 & 1 & 0 \\ 0 & 0 & \frac{1}{2} \end{bmatrix} \begin{Bmatrix} \kappa_{xx} \\ \kappa_{yy} \\ \rho_{xy} \end{Bmatrix} \tag{9.5}$$

Here the *plate flexural rigidity* is $D = Et^3/12$. The rigidity matrix is a diagonal matrix, therefore bending in the x-direction and y-direction are independent of each other. Torsion is another independent transfer mechanism.

The bending behaviour in the x- and the y-direction is modelled with the spring elements derived in Section 9.1. Thus an orthogonal grid of beams is obtained. To account for twisting moments, torsion panels are inserted in the grid in the same way as shear panels were in the membrane model. From Section 5.2 we know that a field of constant m_{xy} can exist if the panel is loaded by two pairs of equilibrating corner point loads occur, and that the edges remain straight. On the basis of this knowledge a torsion panel can be derived with four degrees of freedom as shown in Figure 9.8. The stiffness

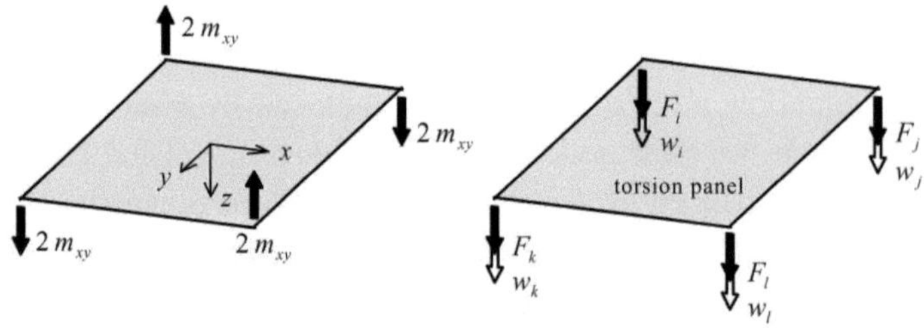

Figure 9.8 Loading for positive twisting moment. Left physical reality, right stiffness method.

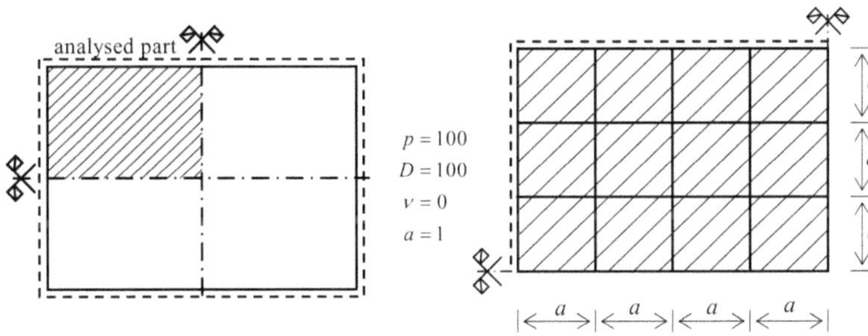

Figure 9.9 Uniformly loaded simply-supported rectangular plate.

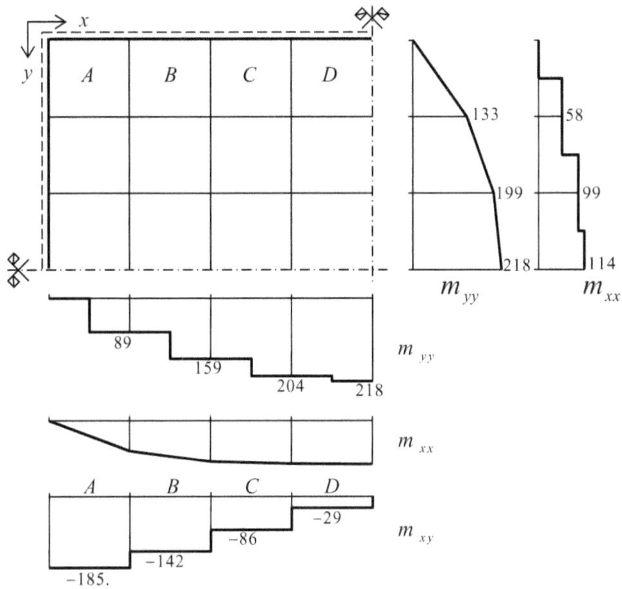

Figure 9.10 Moment distributions in the slab.

matrix of this panel is

$$\begin{Bmatrix} F_i \\ F_j \\ F_k \\ F_l \end{Bmatrix} = \frac{D}{a\,b} \begin{bmatrix} 2 & -2 & -2 & 2 \\ -2 & 2 & 2 & -2 \\ -2 & 2 & 2 & -2 \\ 2 & -2 & -2 & 2 \end{bmatrix} \begin{Bmatrix} w_i \\ w_j \\ w_k \\ w_l \end{Bmatrix} \tag{9.6}$$

Any plate that may be considered a composition of rectangular plate parts can now be modelled with spring elements and torsion panels.

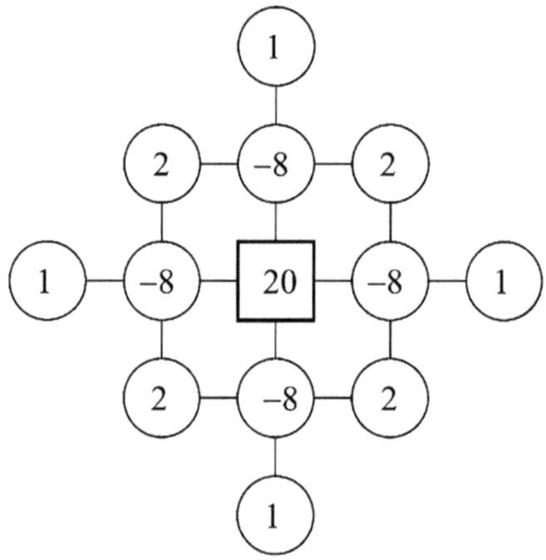

Figure 9.11 Scheme for both FDM and discrete model.

9.2.1 Example 1. Rectangular Simply-Supported Plate

We will perform the discrete analysis for the example of Figure 9.9. For reasons of symmetry only one quarter of the plate needs to be modelled. The computational result is presented in Figure 9.10. As was seen earlier for the normal forces in the membrane solution, we again obtain a continuous distribution for the bending moments in the one direction and a discontinuous one in the other. Like the shear forces in the membrane solution, now the twisting moment is discontinuous.

Advantage of discrete model

In the Finite Difference Method (FDM) the bi-harmonic differential equation is replaced by a set of linear algebraic equations, one for each mesh node. For nodes which are at sufficient distance from the edge the scheme of coefficients is shown in Figure 9.11. The discrete model with flexural springs and torsion panels leads to the same result. The advantage of the discrete modelling is the ease of handling boundary conditions, discontinuities in thickness, and non-square meshes.

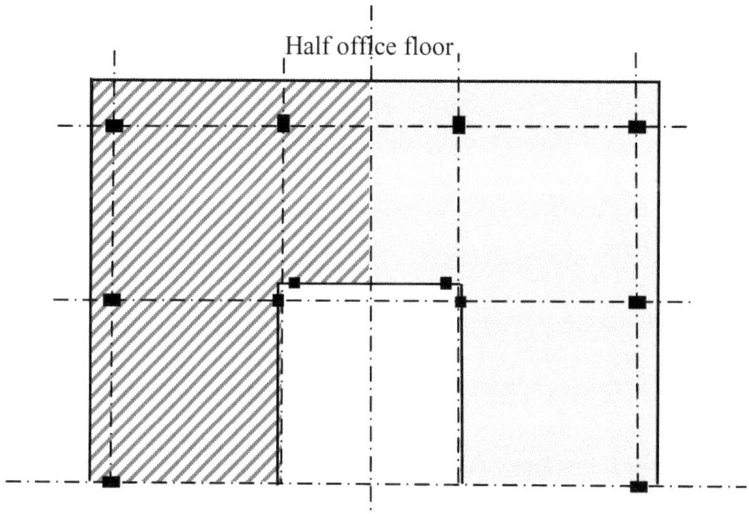

Figure 9.12 Application of discrete model to office building floor.

9.2.2 Example 2. Lift-Slab in Office Building

The model was in use in the 1960s for the analysis of viaducts and floor slabs in office buildings. An example is the floor of the office building in Figure 9.12 as built at Amsterdam Airport. The contractor cast and cured all floors at ground level around a tall central shaft (not shown in Figure 9.12), and then lifted them in place. Therefore they are not clamped to the central shaft, but just connected at discrete points. The hatched quarter of the floor has been considered in the analysis. The bending moments for two directions are shown in Figure 9.13.

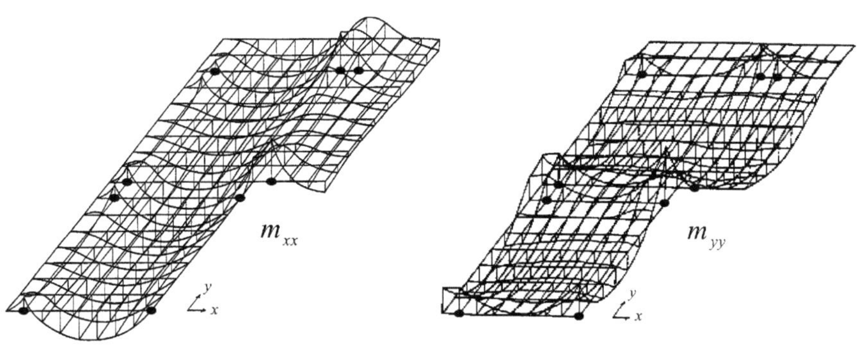

Figure 9.13 Bending moments in quarter of floor slab.

9.3 Didactical Model for Simply-Supported Plate

The discrete model is no longer in use since Finite Element Analysis has replaced it, however it still has great didactical value. This value is illustrated for the simply-supported plate, subjected to a two-way sine load, the exact solution of which we discussed in Section 5.3. Here we recall the most important results of the exact analysis. The deflection w and bending moments m_{xx} and m_{yy} have double-sine distributions, with a maximum in the centre of the plate. The twisting moment has a double-cosine distribution with a value zero in the horizontal and vertical line of symmetry. The maximum value occurs at the four corners. The shear forces v_x and v_y have sine-shaped distributions in the one direction and cosine-shaped in the other. Their maximum value appears at the edges. The distributed support reactions f (positive if directed downward) along the four edges are sine-shaped with a maximum value halfway along the edges. The various maxima are, apart of the sign

$$
\begin{aligned}
w &= \frac{\hat{p}\, l^4}{4\pi^4 D}, \quad m_{xx} = m_{yy} = \frac{\hat{p}\, l^2}{4\pi^2}, \quad m_{xy} = \frac{\hat{p}\, l^2}{4\pi^2} \\
v_x &= \frac{1}{2\pi}\hat{p}\, l, \quad v_y = \frac{1}{2\pi}\hat{p}\, l, \quad f = -\frac{3}{4\pi}\hat{p}\, l
\end{aligned}
\tag{9.7}
$$

where $\hat{p}$ is the maximum value of the load and D is the plate stiffness. The support reaction is 50% larger in absolute value than the maximum shear force at the edge. The negative sign means that it concerns a compressive support reaction in the opposite direction to the load p. Finally it was found that four balancing concentrated corner tensile support reactions R occur with the value

$$
R = \frac{1}{2}\,\hat{p}\, l^2/\pi^2 \tag{9.8}
$$

We now start to explain the discrete model. The coarsest mesh possible is a two-by-two grid with one central node; there is just one degree of freedom. However, a three-by-three mesh leads also to one degree of freedom; There are four free nodes in the plate, but they all have the same displacement. This finer mesh will produce more information, and therefore is chosen. The square plate has edges of length l which are divided in three equal parts a. Figure 9.14 shows the applicable discrete model of four beams, two in each direction.

In a three-by-three grid we need in general nine torsion panels, but in our case five of them occur on lines of symmetry and therefore will have zero twist. So, just the four corner panels need be entered in the analysis,

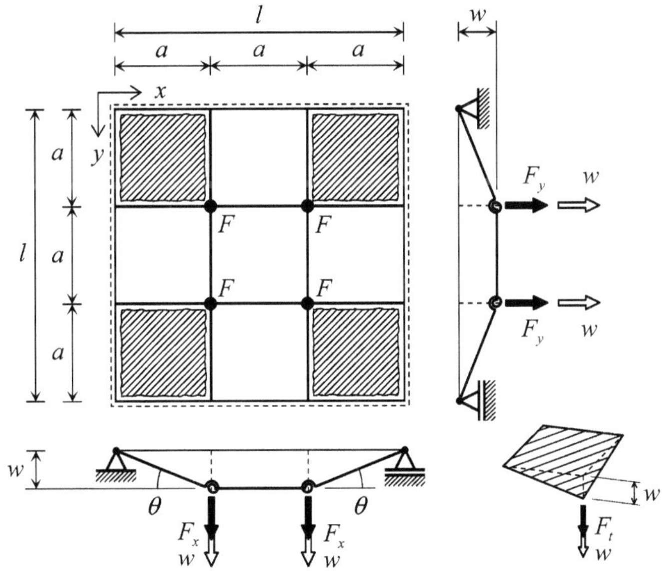

Figure 9.14 Elementary spring-panel model for a square plate.

which confirms the importance of torsion in the corners. The two-way sine load is replaced by four point loads F each pa^2. We restrict the analysis to a zero Poisson's ratio. The model is very simple. The load F at each free point is carried by three elements, a beam in the x-direction, a beam in the y-direction and a torsion panel. The three contributions are F_x, F_y and F_t, respectively. They are derived in an elementary way on the basis of the properties of the spring model and torsion model

$$F_x = \frac{D\,w}{a^2}, \quad F_y = \frac{D\,w}{a^2}, \quad F_t = 2\,\frac{D\,w}{a^2} \tag{9.9}$$

The three contributions are related to the same displacement and are linked in parallel. Therefore they can be summed, which leads to the relation between the displacement w and the load F

$$F_x + F_t + F_y = F, \quad \frac{D}{a^2}(1+2+1)\,w = F, \quad w = \frac{Fa^2}{4D} \tag{9.10}$$

Clearly

$$F_x = \frac{1}{4}F, \quad F_y = \frac{1}{4}F, \quad F_t = \frac{1}{2}F \tag{9.11}$$

This very elementary model effectively confirms what was seen earlier after solving the bi-harmonic equation for a double-sine load in Section 5.3. The

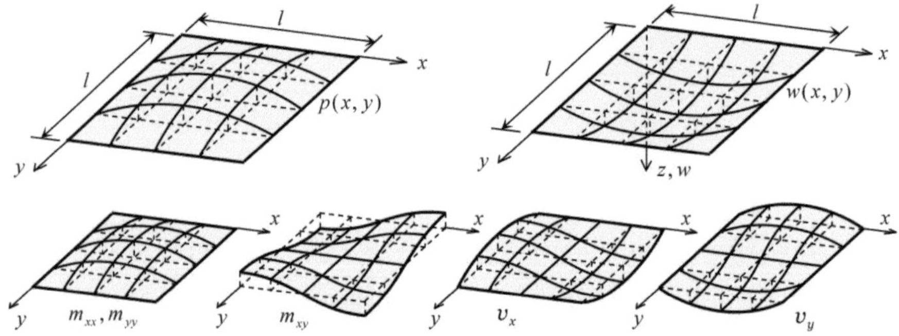

Figure 9.15 Exact results for a two-way sine load.

torsion in the plate carries half the load, and the deflection is a quarter of the value that occurs if one beam had to transfer all the load F. The bending moment at the position of the rotational springs, the twisting moment in the panels, and the shear forces midway between the two nodes become

$$m_{xx} = m_{yy} = m_{xy} = \frac{1}{4}F, \quad v_x = v_y = \frac{1}{2}\frac{F}{a} \tag{9.12}$$

As we found for zero Poisson's ratio in the exact solution, the maximum torsion moment is equal to the maximum bending moment. Figure 9.16 compares the results of the discrete moments to the outcome of the exact analysis for a number of sections over the plate. Figure 9.17 does the same for the shear force and support reaction. Finally, we can calculate the reaction forces in the edge and corner nodes. The simple model leads to $R_{\text{corner}} = \frac{1}{2}F$, $R_{\text{edge}} = -\frac{3}{4}F$. Figure 9.18 presents an overview of these boundary forces. The distributed support reaction $f = R_{\text{edge}}/a = -\frac{3}{4}F/a$ is again 50% larger in absolute value than the shear force v_x. For vertical equilibrium of the total plate it is required that $4R_{\text{corner}} + 8R_{\text{edge}} + 4F = 0$. This is indeed satisfied: $4(\frac{1}{2}F) + 8(-\frac{3}{4}F) + 4F = 0$. All the results that were seen in the solution of the bi-harmonic equation reappear in this elementary model.

Comparison

The results of the discrete model are compared to the exact solution of the square simply supported plate subjected to a double-sine load. Table 9.1 lists the maximum occurring values of the displacement w, the moments m, the shear force v, the distributed support reaction f and the concentrated corner reaction R_{corner} is given. The results of the discrete model and the exact so-

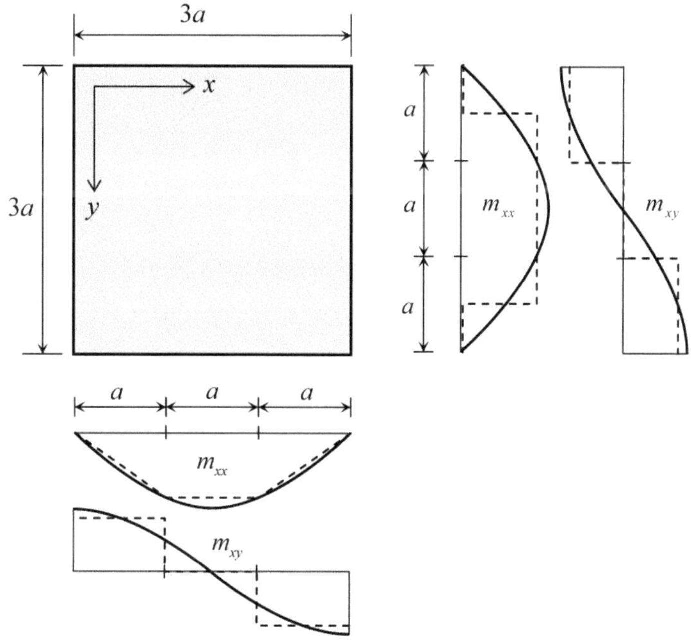

Figure 9.16 Comparison between discrete model and exact solution for moments.

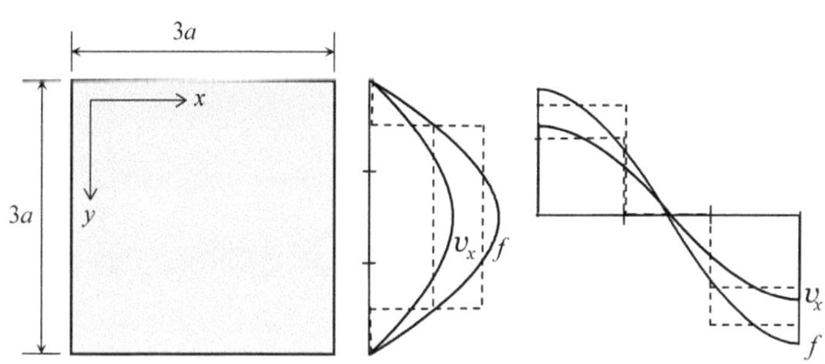

Figure 9.17 Comparison between discrete model and exact solution for shear force and support reaction.

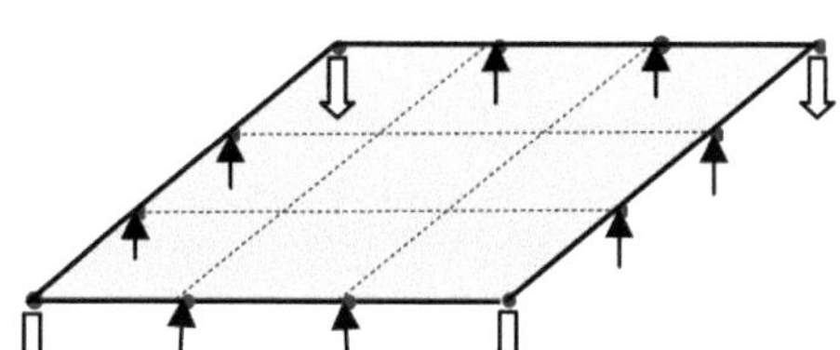

Figure 9.18 Distribution of support reactions.

Table 9.1 Comparison of discrete model with exact values for two-way sine load.

	Exact values divided by pl^2/π^2	Values for discrete model divided by $F = pl^2/9$
w	$l^2/(4\pi^2 D)$	$l^2/(4 * 9D)$
m	1/4	1/4
v	$\pi/2l$	$3/2l$
f	$-3\pi/4$	$-3 * 3/4$
R_{corner}	1/2	1/2

lution are very close to each other. The simple model accurately shows the main aspects in the force transfer of a simply-supported plate:

1. Maximum bending moments m_{xx} and m_{yy} occur at the centre of the plate.
2. Maximum twisting moments m_{xy} occur at the corners, and have the same magnitude as the bending moments.
3. The distributed support reactions are 50% larger than the shear forces.
4. Lumped tensile corner support reactions are twice the size of the twisting moments.

9.4 Discrete Model for Plate on Flexible Edge Beams

In the previous section we modelled the simply-supported plate which had been discussed in Section 5.3 for a two-way sine load. In Section 5.4 we considered another interesting case, a flexible edge beam, which leads to a twist-less plate, at that time for a homogeneously distributed load. Because the two cases have different loads and different boundary conditions, we had to study them with different displacement fields. With the discrete model we can study both cases in one model for the same load. We will do so for the square model of Figure 9.14 and a homogeneously distributed load p.

The new model is shown in Figure 9.19. The grid consists of four beams in the x-direction, four in the y-direction, and four torsion panels. The beams inside the plate represent a plate strip of width a. The beams at the position of the edge represent the actual applied edge beam plus a plate strip of width $\frac{1}{2}a$. Because the edges can deflect, an additional degree of freedom must be introduced. We call the displacement of each inner node w_1 and of each edge node w_2. Now the stiffness matrix of the plate has two rows and two columns.

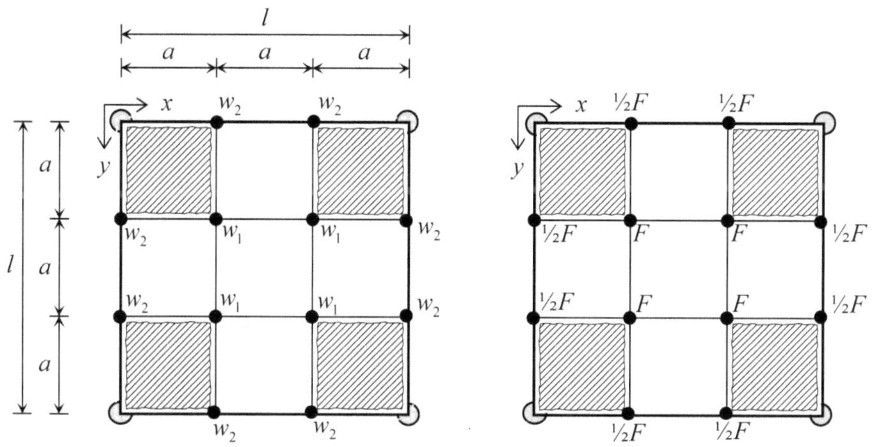

Figure 9.19 Discrete model for plate with flexible edge beams.

We introduce the edge beam flexural rigidity $EI = \lambda(\frac{1}{2}lD)$, where $l = 3a$ is the length of the plate edge and D the plate flexural rigidity; the parameter λ relates the beam rigidity to the rigidity of the half plate width. Now we obtain, on the basis of the properties of the rotational spring elements and torsion panel, the following matrix equation

$$\frac{D}{a^2}\begin{bmatrix} 4 & -6 \\ -6 & 11+3\lambda \end{bmatrix}\begin{Bmatrix} w_1 \\ w_2 \end{Bmatrix} = F\begin{Bmatrix} 1 \\ 1 \end{Bmatrix} \tag{9.13}$$

The force F is equal to pa^2. We found the solution of this matrix equation for three different values of the parameter λ

$$\begin{aligned} \lambda = \infty &\rightarrow w_1 = \frac{1}{4}\frac{Fa^3}{D}, \quad w_2 = 0 \\ \lambda = 1 &\rightarrow w_1 = \frac{Fa^3}{D}, \quad w_2 = \frac{1}{2}\frac{Fa^3}{D} \\ \lambda = 0 &\rightarrow w_1 = \frac{17}{8}\frac{Fa^3}{D}, \quad w_2 = \frac{Fa^3}{D} \end{aligned} \tag{9.14}$$

The computation of the bending and torsion moments from these displacements is a straight-forward procedure. The bending moment along a section at mid-span and the torsion moment along the edge are presented in Figure 9.20. The case of infinite large λ corresponds to the simply-supported case. The same solution is obtained as in the previous Section. The twisting moments and the bending moments are equal. For

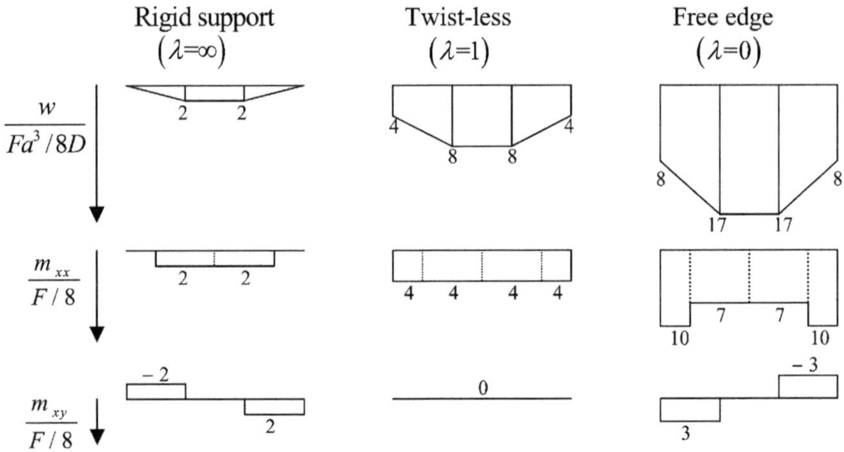

Figure 9.20 Results for different edge beam stiffness.

$\lambda = 1$, the plate becomes twist-less; all twisting moments are zero; the edge beam has the same stiffness as the half plate width. For $\lambda = 0$ the plate has free edges and is supported just by four compressive point loads in the corners. Twisting moments occur, but they have an opposite sign compared to the simply-supported plate. There, the corner reaction was tensile, here it is compressive. When λ decreases from infinity to zero, we see the twisting moments switch sign, and notice a substantial increase of the bending moments. In Figure 9.20, note the large increase of deflection, the substantial increase in the bending moments, and the switch in sign of the twisting moments.

The total bending moment over the full width of the plate at mid-span due to the load is $3Fa$, which is equal to $pl^3/9$. Part of this moment is carried by the plate and part by the edge beams. For the three considered cases of rigid supports, twist-less plate and free edges, the plate part is 1/6, 1/2 and 1, respectively. Two edge beams account for the remaining part, 5/6, 1/2 and 0, respectively. The moment in each edge beam becomes $pl^3/21.6$, $pl^3/32$ and 0, respectively.

It is interesting to compare these values to those arising from the practical approach of structural engineers and recommendations in some codes of practice, in which the load is supposed to flow to the beams as depicted in Figure 9.21, referred to as *envelope approach*. For a square plate this leads to a distributed beam load of triangular shape with maximum value $pl/2$ at mid-span. In its turn, this load leads to a bending moment $pl^3/24$, which is

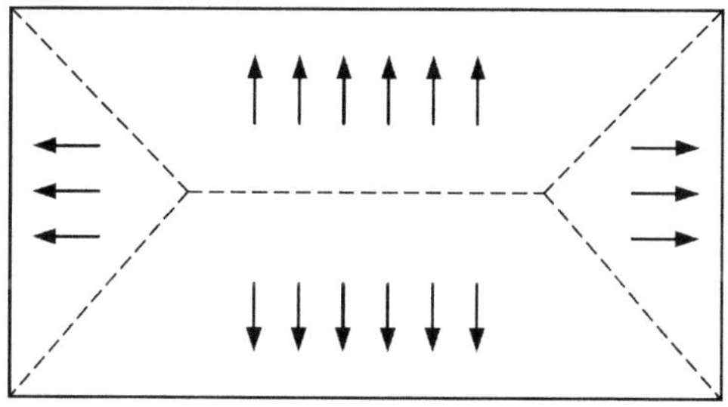

Figure 9.21 Loading of edge beams in 'envelope' approach.

10% too small, compared to the exact value $pl^3/21.6$ for an infinite rigid beam. For flexible edge beams in a twist-less plate with exact moment value $pl^3/32$, it is too large. Then the envelope approach overestimates the moment by 50%.

If a structural engineer is detailing flexible edge beams on the basis of the envelope approach, the reinforcement in the plate itself may be too weak. Even though safety may not be affected, there may be severe cracking in practice.

9.5 Message of the Chapter

- Plate bending can be modeled by a grid of beams filled in by torsion panels. The beams in their turn can be modeled by rotational spring elements.
- The spring-panel model elegantly shows that the central part of a square simply-supported plate under distributed loading is dominated by bending and the corner parts by torsion, and that the maximum bending moment is equal to the maximum twisting moment for zero Poisson's ratio.
- Equations in the global stiffness matrix are similar to those in the classical Finite Difference Method.

- The effect of edge beams can be shown by a simple hand calculation with two degrees of freedom. Very stiff edge beams and very flexible edge beams both lead to twisting moments in the corner regions of the plate, however of opposite sign. For an in between edge beam stiffness the plate is perfectly twist-less.

- For infinitely rigid edge beams, the engineering envelope approach to calculate the maximum edge beam moment is 10% too optimistic. It is 50% too pessimistic for flexible beams with a torsion-less plate.

Part 3
FE-Based Design in Daily Practice

Chapter 10
FEM Essentials

In Chapters 8 and 9 the stiffness method was introduced in the framework of discrete models out of a pre-FEM era. In those models we had to apply different element types to model the complete membrane behaviour: springs for normal forces and panels for shear forces. The same was necessary in modeling plates in bending; there separate rotational springs and torsion panels were used. Compared to this approach the Finite Element Method (FEM) has been a major step forward. One and the same membrane element accounts for normal forces in two directions and shear forces, and one and the same plate bending element accounts for bending in two directions and torsion. The present Chapter is an overview of the main features of FEM codes and comments on its practical use. The aim of this book is not an in-depth presentation of FEM theory. For that purpose we refer to standard text books of Zienkiewicz et al. [16] and Hughes [17].

10.1 Elements and Degrees of Freedom

The Finite Element Method provides an approximation to structural behaviour. The first step is to divide the structure into a large number of *elements*. In this book the elements are plane elements; the dimension in the third direction is small compared to the other two. The elements are joined to each other at element corners; these are called *nodes*. Sometimes also mid-side nodes occur on the edges of elements. In the nodes we choose degrees of freedom. In a membrane plate analysis they are two orthogonal displacements u_x and u_y. In a plate bending element we normally use three degrees

J. Blaauwendraad, *Plates and FEM: Surprises and Pitfalls*, Solid Mechanics and Its Applications 171, DOI 10.1007/978-90-481-3596-7_10,

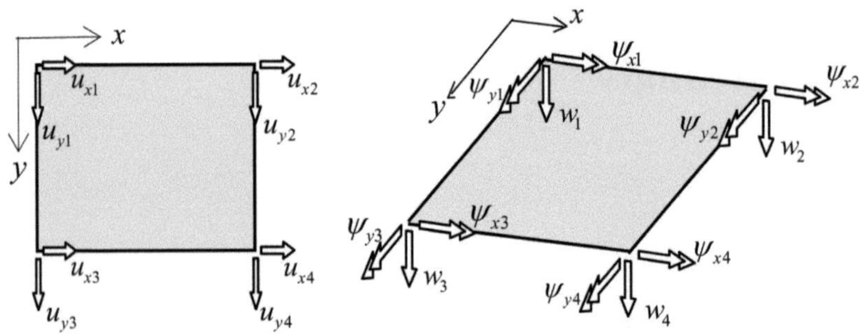

Figure 10.1 Degrees of freedom for membrane and bending analysis.

of freedom per node: the displacement w, the rotation ψ_x about the x-axis and ψ_y about the y-axis. For rectangular examples we refer to Figure 10.1.

Sign definition

We stress that the definition of the rotations in FE codes is different from the definition we used in Chapters 3 and 4 for the derivation of differential equations. To avoid confusion we use another symbol. In Chapters 3 and 4 we had chosen φ, now we use ψ. The sign convention and subscripts are also different. The rotations φ_x and φ_y in Chapters 3 and 4 are positive when they lead to positive displacements u_x and u_x, respectively, for positive z-values in the plate. The rotations ψ_x and ψ_y have a different definition. They are rotations about the x- and y-axis, respectively, and are positive according to the right-hand rotation rule.

Plate elements may be spatially assembled. Examples are box-beams and multi-cell bridges. In such structures, elements are needed both for membrane and bending action; we usually need six degrees of freedom per node. Regretfully, structural engineers call such elements *shell elements*. It is true, FE codes include shell elements for curved surfaces, and in the limit case of zero curvature they are used for combined membrane bending action. We recommend calling such flat elements *membrane-bending elements*. In shell structures there is interaction between the membrane and bending state; this is absent in flat plates.

Commercially available packages usually offer a number of element shapes to be used. Figure 10.2 shows triangles, rectangles and quadrilaterals that can be inserted in the model. If a mid-side node is applied, the edge

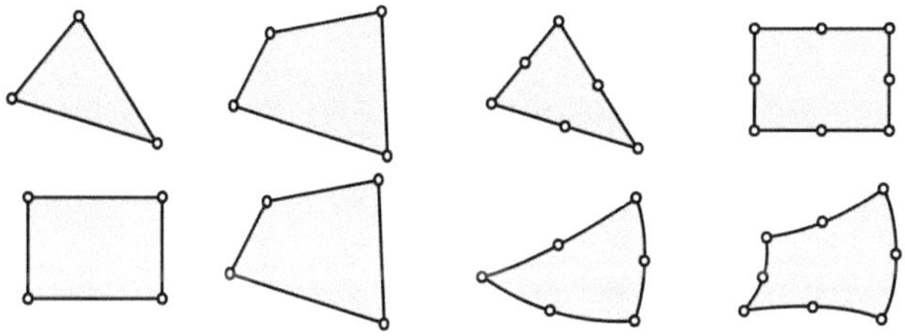

Figure 10.2 Commercially available packages offer various element shapes.

can be curved. For reasons beyond the scope of this chapter, such elements are called *isoparametric* elements. Sometimes a commercial package offers a quadrilateral element which, unknown to the user, is in fact an assemblage of triangles.

In order to make an analysis the structure is divided into elements. Figure 10.3 shows two examples. The left part comes close to the Brazilian test to determine the splitting tensile strength of a concrete cylinder. The right-hand part of the figure shows a shear wall example with a vertical row of openings as may occur in a tall building. In both examples a coarse mesh is drawn. In reality finer meshes will be applied. It is common practise that the software itself makes an appropriate mesh on the basis of the available element types. Usually the user needs to specify just the average size of the elements; the program does the rest. Figure 10.4 shows two examples of spatial structures which are assembled from flat membrane-bending plate elements: a multi-cell bridge and a part of a train carriage.

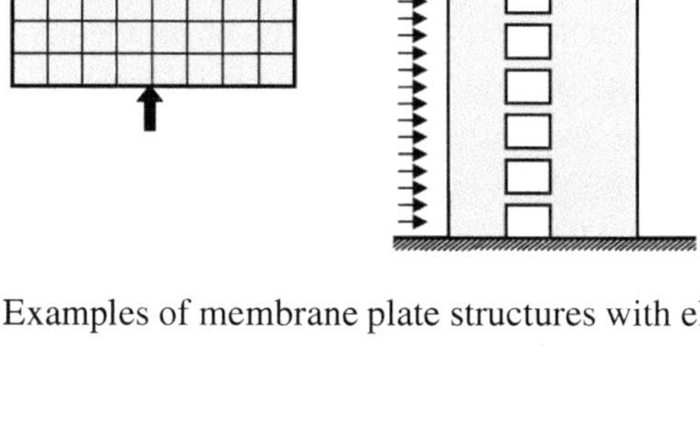

Figure 10.3 Examples of membrane plate structures with element mesh.

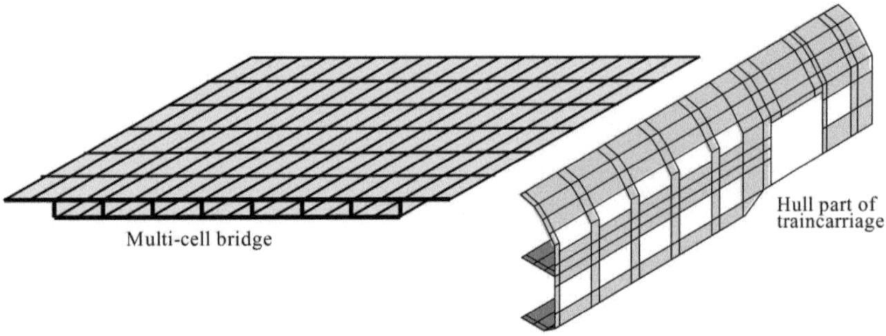

Figure 10.4 Examples of spatial structures composed of plate elements.

10.2 Stiffness Matrix and Constraints

The degrees of freedom at a node are common to all elements that meet at that node. An individual element, in turn, shares the degrees of freedom of different nodes. These degrees of freedom together form the *displacement vector* of the element (this may also contain rotations). A generalized force (which may be also a moment) is associated with each degree of freedom; these forces together form the *force vector* of the element. The element *stiffness matrix* relates the element displacement vector to the element force vector. For a triangular element with corner nodes i, j and k the matrix relation appears as

$$\begin{bmatrix} * & * & * \\ * & * & * \\ * & * & * \end{bmatrix} \begin{Bmatrix} u_i \\ u_j \\ u_k \end{Bmatrix} = \begin{Bmatrix} F_{e,i} \\ F_{e,j} \\ F_{e,k} \end{Bmatrix} \tag{10.1}$$

Herein the vector u_i represents the degrees of freedom in node i and $F_{e,i}$ the vector of generalized element forces. The stiffness matrix governs the structural behaviour of the element. Its derivation is based on the approximation of the displacement field within the element. The higher the degree of polynomials in the field, the more accurate the element will perform. As a rule, the performance of an element is better if it has more degrees of freedom, but it is not a guarantee. The quality of an element is dependent on a number of items: the chosen displacement field, the way numerical integrations are done, the extent to which displacements in adjacent elements are compatible, etc. An element with mid-side nodes is expected to perform better for the same mesh compared to elements without, but sometimes they may do not.

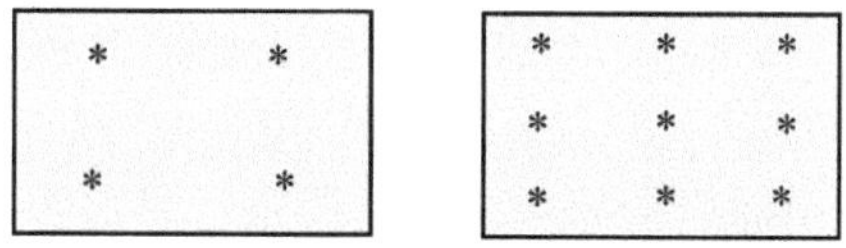

Figure 10.5 Two schemes for Gaussian points.

Programmers use *Gaussian points* in the mathematical integration procedure to construct element stiffness matrices. Gaussian points relate to integrals of polynomials over the area of an element. Gauss showed that one can write such integrals as weighted sums of values of the polynomial at a number of discrete points. The points do not coincide with nodes, but are situated inside elements at some distance from the edges. In rectangles sometimes a two-by-two scheme is used, sometimes three-by-three, see Figure 10.5. Normally the user need not know at all about such integration points, but occasionally programs may refer to the Gaussian points, therefore they are mentioned here. We refer for more details to [11] or [12].

The global stiffness matrix equation of a total structure with N degrees of freedom is similar to that for a discrete model as shown in Eq. (8.2)

$$\begin{bmatrix} \cdot & \cdot & \cdot & \cdot \\ \cdot & \cdot & \cdot & \cdot \\ \cdot & \cdot & \cdot & \cdot \\ \cdot & \cdot & \cdot & \cdot \end{bmatrix} \begin{Bmatrix} u_1 \\ \cdot \\ \cdot \\ u_N \end{Bmatrix} = \begin{Bmatrix} F_1 \\ \cdot \\ \cdot \\ F_N \end{Bmatrix} \tag{10.2}$$

The global stiffness matrix is an assemblage of the individual element matrices. Simple examples of the procedure appeared in Chapters 8 and 9. The right-hand vector holds the load to which the structure is subjected. This load consists of point loads at the nodes. When the user inputs a distributed load, the program will replace it by statically equivalent point loads. With the high speed of computers this is no problem in practice, and the user can apply fine meshes. The assembling procedure has a physical meaning. If M elements join together in node i then the M generalized element force vectors $F_{e,i}$ together balance the applied load F_i at that node.

The set of equations in (10.2) cannot be solved as long as rigid body displacements can occur, for then the global stiffness matrix will be singular. We must specify displacement constraints to prevent such singularity. Therefore, all commercial packages offer features to specify rigid or flexible supports. If a support is rigid, the corresponding displacement in that direction must be zero. So, it is no longer a degree of freedom. Therefore, the row and column in the global matrix equation which correspond with that degree of freedom must be omitted from the set. If a spring support in a node is specified, the program will add the spring stiffness to the main diagonal term which corresponds with the degree of freedom to be supported. This option can also be used to specify a rigid support. Then, a very large spring stiffness must be introduced. Some programs offer the choice for a support type in which only a compressive force can occur. In fact, this makes the analysis nonlinear. The analysis can be made by a linear-elastic analysis in an iterative way. We start including all supports. After the analysis we release the supports with a tensile reaction and restart the analysis. The iterative procedure is stopped when all support reactions are compressive or zero. Some programs do the iterative procedure itself.

10.3 Model Input

The preparation of input for a FE analysis covers all data for assembling a solvable set of algebraic equations. At least the following data must be specified:

- the shape of the structure,
- fineness of the mesh,
- element type and/or section profile,
- supports,
- material properties,
- load cases,
- applicable code of practice,
- combinations and weight factors.

Finite Element packages will always offer the feature of several load cases and the possibility of making various combinations. Codes of practice define which combinations must be considered, and which load factor must be assigned to each load case in a combination. The user can make a choice of point loads, line loads and distributed loads. Usually the load case for the self-weight of the structure is generated automatically on basis of the inputted data for geometry and material properties.

Often the element mesh is automatically generated by the program. However, the user may be asked to indicate the order of magnitude of the average element size. The choice of the element type is a very important decision by the user. For instance, in case of a slab analysis one must decide whether to use Kirchhoff elements or Mindlin ones. In Chapter 15 we will explain the importance of this choice. Many a user is not aware, however, that commercial packages have default options. For truss or beam elements, to be inserted in plate models, the user must make a choice from a section library.

Support data are easy to understand. As we noted earlier, a degree of freedom is either completely restrained, or a spring is introduced. Occasionally a special compression-only support is offered. Spring supports may be point springs, springs distributed along edges or over the area of an element.

10.4 Output Selection

Finite Element packages always offer a wide choice of output options. Standards are

- contour plots,
- trajectories,
- section graphs,
- lists,
- unity check,
- dimensioning/detailing.

The option of coloured contour plots has become very popular. These plots can be selected for displacements and stresses (or stress resultants like membrane forces, plate moments and transverse shear forces). Contour plots for principal stresses, forces and moments and their trajectories are more or less standard. Trajectories for shear forces are an exception, unfortunately.

Plea for graph output in sections

The great popularity of contour plots is to be regretted, the more so when they are used in combination with options for automatic detailing, results of which are also shown as contour plots. It prevents the structural engineer from fully grasping in which way the structure transfers load to supports. A far more valuable facility is the graph option for displacements or forces in sections over the structure. We will return to this in Chapter 14. Software providers should take pride in offering this option. It is very helpful if the structural engineer can output the integral of forces and moments in sections or dedicated part of sections.

Contour plots should at most be used to decide in which sections graphs will be shown. A good alternative to contour plots are 3D pictures of displacements, forces and moments. They are very instructive, because they appeal to the engineer's sense for structural behaviour.

Blameworthy use of envelopes of load cases

A really blameworthy practice is to consider an envelope of load cases and combination results, rather than judging the result of each load case and combination individually. In that way the structural engineer cannot pick up how the structure behaves, and will have no idea about actual safety levels. Apart of this, the approach is not economic. Structural engineers should investigate results of load cases and combinations individually, at least for the most important ones.

When specifying stresses or stress resultants that the engineer wants to be presented, there are several possibilities, such as stress (resultants) at nodes or in the element centres. Stresses at a node differ from element to element. The software may offer the choice of computing the average at a node or the values at the centre of elements. Averaging is a pleasant feature, but is misleading for elements which join in a node and have different thickness. Exceptionally, software may present output at Gaussian points. In such cases the software most probably deals with a specialized subject.

All programs will calculate support reactions. Usually an equilibrium check is done in so far that a check is made to compare the total load to the sum of the support reactions. Strictly speaking it is not proving that the set of equations has been solved correctly, because an ill-conditioned set of equations may lead to a wrong solution even though the test is satisfied. However, in practice it is valuable to test the correctness of the applied load, because ill-conditioned sets of equations are rare in practice.

Finally, the user can usually ask to see at which positions in the structure the permissible stress level is surpassed. This is of great use for steel structures. The stresses can be combined to yield the Von Mises stress which is then compared to the yield stress. In a unity check it is tested whether the ratio of the Von Mises stress and the yield stress is smaller than 1. If not, the design must be changed. For reinforced concrete slabs a FEM package often offers the feature of calculating the wanted reinforcement ratios. Because of its importance, Chapter 16 is fully devoted to this subject.

10.5 Message of the Chapter

- The reliability of a FE analysis highly depends on the competence of the structural engineer. Garbage in is garbage out.

- Seemingly, higher-order elements need not necessarily perform better than simpler elements.

- Programmers sometimes make choices which are hidden from the user. The user should be aware of set defaults.

- FE codes may offer post-processing options for unity-checks and detailing of reinforcement. The user should be aware of the assumptions underlying such design options.

- Contour plots are a popular way to present results, but they prevent the structural engineer from grasping how a structure will transfer load to supports.

- 3D plots and section graphs may appeal much more to the engineer's sense for structural behaviour. Software providers should take pride in offering such options. It must be possible to output the integral of forces and moments in sections and dedicated part of sections.

- The practice of making envelopes for results of load cases and combinations is blameworthy and fundamentally wrong. Structural engineers should investigate results of load cases and combinations individually, at least for the most important ones.

Chapter 11
Handling Membrane FEM Results

11.1 Surprising Stresses

11.1.1 Effect of Poisson's Ratio

We consider a box-shaped steel bridge as shown in Figure 11.1. The box has two horizontal walls, at the top and bottom respectively, and two vertical walls, one at each outer side. There are no inner vertical longitudinal webs in the bridge. The structure is a new bridge on an existing pier that had been used for a narrower structure. Therefore, the two bearings at each bridge end do not coincide with the vertical side walls of the box structure, but are in a position more inward. It makes the end diaphragm behave as a beam with a four-point loading as drawn in Figure 11.1. The diaphragm consists of three parts, two squares outside the bearings and one between the bearings which has a length over depth ratio of two. The outer vertical walls are schematized as vertical stiffeners at the ends of the diaphragm. These walls carry forces F to the two ends of the diaphragm. These point loads cause support reactions F in the bearings. In order to introduce these concentrated sup-

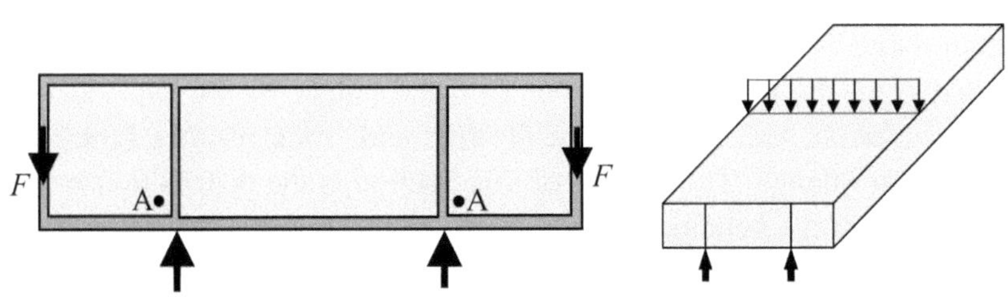

Figure 11.1 End diaphragm of box-shaped bridge.

J. Blaauwendraad, *Plates and FEM: Surprises and Pitfalls*, Solid Mechanics and Its Applications 171, DOI 10.1007/978-90-481-3596-7_11,

port reactions in the web of the diaphragm, two vertical stiffeners have been added at the position of the bearings. The diaphragm will bend such that the upper edge becomes longer and the lower edge shorter. Because of compatibility reasons, parts of the horizontal walls of the box will then participate in the transfer of the forces. In Figure 11.1 these parts have been modeled as horizontal stiffeners over the full length of the diaphragm, top and bottom.

Now we draw attention to the spots A in the diaphragm. The normal force in the vertical stiffener will be $N_{\text{vert}} = -F$. The normal force in the horizontal stiffener at that position will also be $N_{\text{hor}} = -F$. This result means that the strains are equal, to ϵ, in the vertical and horizontal stiffeners at position A. The stress σ in the vertical stiffener is $\sigma = E\varepsilon$. Because of compatibility the strains in the web material adjacent to the stiffener must be of the same size: $\varepsilon_{xx} = \varepsilon_{yy} = \varepsilon$. The stress in the web plate must be calculated with Eq. (1.13):

$$\sigma_{yy} = \frac{E}{1-\nu^2}\left(\varepsilon_{yy} + \nu\varepsilon_{xx}\right) = \frac{E}{1-\nu}\varepsilon \tag{11.1}$$

For Poisson's ratio $\nu = 0.3$ we find a stress is $\sigma_{yy} = E\varepsilon_{yy}/0.7 = 1.43E\varepsilon$, 43% larger than the stress $\sigma = E\varepsilon$ in the adjacent stiffener with the same strain. Accounting for Poisson's ratio in this way can be important, particularly if buckling must be considered. The web is compressed, both in the horizontal and vertical direction.

11.1.2 Effect of Kink in Beam Flange

Now consider a clamped beam of I-section subjected to a point load at its free end, as shown in Figure 11.2. The cross-section of the beam consists of two flanges and a web. The width of the flanges and depth of the web are equal. The top flange is straight over the full length of the beam. The bottom flange is parallel to the top flange over about two-third of the beam counted from the clamped end, but then the height of the web decreases linearly to about half its height at the free end. As a result a kink occurs in the bottom flange. The subject of this Section is the distribution over the beam length of the bending stress σ in the connection line between the web and the bottom flange. Classic beam theory predicts $\sigma = M/W$, where M is the bending moment and W the section modulus. The bending moment M varies linearly over the length of the beam; W is constant between the clamped end and kink, and decreases in a nonlinear way to about one quarter of the constant value at the free end. So the expected stress distribution will

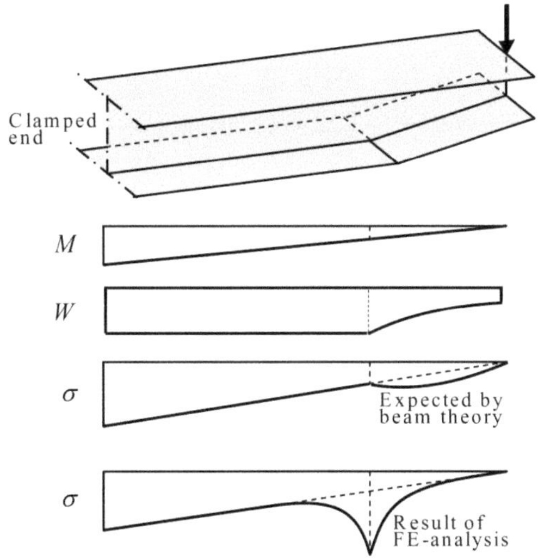

Figure 11.2 Unexpected effect of kink in bottom flange.

be linear between the clamped edge and kink. Between the kink and free end the stress will be larger than for a constant W, but it will become zero at the free end.

As an alternative we can perform the calculation with the aid of a FEM program in which we choose membrane elements. We model the flanges by rectangular elements, say four over the half width. Choosing an aspect ratio of about two, we obtain the wanted number of elements over the length. For the web the same number of elements over length and depth is used. A sketch of the result to be expected from the FE analysis is included in Figure 11.2. It is completely different from the expectation on basis of the classical beam theory.

The explanation for the unexpected FEM result has to do with the fact that the classical beam solution is not admissible at the location of the kink. That solution predicts that a stress $\sigma = M/W$ acts in the horizontal flange at the position of the kink and that the same stress acts in the inclined flange at that position. The two membrane forces meet each other at an angle, and therefore equilibrium can only exist if a third force in another direction acts in the same point to balance them. In reality there is no third force, so the membrane stresses in the bottom flange can be zero only at the location of the kink. In fact, there the I-section beam behaves as a T-section beam, consisting of the web and the top flange, shown in Figure 11.3. The section modulus reduces substantially, the neutral line shifts upward, and the bottom stress

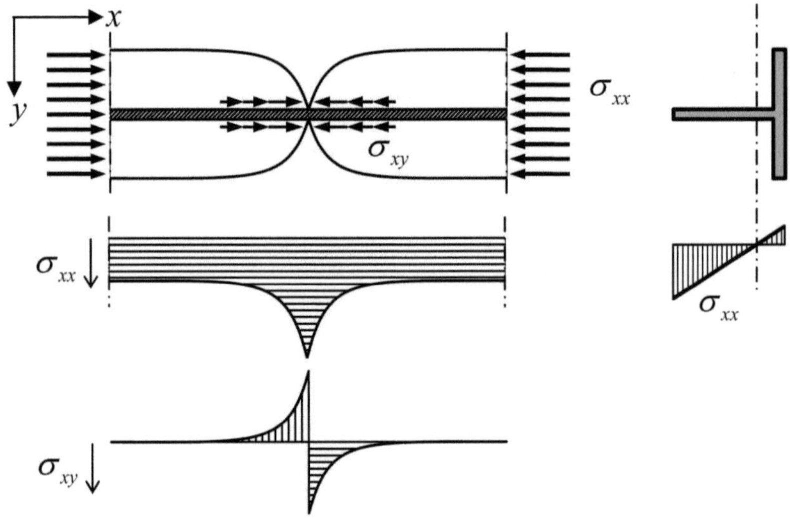

Figure 11.3 Explanation of unexpected high bending stress.

increases. At some distance from the kink, the flange will contribute again. Shear stresses σ_{xy} between the web and each flange half must develop to obtain this effect. These shear stresses abruptly change sign and direction at the kink, and the bending stress σ_{xx} correspondingly decreases in two directions.

FE analyses bring to light omissions in design

FE analyses show stress concentrations where classic beam calculations suppose smooth stress distributions. Sometimes they bring merciless to light omissions in design.

Those who are familiar with bond stresses in a cracked reinforced concrete bar under tension will see a similarity. At a crack the reinforcement bar has to carry all the tensile force, and at some distance from the crack the concrete and bar carry the force together. A transition length occurs in which high bond shear stresses transfer part of the high tensile force in the crack to the concrete.

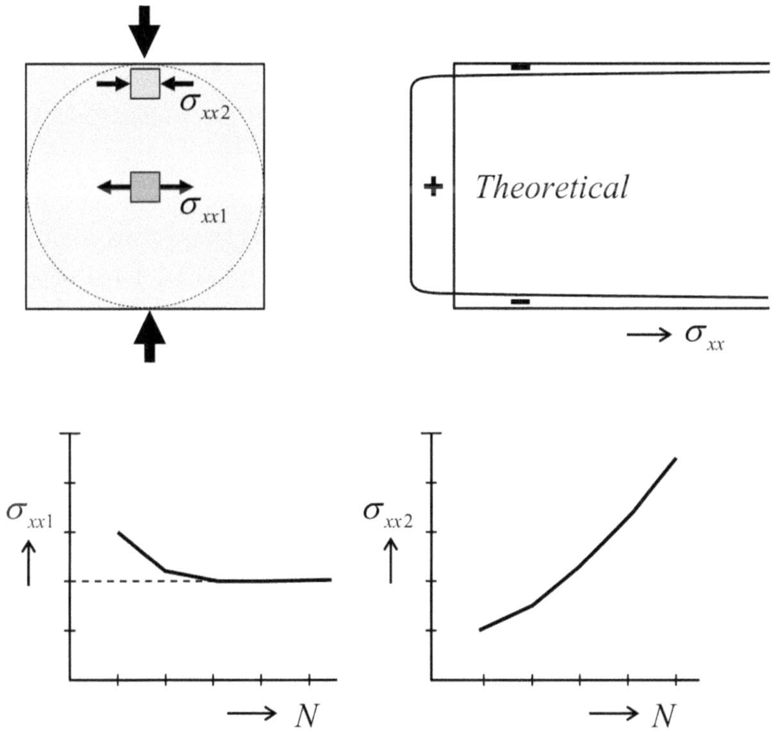

Figure 11.4 FE analyses do not converge in case of stress singularities.

11.2 Stress Singularities in FEM

In Section 6.2.2 we derived the stress state in the Brazilian splitting test. A cylindrical body is compressed by two opposed line loads. We found a constant horizontal tensile stress over the vertical plane of the cylinder and concentrated horizontal compressive point loads, one on the top and the other at the bottom, to balance the tensile stresses. In Figure 11.4 this test is simulated by a FE analysis. Instead of the circular cross-section, a square is chosen. The four corner areas outside the inner circle are relatively low-stress regions which hardly influence the stress state in the vertical plane of symmetry. The theoretical solution of the horizontal stress σ_{xx} is shown, and the stress is computed by FEM in two positions, point 1 at the top edge of the square and point 2 in the centre. The analysis is done for different mesh finenesses, indicated by the number of elements N over the width and height of the cross-section. It is expected in a FE analysis that mesh refinement would make the stress result converge to its final correct value. We notice that this is indeed the case for the stress at point 2, but not at point 1.

Lesson on singularities

Convergence is obtained for stresses of finite value, but not at locations where the membrane plate theory predicts a singularity. Then no convergence will occur. The same is true for an infinitely large bending moment in plate bending, the subject of Chapter 14. There, too, mesh refinement does not make sense.

11.3 FEM-Supported Strut-and-Tie Modeling

If we have to design the reinforcement in concrete walls, it may be helpful for understanding force transfer to draw a Strut-and-Tie Model (STM). It is a truss-type model in which the load is represented by a set of well-chosen lumped forces, and the transfer to the supports occurs through a system of compressed struts and tensioned ties. In simple statically determinate problems the structural designer easily knows how to choose the strut and ties, but statically indeterminate structures may be puzzling, because more than one possibility exists. Figure 11.5 shows a simple example. A square silo is supported at the four corners and is subjected to a homogeneously distributed vertical load over the area of the walls. This load is due to own weight and possible friction of bulk material in the silo.

Each silo wall is a deep beam, of which the maximum bending moment occurs in the vertical line of symmetry at mid-span. The distribution of the horizontal stresses in this section must be known in order to design the reinforcement properly. The strut-and-tie model for an individual wall is simple, as appears from Figure 11.5. Green is compression, red is tension. In the line of symmetry at mid-span of the wall there is a compressive horizontal force and a tensile horizontal force of equal size. The product of the distance between these two forces times the value of the forces is the bending moment. The problem for the structural engineer is to make a fair guess about the distance, then calculate the tensile force and decide on the amount of horizontal reinforcement. Because the structure is highly statically indeterminate, the choice is hard to make. A FE analysis may help to make a good estimate.

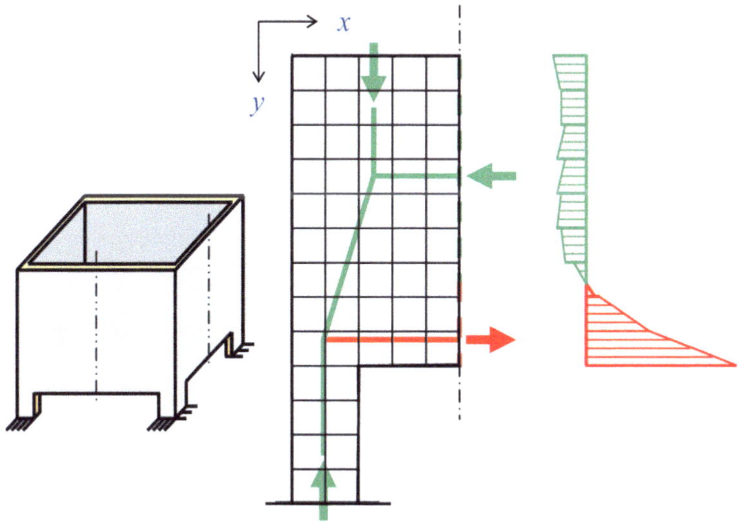

Figure 11.5 FEM as support for Strut-and-Tie Model

We have analyzed the wall with a course mesh, which is appropriate. In fact the wall behaves as a deep beam, as discussed in Section 2.2.2. There we assumed that the wall has a distributed support over the full height of each vertical edge. The result for deep walls will be that the vertical support reaction is very much concentrated in the lower part of the edges. This distribution of the support reaction is close to the real support of the deep wall. The difference is then not significant. And for the rest, it anyhow is not relevant where the support occurs along the vertical edge. This does not influence the total beam moment which must be transferred in the vertical line of symmetry of the wall. The result of the FE analysis has been included in Figure 11.5. First, we see that the computed result agrees well with the theoretical expectation in Figure 2.13 for a short, tall wall. Second, the stress distribution along the line of symmetry nicely shows where we can choose the center of gravity of the tensile stresses and of the compressive stresses, which determine the positions of the horizontal tie and strut, respectively.

FE analysis as support for strut-and-tie modeling

FE analysis supports the structural engineer in making strut-and-tie decisions. It helps positioning the reinforcement and determining lever arms.

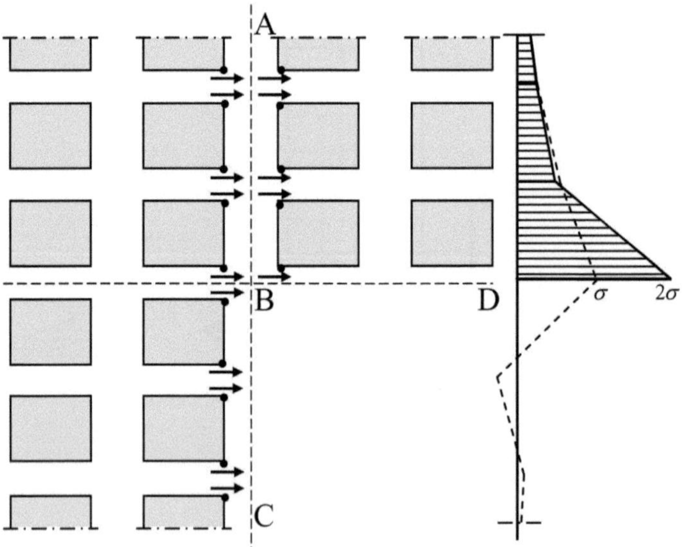

Figure 11.6 Exploded view of elements around re-entrant corner.

11.4 Re-entrant Corner

The silo wall of Section 11.3 offers a good occasion to draw attention to a typical aspect which accompanies FE analyses for structures with a re-entrant corner. An example is displayed in Figure 11.6, which could be the corner between the column and the bottom edge of the wall. We have drawn in an exploded view rows and columns of rectangular elements around the corner. The re-entrant corner has edges BC and BD, so the corner is at node B. For ease of explanation it is assumed that the load is applied outside the considered part of the element mesh. We consider the vertical line which runs from node A over B to C, and focus on the horizontal forces at the nodes and the equilibrium of these forces. The two nodes between A and B can be seen as regular mesh nodes. Here four elements join together in a node and each element brings in an element force to the equilibrium equation for that node. The two forces coming from the left are more or less equal, and the same holds true for the two forces coming from the right. Equilibrium requires that the sum of the four forces is zero. The result is that the horizontal stresses in the four elements will be almost equal. An averaging procedure makes sense in such points. In corner node B a different situation exists. Now there are three element forces and they must sum to zero. The two left forces, more or less of equal size, must balance those coming from the right. Therefore, the horizontal stress in the element to the right of node B is about twice the stress in the elements to the left of the node. If the stress to the left of the node is about σ in both elements, then it will be 2σ to the right of the node. This latter size must be considered as the most realistic one. The average value is 1.33σ, which is only two-third of the maximum stress 2σ, the realistic one. Below corner B, only two element forces remain, which must be equal and opposite.

The stress state in the re-entrant corner is singular. In fact infinitely large shear stresses must occur in the horizontal line through corner node B to achieve that the stress 2σ in the right plate part is spread over the two plate parts left of the vertical line. A disturbance occurs in the FE analysis which is still noticed in the edge nodes below the corner B. The stress distributions over sections just left and right of the vertical line through the corner node, sketched in Figure 11.6, visualize this. The dotted line represents the stresses to the left of the line ABC. It takes a couple of elements before the stress really is zero on the free edge BC. The explained phenomenon is inherent to the finite element method. The finer the element mesh, the smaller the region will be where the disturbance is seen. Similar effects are seen in plate bending near re-entrant corners. What we have explained here for plate forces, will occur there for plate moments.

Note on averaging procedures

Averaging procedures can be misleading at re-entrant corners. The average value can be far less than the real maximum stress.

11.5 Tall Wall with Openings

Multi-storey buildings often transmit wind loads to the base through shear walls of reinforced concrete. Shear walls may be positioned as end walls in building plans or as inner walls. At each storey an opening may occur; in end walls they allow for windows, and in inner walls they are necessary if a corridor crosses the wall, see Figure 11.7. We can view the tall wall with openings as a structure of two slender walls, left and right respectively, which are connected by horizontal cross-beams at each floor. Walls and beams together form one monolithic structure. If the shear wall is subjected

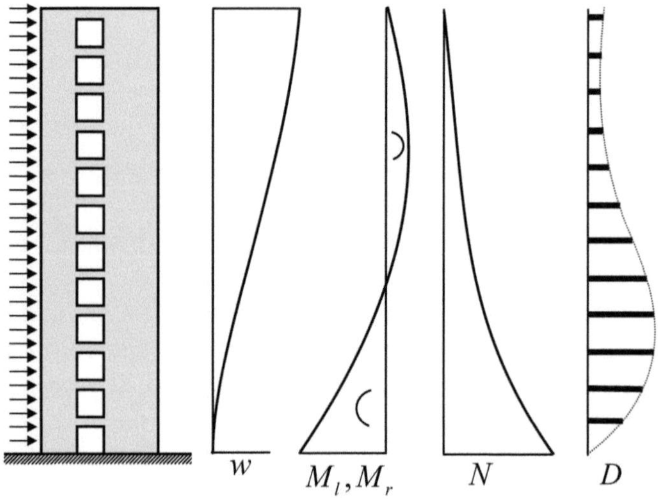

Figure 11.7 Expected behaviour of tall wall with openings due to wind.

to wind load, the cross-beams will act as dowels, and the wall will deflect (w) as sketched in Figure 11.7. The structural engineer is interested in the size of the transverse dowel forces (shear forces) D in the cross-beams, in order to detail the reinforcement in an adequate way. As a consequence of the dowel forces D, a tensile base force N will act in the left wall and a compressive base force N in the right. These normal forces carry part of the total wind moment at the base. The other part consists of bending moments in the left and right wall at the base, M_l and M_r respectively. The total horizontal force due to the wind is the base shear, which is divided over the left and right wall V_l and V_r respectively. The base information is of interest to the structural designer in order to properly design the foundation. The expected diagrams for normal forces N, moments M, and beam dowel forces D are included in Figure 11.7. Of course, the analyst must superimpose the load due to self-weight.

In the 1960s and 1970s a lot of research on the force distribution in this type of structure was published on the basis of differential equations, among which contributions of Rosman [18] are well known. Today structural engineers will most probably apply FE analysis. In this section we comment on the FE modeling of this type of structure. We will do this in two ways: in Section 11.5.1 the focus is on modeling with membrane elements; in Section 11.5.2 it is on modeling as a frame structure.

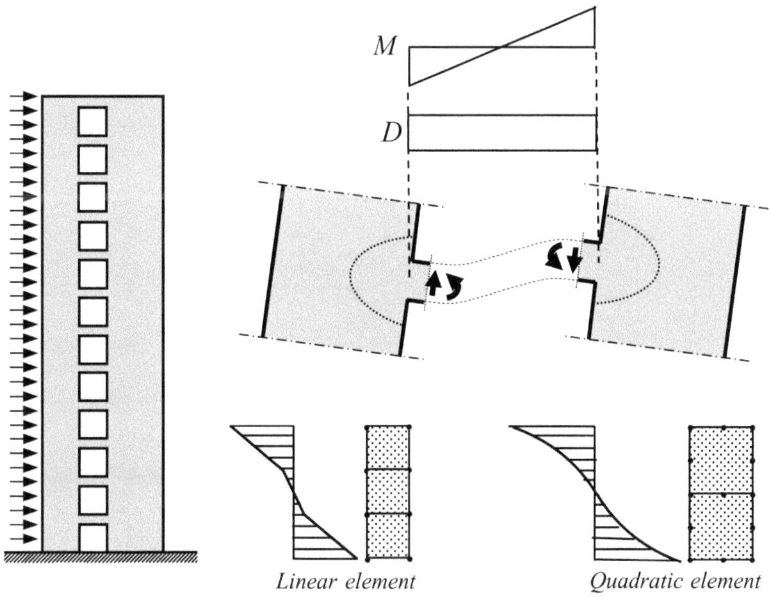

Figure 11.8 Tall wall with row of openings.

11.5.1 Modeling with Membrane Elements

In Figure 11.8 we consider in more detail the force transfer in an individual cross-beam. The midpoint of beams is a point of counter-flexion. A constant shear force D occurs in the cross-beams, and the diagram for the bending moment M is linear with a zero value at mid-span.

At the connection between the cross-beams and the vertical walls, there will be high stress concentrations. Therefore, a sufficiently fine mesh must be chosen around beam-wall connections. Further away in the vertical walls a courser mesh will be adequate. However, a coarse mesh is not really necessary in view of the speed and the mass storage of current computers. If the structural engineer chooses the membrane element with mid-side nodes (quadratic displacement) it suffices to use two elements over the depth of the cross-beam. For the element with only four corner nodes (linear displacements) three or more elements over the depth must be applied. Additionally, we recommend choosing the constant shear element because it performs better in situations where the element must reproduce bending states. The expected distribution of bending stresses in the cross-beams for linear and quadratic elements is included in Figure 11.8.

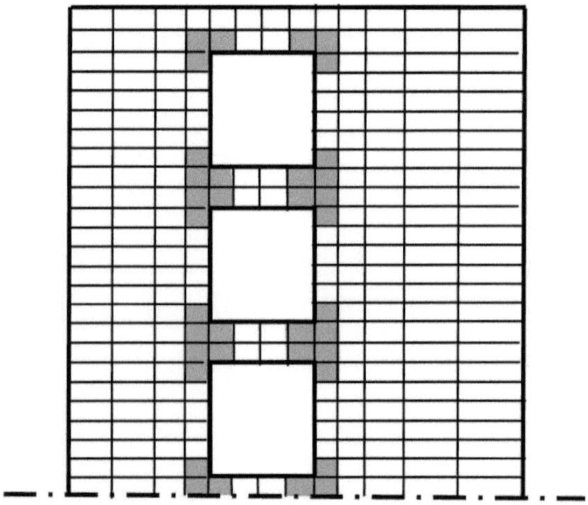

Figure 11.9 The shaded elements will be seriously cracked.

Figure 11.9 shows the mesh with two elements over the depth. Because the largest moment occurs at the end of the cross-beam, and there is stress concentration at the connection, we expect substantial cracking. We can account for that by reducing the modulus of elasticity. Compared to the middle part of the cross-beams the actual stiffness might be halved. In Figure 11.9 we have marked these elements by shading. Reduction of the stiffness is important if the structure is sensitive to geometrical nonlinearity (second-order effects) and an increase of bending moments must be considered in stability checks.

11.5.2 Modeling as Frame

Another way to investigate the force distribution in a tall wall with openings is to model the structure as a frame. In this case we must pay attention to a number of things. The frame consists of two vertical members (line elements), which coincide with the centre lines of the two slender walls. A horizontal member is placed in the centre line of each cross-beam. The modeling of the vertical members does not raise problems. The engineer must just make sure that the computer program accounts for deformation due to both normal forces and bending moments. Shear deformation need not be considered for slender walls. In the frame model, we introduce horizontal members with rigid end parts. The length of these rigid parts must be chosen

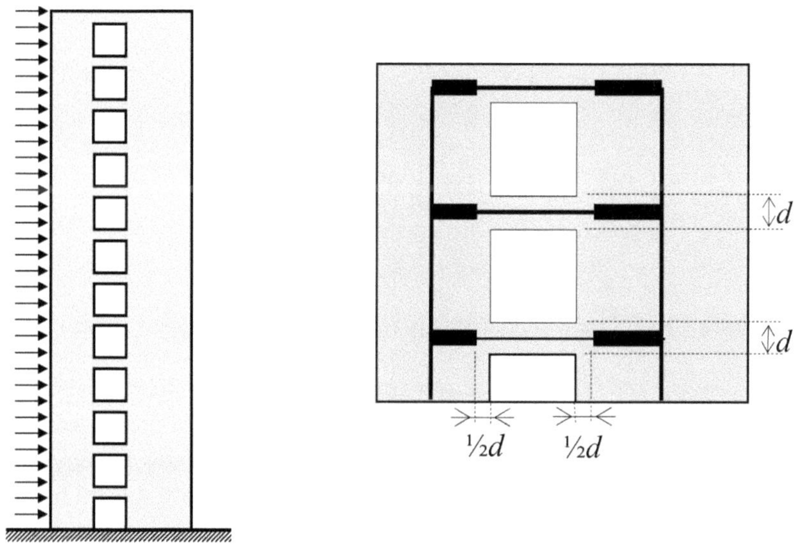

Figure 11.10 Frame model of tall wall with openings.

with care. At first glance we may make the length half the width of the vertical walls, but then we make them too long. At the junction of the cross-beam and the walls, there will be deformation of the wall. Therefore, we recommend working with a fictitious length of the cross-beams larger than the real length. At each end the cross-beam may be extended by a length equal to the half depth of the beam, see Figure 11.10. So, the length of the rigid parts become smaller than the half width of the vertical walls.

Next to the choice of the length of the cross-beams, we must decide whether they behave as Bernoulli beams (flexural deformation only) or Timoshenko beams (flexural and shear deformation). Even after we increase the length of the cross-beam, it is still not slender. We must consider the half length of the beam, because of the zero value of the moment at mid-span. The half cross-beam is a cantilever beam with a point load at the end (point of counter-flexion), as discussed in Section 2.1.3. The length over depth ratio of the cantilever beam may become of the order of magnitude two and then, according to Eq. (2.57) stiffness reduction occurs. One can account for this either by using a program that accounts for shear deformation or by introducing a judiciously decreased bending stiffness.

Until now, we have assumed that the cross-beams have rectangular cross-section with the same width as the wall thickness. It is probably more com-

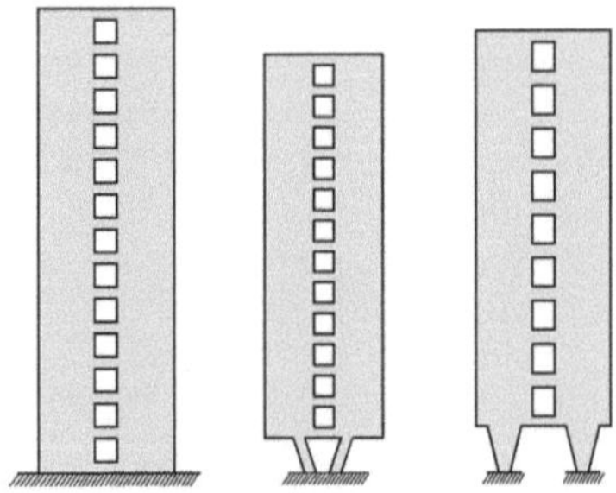

Figure 11.11 Frame models easily include deviating support structures.

mon that floor slabs are fixed to the cross-beams. Then the engineer has to increase the bending stiffness of the cross-beam in the model, because of two reasons. The first reason is that the cross-beam becomes a T-shaped beam because the floor behaves as a flange of the cross-beam. It is not likely that this contribution is substantial. Participation of the floor as flange presupposes that a normal force can operate in the flange, however this requires a sufficient length to build up the normal force, and the half cross-beam is too short to obtain this. The other reason to include participation of the floor slab is the bending of the slab. It has to follow the curvatures of the cross-beam, so an effective slab width must be chosen, the bending stiffness of which is added to the bending stiffness of the cross-beam. To choose this effective slab width we could draw a line from the cross-beam centre with an angle of 45 degrees to the cross-beam. This implies, at the connection with the wall, a slab width at each side of the cross-beam of half the length of the beam.

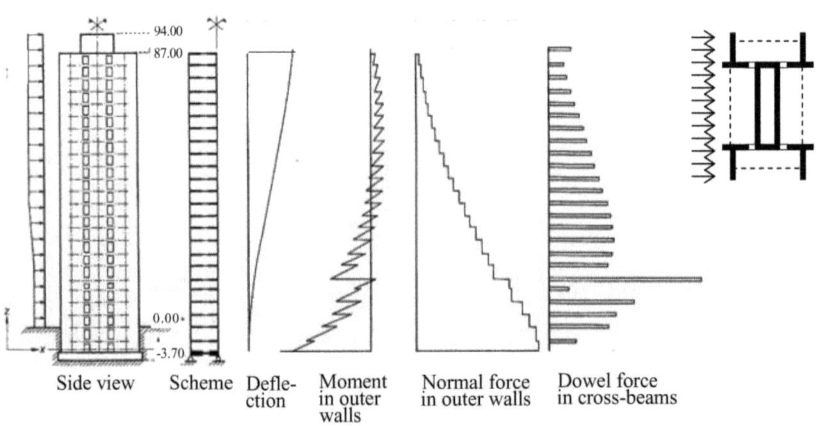

Figure 11.12 Example of three coupled walls with two rows of openings.

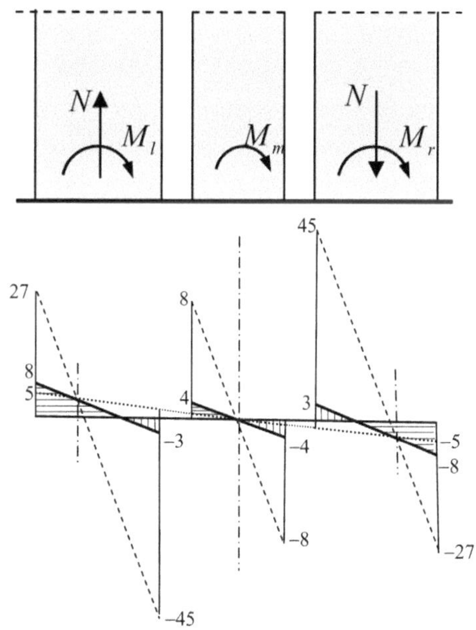

Figure 11.13 Stress distribution in walls at base level. Dashed line for pin-connected cross-beams. Dotted lines for infinitely rigid cross-beams. Full line for real cross-beam stiffness.

Unlike the first, this second contribution to the beam bending stiffness may be substantial.

The frame method has a number of advantages. First, it provides results in a way structural engineers appreciate. The analysis leads to normal forces, shear forces and bending moments, for which code checking is familiar. Second, special supporting structures can easily be included in the model. Figure 11.11 demonstrates this. Third, it is easy to handle irregular structures with different storey heights and locally deviating cross-beam stiffness, and walls that are hard to model as a membrane plate fit easily in a frame model. Figure 11.12 shows an example. The building of about 90 meters height houses the administration offices of the government of a province in the Netherlands. The plan form is shown in the right-hand top part of the figure. The dashed lines are glass facades. In the plan, four T-shaped columns are connected with the corners of a central rectangular shaft. The connection of the columns to the shaft is by cross-beams at each floor level. A cross-beam of exceptional stiffness occurs at all four corners between the sixth and seventh storey. The building is subjected to wind at the long edge of the plan form.

The structure has been modeled as a system of three vertical walls and two rows of openings, as depicted in Figure 11.12. Because of symmetry, normal forces occur only in the two outer walls. The results of the analysis are included in Figure 11.12. At the position of the cross-beam with exceptional stiffness, there is a large dowel force D, which causes discontinuities in the bending moments and in the normal forces of the outer walls. Figure 11.13 displays the base stresses in the three walls. Note that the stress diagrams have the same gradient; this must hold because the three walls share the same deflection and therefore curvature. In the plot also the stresses are shown which would occur for two extreme situations, one in which the cross-beams are pin-connected and just act as trusses, and another in which the cross-sections are perfectly rigid such that all three walls act together as one wide beam, respectively. The actual maximum stress is about twice the stress for the ideal stiff case, however less than one-third of the pin-connected case.

11.6 Checking and Detailing

Commercial packages offer program features to check whether the design of structures is in accordance with codes of practice, or to dimension structural components. Here we just touch the subject to give an impression.

11.6.1 Steel

After a stress analysis has been performed for a steel structure, a so called unity check is made to confirm that the stress state in the structure is sufficient remote from the state of yielding. For that purpose the Von Mises stress σ_{VM} is calculated. In two-dimensional states this stress is

$$\sigma_{VM} = \sqrt{\sigma_1^2 + \sigma_2^2 - \sigma_1 \sigma_2} \tag{11.2}$$

where σ_1 and σ_2 are principle stresses. The condition is that this stress is smaller than the yield stress σ_y:

$$\gamma_{\text{mat}} \, \sigma_{VM} \leq \sigma_y \tag{11.3}$$

Here γ_{mat} is the partial safety factor to be taken into consideration. Figure 11.14 visualizes the yield condition in the two-dimensional diagram for principal stresses. The actual combination of stresses must remain within the yield contour.

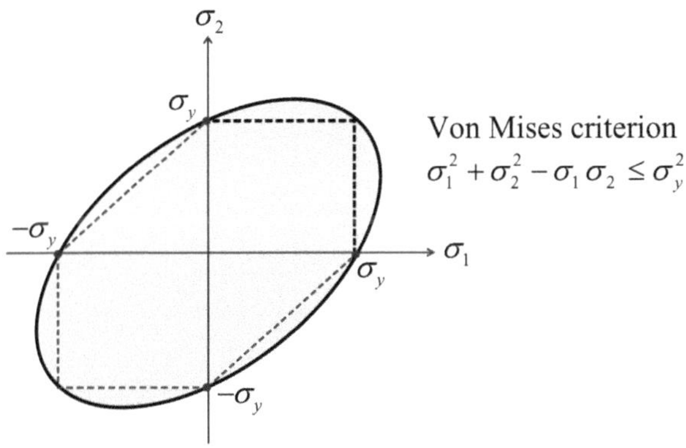

Figure 11.14 Von Mises yield criterion. Dashed line Tresca.

11.6.2 Reinforced Concrete

For reinforced concrete structures, programs offer options to design the reinforcement automatically. We must solve a problem if we want to apply a two-way orthogonal reinforcement, because we have three membrane forces n_{xx}, n_{yy} and n_{xy}. The subject will come up in detail in Chapter 16, but we will give a taste of the approach here. Consider a square panel with tensile normal membrane forces and a non-zero positive membrane shear force. In its most simple form the procedure is to replace the three membrane forces by two steel forces per unit length, n_{sx} and n_{sy}.

$$\begin{aligned} n_{sx} &= n_{xx} + n_{yx} \\ n_{sy} &= n_{yy} + n_{xy} \end{aligned} \tag{11.4}$$

and dimension the reinforcement in the x- and y-direction on the basis of these two forces. The idea behind this procedure is the assumption that the concrete is cracked. For the direction of the cracks only the shear membrane force n_{xy} is considered, which leads to principle forces in the direction of the diagonals of the panel, one tension and one compression respectively. The cracks are supposed to develop normal to the tensile principal force. This situation is shown in the top part of Figure 11.15. There the lower right triangular part represents concrete, and the upper left triangle shows the reinforcement bars. Equation (11.4) follows from the equilibrium conditions in the x- and y-direction of the concrete triangle. If the shear force has an opposite sign, the crack direction reverses. The relations in Eq. (11.4) still hold true if we use the absolute value of the shear forces, as follows from the

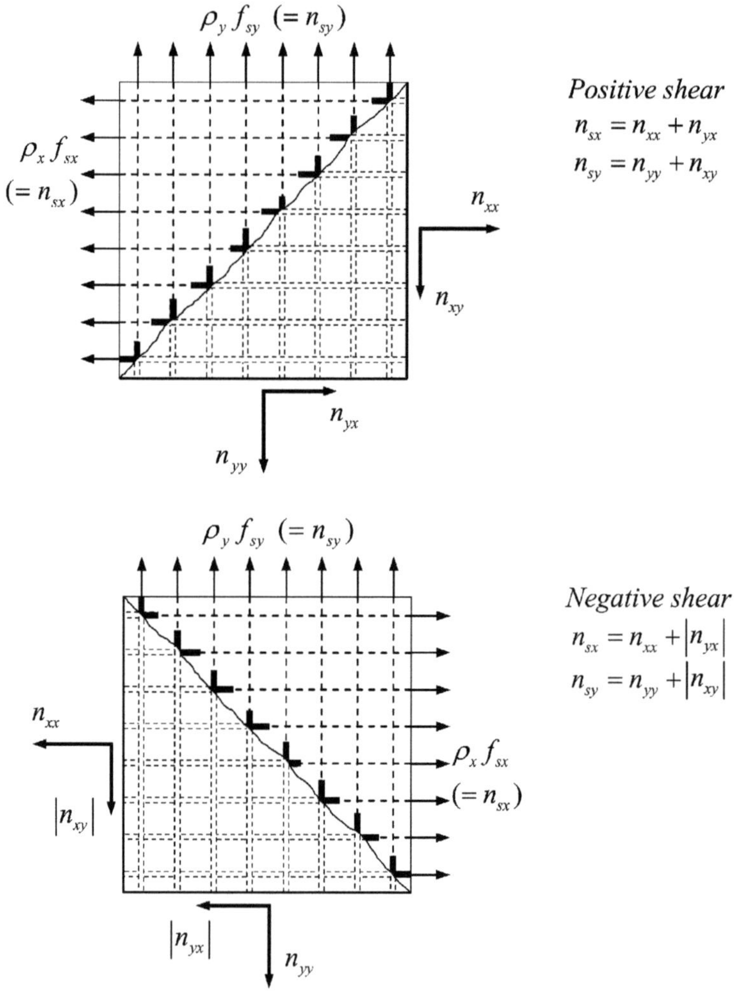

Figure 11.15 Determination of the reinforcement forces n_{sx} and n_{sy}.

bottom part of Figure 11.15. The concrete struts are now in the direction of the other diagonal. It is a consequence of the assumed cracked state that compressive stresses occur in the concrete. A shear membrane stress σ causes a compressive stress of 2σ in the concrete struts, as is seen in Figure 11.16. This follows from the equilibrium of a strut of width $s\sqrt{2}$ if s is the spacing between the reinforcement bars.

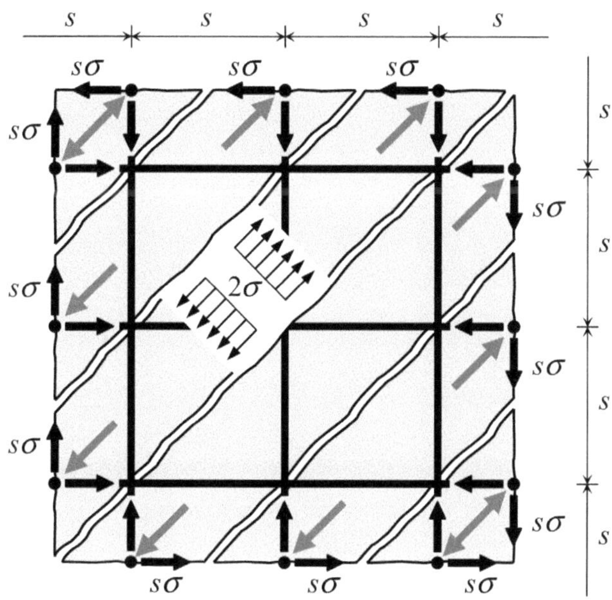

Figure 11.16 Cracked concrete, tensioned reinforcement, diagonal compressive struts.

11.7 Message of the Chapter

- Lateral contraction due to Poisson's ratio can be responsible for unexpected local increases of stress.
- FE analyses show stress concentrations where classic beam calculations suppose smooth stress distributions. Sometimes they reveal omissions in design.
- If the theory of elasticity predicts a singular stress, refinement of mesh in the FE analysis will lead to ever increasing stress values. No convergence will be obtained.
- Linear-elastic FE analysis can be a help to structural engineers who like to use strut-and-tie models in detailing reinforced concrete.

- Re-entrant corners are singular stress spots in membrane structures. Here, averaging of discontinuous stress values must be done with care.

- The structural engineer can account for severe cracking in a linear-elastic FE analysis by judiciously reducing Young's modulus in cracked regions. An example is a tall wall with a row of rectangular openings. This structure can alternatively be handled as a frame.

Chapter 12
Understanding FEM Plate Bending

12.1 Intended Goal and Chosen Structure

This is the first chapter on FE-based analysis of plate bending. We include this for in-depth understanding of the nature of the approximation. The Finite Element Method does not violate the kinematic and constitutive relationships, but satisfies equilibrium conditions only in an average sense. We will investigate to what extent equilibrium is maintained. The shape of the plate is purposely chosen so that all kinds of boundary conditions can be included. For the goal of this Chapter we choose a coarse mesh. When discussing real structural problems in subsequent chapters, we will apply finer meshes. The chosen concrete structure is shown in Figure 12.1, with a set of axes x,y in the horizontal plane, shown in the left-hand top corner. The slab has 40 elements and 55 nodes. The nodes and elements are numbered. Element numbers are put in a frame throughout the whole chapter, to distinguish them from node numbers. All elements have the same size, with a length $a = 0.9$ m in the x-direction and a length $b = 1.3$ m in the y-direction. The plate thickness is 200 mm. The edge between the nodes 1 and 8 is clamped, and the edges between nodes 8 and 55 and between nodes 51 and 55 are simply-supported. The remaining edges are free. A ball support is placed in the corner node 33, which is a constraint only to the vertical displacement. The slab is loaded by a homogenously distributed load $p = 800$ N/m^2. The material Young's modulus is $E = 30.000$ N/m^2 and Poisson's ratio $\upsilon = 0.2$.

The finite element program is based on the frequently applied rectangular elements with four corner nodes and three degrees of freedom per node, the vertical displacement w and two rotations ψ_x and ψ_y. As stated in Chap-

J. Blaauwendraad, *Plates and FEM: Surprises and Pitfalls*, Solid Mechanics and Its Applications 171, DOI 10.1007/978-90-481-3596-7_12,

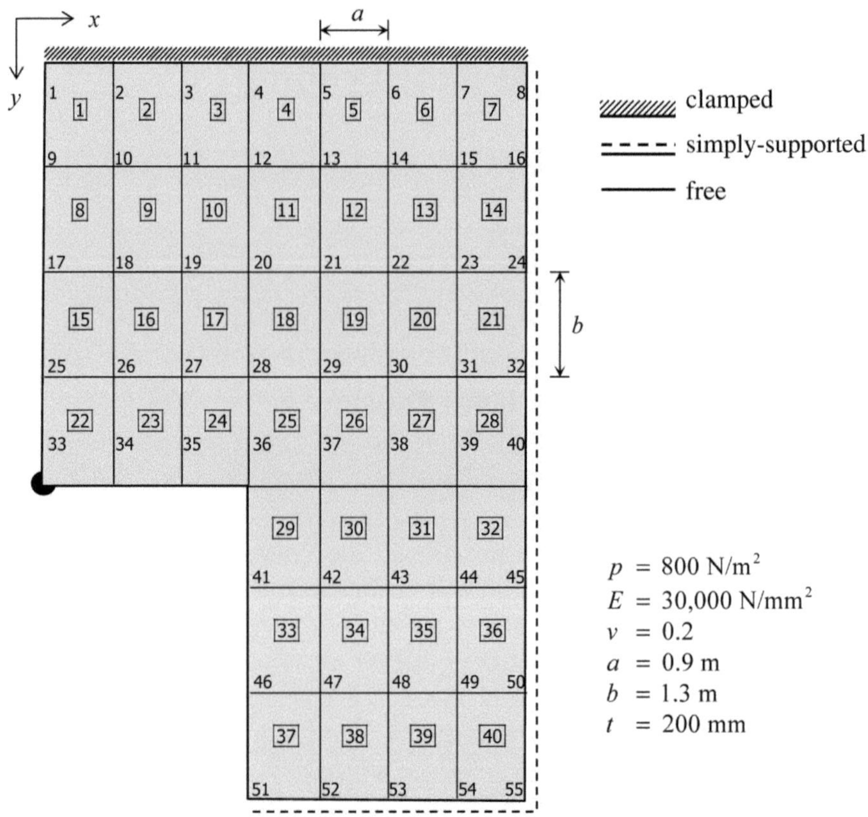

Figure 12.1 Mesh of a plate with node and element numbering.

ter 10, rotations in FE codes are defined about the x- and y-axis, respectively. In this element type the two bending moments m_{xx} and m_{yy} are linear in both directions x and y, so will in general have different values in the four element corners. The twisting moment m_{xy} has a distribution which is slightly parabolic, and also may take different values in the four corners. The shear force v_x is constant in the x-direction and varies linearly in the y-direction, whereas the shear force v_y is constant in the y-direction and varies linearly in the x-direction. The program replaces the distributed load over the element area by lumped vertical forces at the four corners. No torques, associated with the rotations ψ_x and ψ_y, are generated.

We need to discuss the way in which the boundary conditions are chosen. At the clamped edge, we have chosen $w = 0$ and $\psi_x = 0$ at each node. We also could have added $\psi_y = 0$ in order to enforce the edge to remain a straight line, but we decided not to do so. In reality meshes are much finer

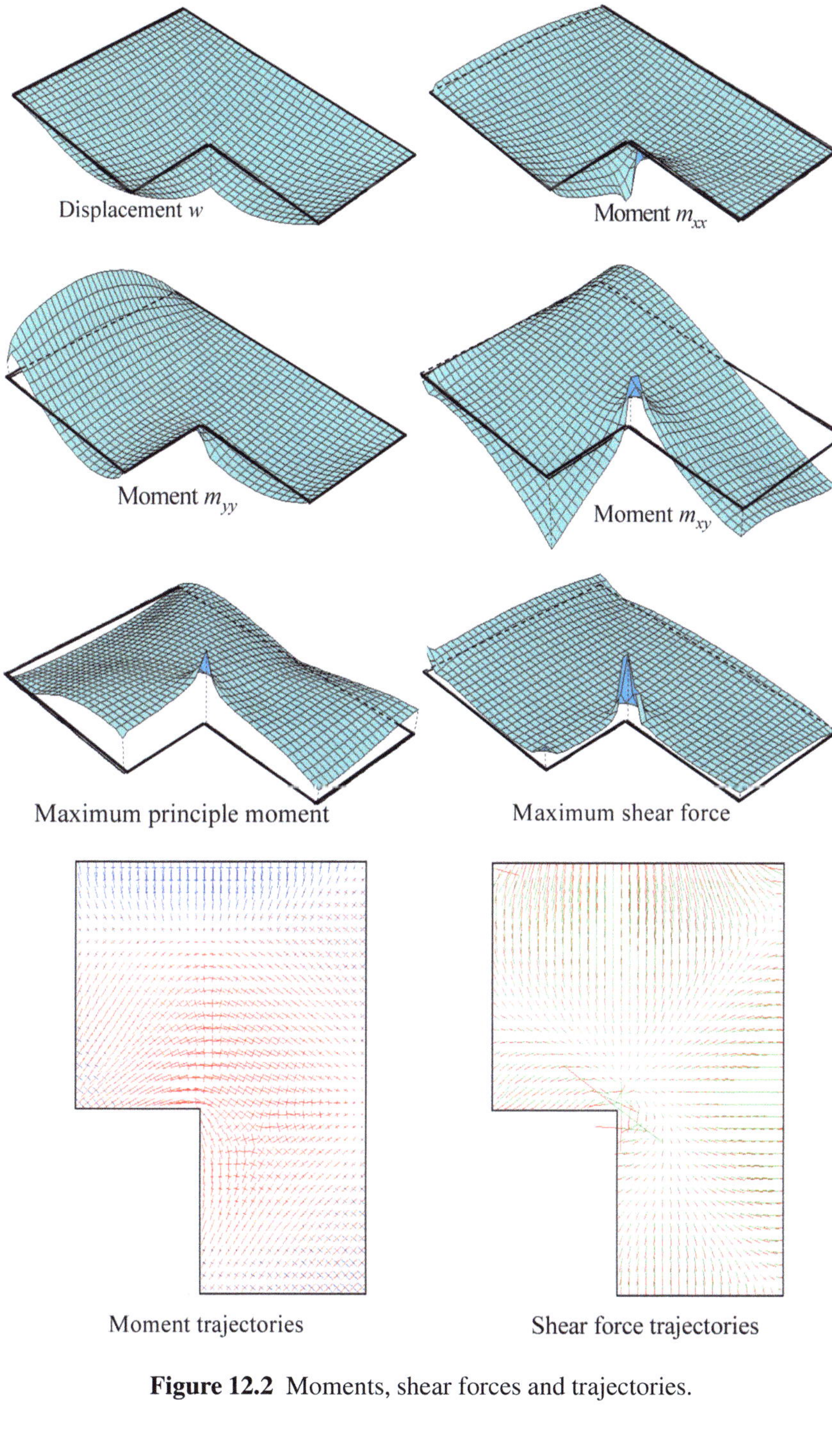

Figure 12.2 Moments, shear forces and trajectories.

than those chosen in this chapter, and then the requirement to model a straight line is met automatically. For the simply-supported edge we can argue in the same way. There we have just specified $w = 0$. However, there is a more serious argument. In case of the thin plate theory (Kirchhoff) the support reaction in w-direction will not change when we introduce the additional rotation constraint. This is inherent to the presuppositions of the theory, for which we refer to Section 4.4. Things are different for the thick plate theory (Mindlin). If we put additionally $\psi_x = 0$ for the simply-supported vertical edge in Figure 12.1, then the twisting moments lead to support torques T_x associated with ψ_x. And the support reactions in w-direction will become different. Putting $\psi_x = 0$ at the clamped edge will not make much difference because there the twisting moment is zero.

Contrary to practitioner's practice, we have listed the computation results for the moments and shear forces in Tables 12.1a and 12.1b for each element and all four nodes in the element. This is done because we want to do checks on detailed scale. In these tables we indeed see that the moments take different values in each corner of the element, but the shear forces do not. For the latter, only two different values per element occur, each value holding true in two nodes (constant in the direction from the one node to the other). In Table 12.2 we present the computed support reactions. This table reflects the way we have specified the boundary conditions. In the nodes 1 to 8 we obtain two support reactions, one for $w = 0$ (direction z in the list) and one for $\psi_x = 0$ (direction x in the list). The nodes 16 to 55 in the table are nodes on the simply supported edges, with $w = 0$, and show just a support reaction in the direction z. The same holds for the ball support in node 33.

In order to give the reader an impression of the real moment and shear force distribution, we have calculated the slab with a finer mesh as well, and we present a number of plots in Figure 12.2. Hereafter, however, we will just work with Tables 12.1 and 12.2 for the coarse mesh. Figure 12.2 shows the displacement w, the moments m_{xx}, m_{yy} and m_{xy} and trajectories for the moments and shear force. The length of a trajectory is an indication of the size of the principal moment or shear force. The bending moment m_{yy} has an opposite sign at the clamped edge compared to the moment remote from the edge. We recognize the irregularity at the inner corner between returning edges. Locally high twisting moments occur there. Also high shear forces happen, as is seen from the large trajectories at that position. The bending moment is highly irregular also.

Table 12.1a. Output for the elements 1 to 20 of the finite element program.

el-em	no-de	m_{xx} (N)	m_{yy} (N)	m_{xy} (N)	v_x (N/m)	v_y (N/m)	el-em	no-de	m_{xx} (N)	m_{yy} (N)	m_{xy} (N)	v_x (N/m)	v_y (N/m)
1	1	278	-2090	77	-2021	910	11	12	264	-480	-1	-115	1416
	2	-896	-3138	2	-2021	2423		13	131	-489	369	-115	1631
	9	-126	-456	-307	443	910		20	1036	985	-42	110	1416
	10	159	-223	-381	443	2423		21	1220	937	329	110	1631
2	2	-763	-3185	-81	-310	2578	12	13	426	-398	342	-320	1156
	3	-712	-3556	53	-310	2491		14	47	-395	694	-320	1522
	10	180	-145	-443	-337	2578		21	1015	863	338	64	1156
	11	49	-391	-309	-337	2491		22	1170	808	690	64	1522
3	3	-728	-3576	-77	-220	2631	13	14	604	-209	695	-669	671
	4	-811	-3734	85	-220	2758		15	-101	-271	957	-669	1237
	11	143	-355	-205	-20	2631		22	832	666	679	-39	671
	12	188	-418	-42	-20	2758		23	923	578	941	-39	1237
4	4	-779	-3734	-87	218	2681	14	15	538	-34	855	-272	146
	5	-777	-3514	114	218	2760		16	-197	361	962	-272	-19
	12	131	-422	79	234	2681		23	573	399	865	-1042	146
	13	306	-332	279	234	2760		24	111	-218	973	-1042	-19
5	5	-709	-3486	-78	642	2419	15	17	42	1465	-706	238	32
	6	-677	-2855	138	642	2534		18	468	1254	-630	238	460
	13	76	-393	409	560	2419		25	-122	1761	-1248	627	32
	14	450	-132	626	560	2534		26	981	1377	-1172	627	460
6	6	-585	-2790	-54	1294	1887	16	18	555	1203	-718	255	415
	7	-356	-1661	134	1294	1793		19	896	1120	-407	255	481
	14	29	-263	728	719	1887		26	1019	1452	-1067	196	415
	15	529	53	916	719	1793		27	1566	1135	-755	196	481
7	7	-375	-1587	-9	2042	996	17	19	832	1064	-493	260	814
	8	45	162	83	2042	-45		20	1196	1016	-98	260	324
	15	111	-108	853	-291	996		27	1884	1242	-568	-308	814
	16	73	-348	945	-291	-45		28	1580	1177	-173	-308	324
8	9	-39	-112	-325	68	949	18	20	1012	936	-122	86	1090
	10	137	-245	-383	68	1255		21	1221	870	272	86	444
	17	-76	1460	-651	353	949		28	2234	1350	-126	-635	1090
	18	579	1216	-710	353	1255		29	1536	1312	268	-635	444
9	10	7	-349	-437	150	1338	19	21	1167	882	340	-337	719
	11	183	-381	-250	150	1450		22	999	696	607	-337	530
	18	519	1282	-656	175	1338		29	1826	1347	135	-597	719
	19	938	1070	-468	175	1450		30	1436	1083	402	-597	530
10	11	127	-433	-277	-3	1451	20	22	1114	755	683	-826	403
	12	173	-495	48	-3	1604		23	615	359	853	-826	441
	19	880	1099	-409	153	1451		30	1447	1049	330	-773	403
	20	1153	1005	-84	153	1604		31	994	669	501	-773	441

12.2 Bending Moments and Equilibrium

12.2.1 Discussion of Moment Diagrams

Figure 12.3 shows the plan of the slab with four sections I, II, II, and IV. Moment diagrams for these sections are shown in Figure 12.4. First, they make clear that moments are discontinuous at the boundaries between elements. This implies that moments which should be zero at an edge may

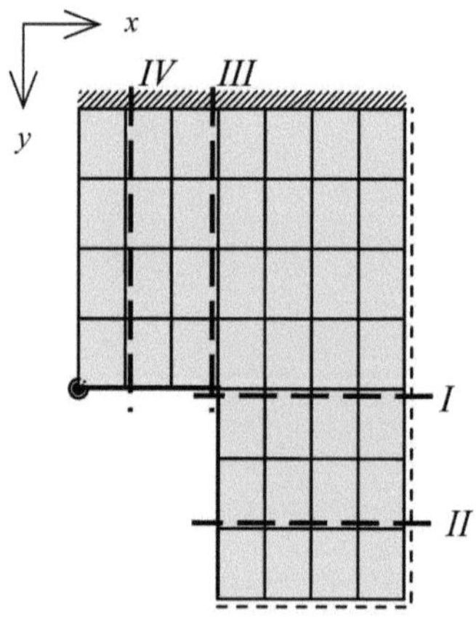

Figure 12.3 Sections where moment diagrams will be shown.

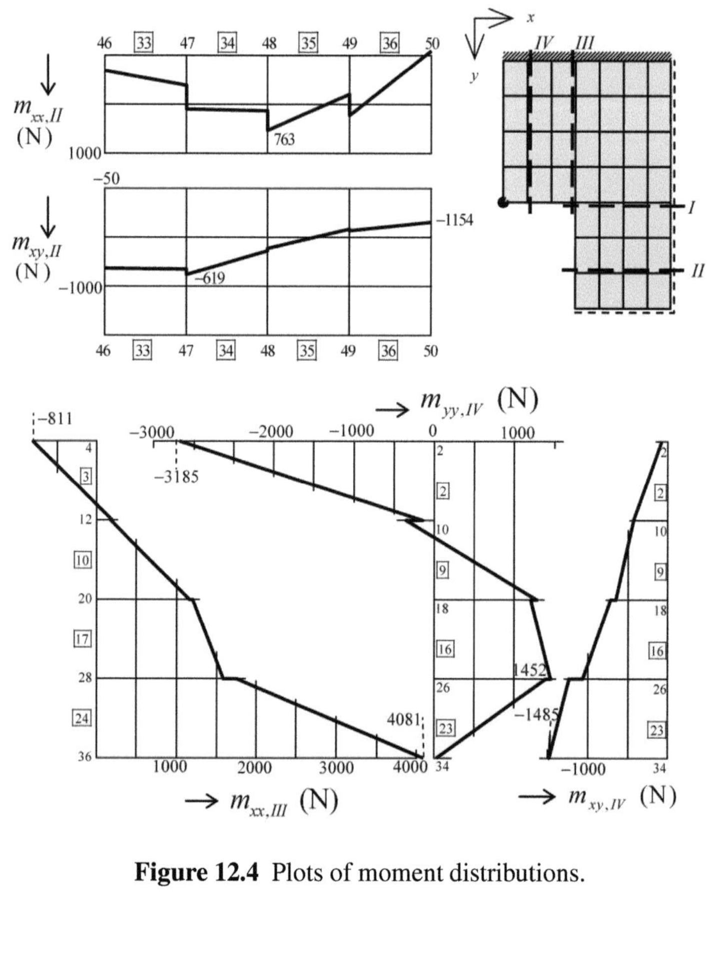

Figure 12.4 Plots of moment distributions.

Table 12.1b. Output for the elements 21 to 40 of the finite element program.

el-em	no-de	m_{xx} (N)	m_{yy} (N)	m_{xy} (N)	v_x (N/m)	v_y (N/m)	el-em	no-de	m_{xx} (N)	m_{yy} (N)	m_{xy} (N)	v_x (N/m)	v_y (N/m)
21	23	793	446	850	-1097	149	31	38	1529	1142	88	-940	-257
	24	-38	96	905	-1097	-21		39	961	678	-60	-940	-95
	31	930	604	443	-1468	149		43	1137	1166	-411	-798	-257
	32	16	-62	498	-1468	-21		44	831	627	-560	-798	-95
22	25	-22	1851	-1434	535	-1000	32	39	1006	656	-17	-1554	-113
	26	916	1458	-1005	535	-447		40	-17	-9	-61	-1554	40
	33	65	215	-2123	1254	-1000		44	835	658	-577	-1363	-113
	34	1691	-25	-1694	1254	-447		45	9	22	-621	-1363	40
23	26	873	1388	-1235	860	-397	33	41	-438	2117	32	762	60
	27	1757	1426	-611	860	-408		42	1079	1508	37	762	-798
	34	1645	28	-1485	773	-397		46	156	1448	-681	-7	60
	35	2578	-63	-861	773	-408		47	306	1206	-677	-7	-798
24	27	1687	1301	-769	-209	-195	34	42	876	1518	-166	-31	-397
	28	1749	975	177	-209	1506		43	1197	1180	-405	-31	-623
	35	2428	18	-443	2617	-195		47	552	1205	-619	-294	-397
	36	4081	1229	503	2617	1506		48	565	858	-858	-294	-623
25	28	1999	1432	113	-470	-806	35	43	1091	1134	-404	-586	-348
	29	1759	1171	118	-470	146		44	888	711	-613	-586	-485
	36	1594	325	583	1387	-806		48	763	922	-878	-775	-348
	37	2009	1402	588	1387	146		49	397	444	-1086	-775	-485
26	29	1735	1178	298	-352	24	36	44	806	637	-570	-1394	-147
	30	1484	1040	278	-352	113		45	15	-75	-652	-1394	54
	37	1651	1318	146	-277	24		49	612	545	-1072	-993	-147
	38	1547	1133	126	-277	113		50	-50	132	-1154	-993	54
27	30	1496	1038	415	-931	35	37	46	-130	1494	-682	362	-685
	31	928	594	423	-931	106		47	603	1217	-635	362	-1310
	38	1474	1123	-1	-855	35		51	189	-28	-880	-170	-685
	39	1011	669	8	-855	106		52	-98	10	-833	-170	-1310
28	31	981	611	466	-1455	15	38	47	399	1232	-652	20	-984
	32	-16	32	473	-1455	2		48	741	943	-854	20	-1170
	39	980	657	-48	-1523	15		52	27	-21	-859	-24	-984
	40	9	-12	-41	-1523	2		53	-31	-20	-1061	-24	-1170
29	36	2657	1611	1262	-1370	-2017	39	48	571	906	-863	-234	-904
	37	1202	1399	298	-1370	279		49	620	615	-1109	-234	-883
	41	-617	2013	790	1090	-2017		53	22	-7	-1088	-88	-904
	42	1239	1522	174	1090	279		54	-6	-85	-1334	-88	-883
30	37	1814	1378	343	-536	-524	40	49	505	542	-1040	-1149	-626
	38	1402	1164	4	-536	-90		50	-8	-190	-1174	-1149	173
	42	883	1594	-53	-184	-524		54	-60	-47	-1284	325	-626
	43	1194	1130	-392	-184	-90		55	50	199	-1418	325	173

deviate from the correct value. This is seen for moment m_{xx} in section II. For realistic mesh fineness this has no practical meaning. The finer the mesh, the closer the moments become to zero. Secondly, we see in section II that the twisting moment m_{xy} is non-zero, but now clearly not as a consequence of the coarse mesh. In Kirchhoff theory a twisting moment is permitted at free edges. It will also be allowable for fine meshes. Otherwise, the twisting moment in section IV must be zero at the clamped edge. This is satisfied apart from the small discontinuities at each element edge. At the opposite

Table 12.2 Support reactions in the boundary nodes.

node	direction	support reaction (N)	node	direction	support reaction (N)
1	z	217	8	z	1262
1	x	-898	8	x	86
2	z	-3154	16	z	-69
2	x	-2856	24	z	-2142
3	z	-2904	32	z	-2807
3	x	-3230	40	z	-2979
4	z	-3161	45	z	-2815
4	x	-3390	50	z	-2087
5	z	-3063	55	z	2413
5	x	-3181	54	z	-1221
6	z	-2764	53	z	-1566
6	x	-2567	52	z	-1582
7	z	-1760	51	z	-2257
7	x	-1397	33	z	-5004

end of section IV, a free edge, we would expect a twisting moment, and there is one. The bending moment m_{yy} in section IV shows the expected shape for a beam simply-supported at the one end and clamped at the other. Section III shows the bending moment m_{xx}. The bending moment m_{yy} in the lateral direction is expected to have a value, which is Poisson's ratio times the clamped moment $m_{yy} = -3185$ N, so $m_{xx} = -0.2 \times 3185 = -637$ N. In Table 12.1a we read the value -763 N, the difference being due to the coarse mesh. At mesh refinement the requirement will be fulfilled better. The distribution of the lateral moment m_{xx} for section III is also included in Figure 12.4. Indeed a negative value occurs close to the clamped edge, of the order of 20% of the clamped moment m_{yy}. The other end of section III coincides with the corner between the returning edges, which is a singular point in the plate. There a peak value occurs of size 4081 N. If we shift section III a little to the right, it will lie in the row of adjacent elements. Then the peak value is 2126 N (average of the value in elements 25 and 29), which is about half of the peak value 4081 N in section III. This verifies the expectation expressed in Section 11.4 for corners with re-entrant edges.

12.2.2 Equilibrium Check for Moments

In a FE analysis we must distinguish two types of moments and forces, as stated in Chapter 10. First, we have the *generalized* element forces and moments, which act between elements at element nodes. These forces and moments correspond with, and have the same sign convention as, the displacement w and rotations ψ_x and ψ_y. Usually they are hidden for the user. Secondly there are the *engineering* bending and twisting moments m_{xx}, m_{yy}, m_{xy} and shear forces v_x and v_y within the elements, with another sign convention. These are available to the users and can be shown in sections through the elements or as contour plots.

In principle, FE analyses produce an approximation; however, equilibrium is always satisfied on the level of the generalized element forces and moments, regardless of the fineness of the mesh. Because support reactions are computed from these generalized forces and moments, they are always in equilibrium with the applied load. The sum of all support reactions in the z-direction in Table 12.2 is 37,443 N, for a total distributed load p of 37,440 N, a difference less than 0.01%. This difference is due to rounding off errors rather than inaccuracy of the method.

The story is different for the engineering forces and moments. No equilibrium is automatically ensured, and we have to inspect to what extent the engineering bending moments and forces satisfy equilibrium. The check for moments is the subject of this section; we do the check on shear forces in the next section.

We will make an equilibrium check of the diagram for the bending moment m_{yy} in section I shown in Figure 12.5. This is the bending moment in the direction normal to the section. We know that the stress state near the corner between returning edges is highly irregular. Therefore, we lower our expectations. We will perform the check on equilibrium as follows. We integrate the internal moment m_{yy} over the length of section I, and compare this total moment with the total moment due to the external load. There is a difficulty: the plate part below section I is not free hanging. The total moment in the section cannot be determined on the basis of load only; we also must consider the support reactions in the edge nodes of this plate part; we know that they are a set of forces and moments equilibrating the load. We calculate the integral of the bending moments m_{yy} by determining the average, which we multiply by the length $4a$ of the section. The values of m_{yy} are drawn from Table 12.1, and we find

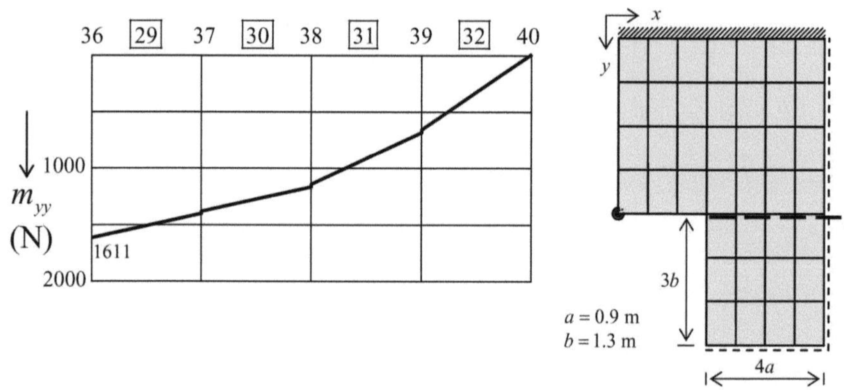

Figure 12.5 Plot of the distribution of m_{yy} in section I.

$$m_{\text{ave}} = \frac{1}{8}\left[\overbrace{m_{36} + m_{37}}^{\boxed{29}} + \overbrace{m_{37} + m_{38}}^{\boxed{30}} + \overbrace{m_{38} + m_{39}}^{\boxed{31}} + \overbrace{m_{39} + m_{40}}^{\boxed{32}}\right] = 1{,}002.4$$

Length of line is $4a$ $= 3.6$ ×

Integral of moments m_{yy} $= 3{,}609$ Nm

The load on the plate part below section I consists of distributed load p and support reactions R at the nodes 45, 50, 51, 52, 53, 54 and 55. The total external moment with respect to section I is calculated.

$$\begin{aligned}
4a \times 3b \times p \times 3b/2 &= -21{,}902 \\
R_{45} \times b &= 3{,}659 \\
R_{50} \times 2b &= 5{,}427 \\
(R_{51} + R_{52} + R_{53} + R_{54} + R_{55}) \times 3b &= 16{,}425 \quad + \\
\text{Total external moment} &= 3{,}611 \text{ Nm}
\end{aligned}$$

We conclude that the sum of the engineering moments m_{yy} in section I is in surprising good equilibrium with the external load and support reactions. The difference of 3,609 and 3,611 N is less than 0.1%. Considering the very coarse mesh, we can be satisfied; our low expectations were unjustified.

12.3 Shear Forces, Support Reactions and Equilibrium

We will now discuss the distribution of the shear forces, compare them to the support reactions, and check their equilibrium with the external load.

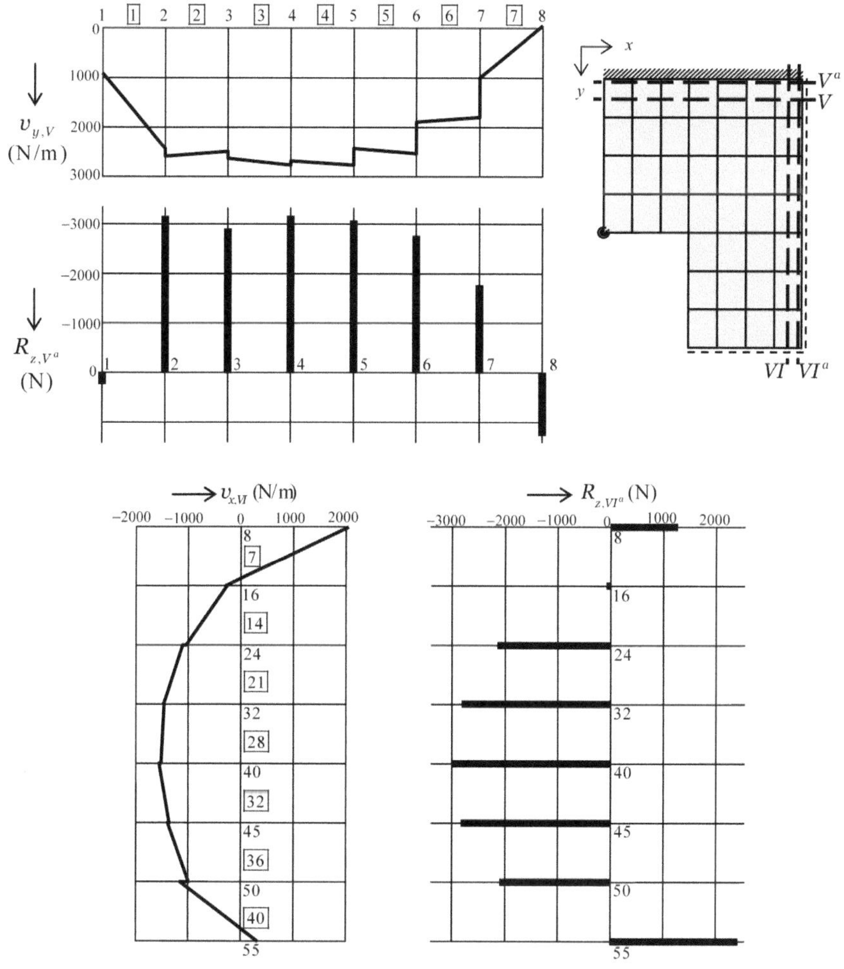

Figure 12.6 Distribution of shear forces and support reactions.

12.3.1 Discussion of Shear Force Diagrams

In Figure 12.6 we show the distribution of shear forces in sections V and VI close to the clamped and simply-supported edge, respectively. These sections are drawn through the centre of elements because the shear force is constant in the direction it acts (v_x in the x-direction and v_y in the y-direction). The distance of the sections to the edge is a half element width. For the same edges we also have depicted the vertical support reactions (z-direction) over the edges V^a and VI^a. The distribution of the shear force v_x in section VI corresponds reasonably with the support reaction in edge VI^a. A tensile force

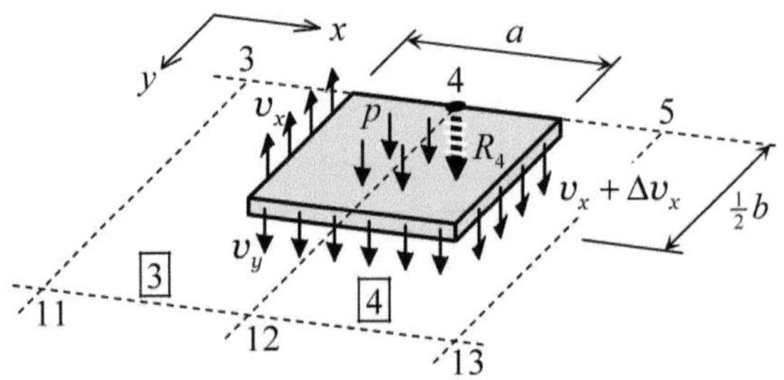

Figure 12.7 Calculation of reaction force R_4 from the shear forces.

appears to occur at the clamped edge. This is not due to the twisting moment which is supposed to be zero at a clamped edge, but rather to the shear force v_x. There is a good agreement between shear force v_y and support reactions along the clamped edge.

12.3.2 Equilibrium Check for Shear Forces

In this section we check to what extent shear forces satisfy equilibrium at four nodes of the slab. We select node 4 on the clamped edge, node 40 on the simply-supported edge, node 55 in the corner between the two simply-supported edges, and node 33 in the corner between the two free edges at the ball support.

Clamped Edge

We start with node 4 at the clamped edge. We expect that the support reaction at the node is due to the shear force v_y over a width $a/2$ to the left and $a/2$ to the right of the node. No effect of twisting moments can occur because this moment must be zero at a clamped edge. From the data in Table 12.1 we calculate that the average value of v_y over the considered length a is $-2{,}714$ N/m. So $av_y = -0.9 \times 2{,}714 = -2{,}441$ N. In Table12.2 we read for the support reaction in z-direction $R_4 = -3{,}161$, so the value on basis of the shear force is 23% too small. There is a simple explanation for this mismatch. We should remind ourselves that the shear force v_y is constant in y-direction. The most probable position where the shear force applies is in a line parallel to the clamped edge, through the centre of the element. Therefore, we should consider a rectangular plate part as shown in Figure 12.7, of sizes a and

$b/2$. This plate part is loaded by the support reaction R_4, distributed load p, shear forces v_x and $v_x + \Delta v_x$, and the shear force v_y. In Figure 12.7 these components are depicted for positive values. The equation for the vertical equilibrium of the plate part is:

$$\boxed{R_4 + \frac{1}{2}abp + \frac{1}{2}b\Delta v_x + av_y = 0}$$

This equation leads to a formula for computing the support reaction:

$$R_4 = -\frac{1}{2}abp - \frac{1}{2}b\Delta v_x - av_y.$$

From the data in Table 12.1 we calculate

$$\left.\begin{array}{l} v_x = -170 \\ v_x + \Delta v_x = 222 \end{array}\right\} \quad \rightarrow \quad \Delta v_x = 222 - (-170) = 392 \text{ N/m}$$

$$v_y = 2{,}714 \text{ N/m}$$

Using $a = 0.9$ m, $b = 1.3$ m and $p = 800$ N/m^2, we obtain

$$\begin{array}{lll} -\frac{1}{2}abp & = & -468 \text{ N} \\ -\frac{1}{2}b\Delta v_x & = & -255 \text{ N} \\ av_y & = & -2{,}442 \text{ N} \\ \hline & & + \\ R_4 & = & -3{,}165 \text{ N} \\ R_4 & = & -3{,}161 \text{ N} \quad \text{(Output Table 12.2)} \end{array}$$

The computer output delivers a practically equal value. The difference is in the order of 0.1%. At mesh refinement the terms, in which b appears will vanish, and R_4 will follow from v_y only. The sign of the support reaction is negative, which implies – due to the sign convention – a compressive force.

Simply-Supported Edge

We do a similar check for node 40 at the simply-supported edge. The relevant plate part is shown in Figure 12.8. Compared to Figure 12.7 for the node at the clamped edge, now two concentrated edge forces are added, due to the non-zero twisting moment, and equal in value to the twisting moment at the actual position. The drawn direction applies for a positive twisting moment. These forces need not be considered at the clamped edge, because there the

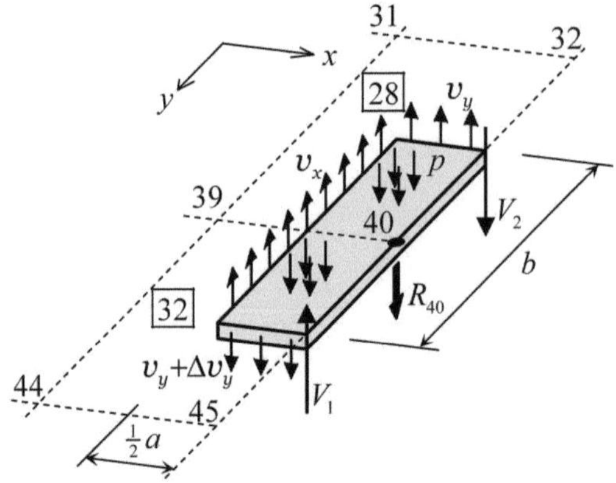

Figure 12.8 Calculation of reaction force R_{40} from the shear forces.

twisting moment is necessarily zero. Now the equation for the vertical equilibrium of the plate part is

$$\boxed{R_{40} + \frac{1}{2}abp - bv_x + \frac{1}{2}a\Delta v_y - V_1 + V_2 = 0}$$

From this relation we find the contributions to the support reaction

$$\begin{aligned}
-\tfrac{1}{2}abp &= -\quad 468\ \text{N} \\
+bv_x &= -1{,}958\ \text{N} \\
-\tfrac{1}{2}av_y &= +\quad\ \ 2\ \text{N} \\
&\qquad\qquad\qquad -2{,}424\ \text{N} \\
+V_1 &= -\ 341\ \text{N} \\
-V_2 &= -\ 216\ \text{N} \\
&\qquad\qquad\qquad -\ \ 557\ \text{N} \\
R_{40} &= \qquad\qquad\ -2{,}981\ \text{N} \\
R_{40} &= \qquad\qquad\ -2{,}979\ \text{N}\ \ \text{(Output Table 12.2)}
\end{aligned}$$

The computed compressive support reaction is again very close to the outputted support reaction in Table 12.2. We have purposely separated the contribution of the concentrated edge shear forces in order to show their contribution to the total balance. Without these forces there would be an unbalance of about 20%.

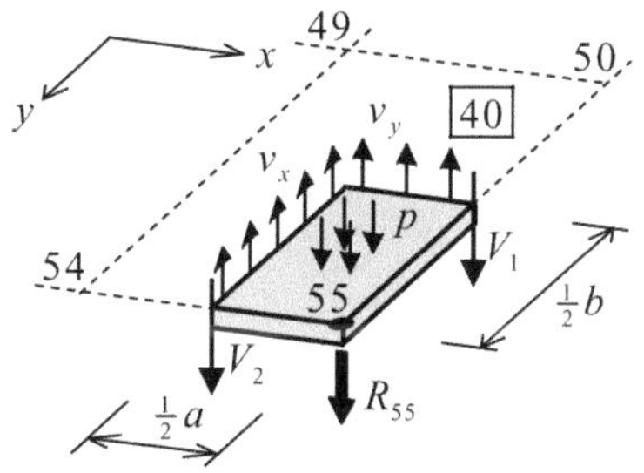

Figure 12.9 Calculation of reaction force R_{55} from shear forces.

Corner between Two Simply-Supported Edges

Figure 12.9 shows the relevant plate part at the corner node 55, and its loading. No further explanation is needed. The equilibrium equation is

$$\boxed{R_{55} + \frac{1}{4}abp - \frac{1}{2}bv_x - \frac{1}{2}av_y + V_1 + V_2 = 0}$$

$$\begin{array}{lcrr}
-\frac{1}{4}abp & = & -\ 234\ \text{N} & \\
+\frac{1}{2}bv_x & = & -\ 28\ \text{N} & \\
+\frac{1}{2}av_y & = & -\ 12\ \text{N} & \\
\cline{3-3}
 & & & -\ 274\ \text{N} \\
-V_1 & = & +1{,}296\ \text{N} & \\
-V_2 & = & +1{,}351\ \text{N} & \\
\cline{3-3}
 & & & +2{,}647\ \text{N} \\
\cline{4-4}
R_{55} & = & & +2{,}373\ \text{N} \\
R_{55} & = & & +2{,}413\ \text{N}\ \text{(Output Table 12.2)}
\end{array}$$

This time the difference is in the order of 2%. Completely in agreement with theory, the corner support force is a tensile force (positive sign) and is practically due to the twisting moment only. The two concentrated forces (each having the size of the twisting moment), are almost equal, so the tensile force is close to twice the twisting moment. At mesh refinement the factor two will be attained even better. The lesser accuracy (though still good) may be due to the slightly parabolic distribution of the twisting moments within an element. For determining the value of V_1 and V_2 at the mid of the respective element edges, we applied linear interpolation between corner values.

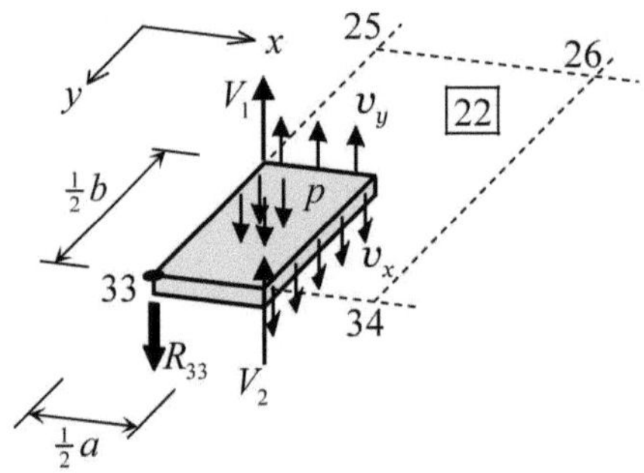

Figure 12.10 Calculation of reaction force R_{33} from the shear forces.

Corner with Ball Support between Two Free Edges

The plate part at node 33 and its loading is shown in Figure 12.10. The equilibrium equation is

$$\boxed{R_{33} + \frac{1}{4}abp + \frac{1}{2}bv_x - \frac{1}{2}av_y - V_1 - V_2 = 0}$$

$$\begin{aligned}
-\tfrac{1}{4}abp &= -\ 234\ \text{N} \\
-\tfrac{1}{2}bv_x &= -\ 698\ \text{N} \\
+\tfrac{1}{2}av_y &= +\ 388\ \text{N} \\
& \qquad\qquad -1{,}320\ \text{N} \\
-V_1 &= -1{,}779\ \text{N} \\
-V_2 &= -1{,}909\ \text{N} \\
& \qquad\qquad -3{,}688\ \text{N} \\
R_{33} &= \qquad\quad -5{,}008\ \text{N} \\
R_{33} &= \qquad\quad -5{,}004\ \text{N}\ \text{(Output Table 12.2)}
\end{aligned}$$

The computer output delivers a practically equal value. At mesh refinement the lengths a and b tend to zero. Then the lumped edge shear forces V_1 and V_2 become equal and have the value $R_{33}/2$.

12.4 Message of the Chapter

- Moments and shear forces are discontinuous in boundaries between elements, shear forces more so than moments.

- Moments in a FE analysis are in accordance with expectations from theory. Twisting moments are zero at clamped edges. Non-zero twisting moments occur at simply-supported and free edges.

- Support reactions need not necessarily be exact, but they are always in perfect equilibrium with the load.

- Bending moment distributions will become more exact for refined meshes. The total moment in a section through the plate is always in good equilibrium with load and support reactions, even for a coarse mesh.

- There is good correspondence between shear forces and support reactions when we consider the right plate part, and include the concentrated edge shear force.

Chapter 13
FE Analysis for Different Supports

In Chapter 4 we became acquainted with various edge conditions in thin plate theory; in Chapter 5 we applied this knowledge to square plates with three different support conditions. Here we meet the three cases again, now they appear in a FE analysis. In Chapter 5 we considered a two-way sine load and a homogeneous distributed load. That was done in order to be able to solve the differential equation. Here we need not make that difference in load type and will subject the plate to a homogenous load in all cases, as we did earlier for the discrete model in Section 9.4. In Chapter7 we became acquainted with the behaviour of circular plates subjected to both distributed load and a point load. Here we consider the behaviour of a square plate due to a central point load. It will appear that the response near the point load is of the same nature as occurs for the point load on a circular plate. In all analyses we choose Kirchhoff theory.

13.1 Simply-Supported Plate

We consider a square simply-supported thin plate and choose a mesh of 20 elements in each direction. In thin plates the span is about 25 times thickness, so the chosen mesh size is in the order of the thickness of the plate. Poisson's ratio is 0.2. Section 13.1.1 is devoted to distributed load and Section 13.1.2 to a point load.

J. Blaauwendraad, *Plates and FEM: Surprises and Pitfalls*, Solid Mechanics and Its Applications 171, DOI 10.1007/978-90-481-3596-7_13,

13.1.1 Distributed Load

Figure 13.1 depicts FE results for a homogeneous distributed load. The distributions are very similar to, and a confirmation of, theoretical results in Fig-

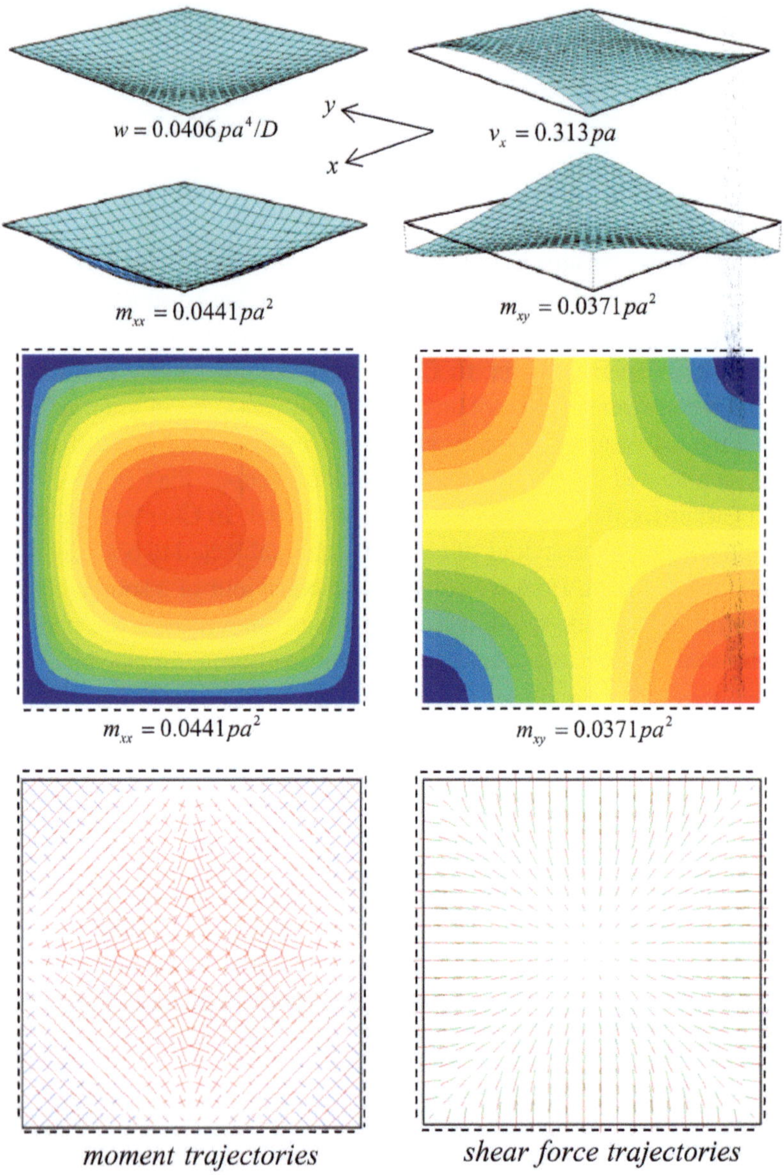

Figure 13.1 FE results. Square plate. Simply-supported. Distributed load.

ures 5.6 and 5.7 for the two-way sine load. In Figure 13.1 we show moments twice, as three-dimensional plots and as contour plots. Three-dimensional representation is preferred, for it highlights areas needing special care. This is particularly important when moment peaks occur. FE codes should offer this output option in any case.

Here we are not interested in values but rather in distributions and ways of load transfer. Therefore we skip legend scales. Maximum values are expressed in terms of the homogeneous load p and the span a. Figure 13.1 shows that the maximum values of m_{xx} occur in the centre of the plate, and of m_{xy} in the corners. These values are $0.0441pa^2$ and $0.0371pa^2$ respectively. They are identical to the theoretical values $0.0442pa^2$ and $0.0371pa^2$, borrowed from Timoshenko and Geere, converted from Poisson's ratio 0.3 to 0.2 [16]. Most plots for the moments and shear forces need no comment. We draw attention to the trajectories of the moments and shear force. For the support condition and load under consideration the shear trajectories approach the plate edges normal to the edge, because the shear force parallel to the edge is zero; Eq. (4.23) leads to a trajectory angle $\beta_o = \pi/2$.

13.1.2 Point Load

Figure 13.2 shows the results for a point load on a square simply-supported plate for the 20×20 mesh. The plots for the trajectories of the principal moments and shear force show that the state of moments and shears is almost perfectly axisymmetric near the point load. Note that the trajectories for the shear force are depicted for a part of the plate with sizes $a/2$ around the point load. This is done because of the large gradient in that region. In the contour plot of the twisting moment we recognise the horizontal and vertical lines of plate symmetry. In the considered example the twisting moments increase in the direction of the corners. The maximum value is of red colour. In other examples the contour plot of the twisting moment may appear as a cloverleaf. At the lines of symmetry m_{xy} is zero (yellow zones). This always holds true for twisting moments and shear forces for symmetrical loads. They are antisymmetric quantities in contrast to bending moments which are symmetric, and have maximum values on lines of plate symmetry.

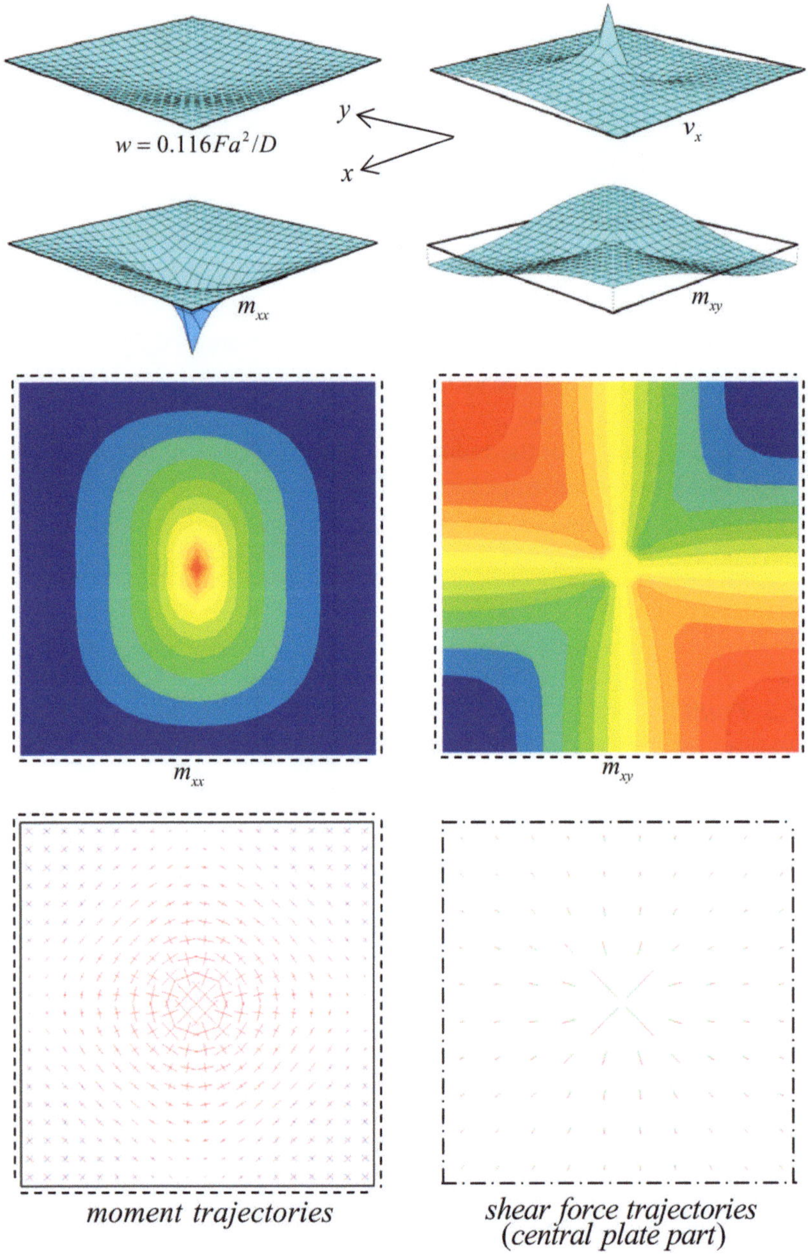

Figure 13.2 FE results. Square plate. Simply-supported. Central point load.

13.2 Corner Supports

We repeat the analyses for a plate with free edges and corner supports. In Section 13.2.1 we consider distributed load, in Section 13.2.2 a point load.

13.2.1 Distributed Load

Figure 13.3 shows the results for the distributed load. The corners are supported by balls, which permit rotations but prevent vertical displacements. Now the maximum moment does not occur in the plate centre, but mid-span of the free edge. The maximum twisting moments again occur in the corners, however with opposite signs compared to the simply-supported case. There a tensile corner reaction occurs, here a compressive one.

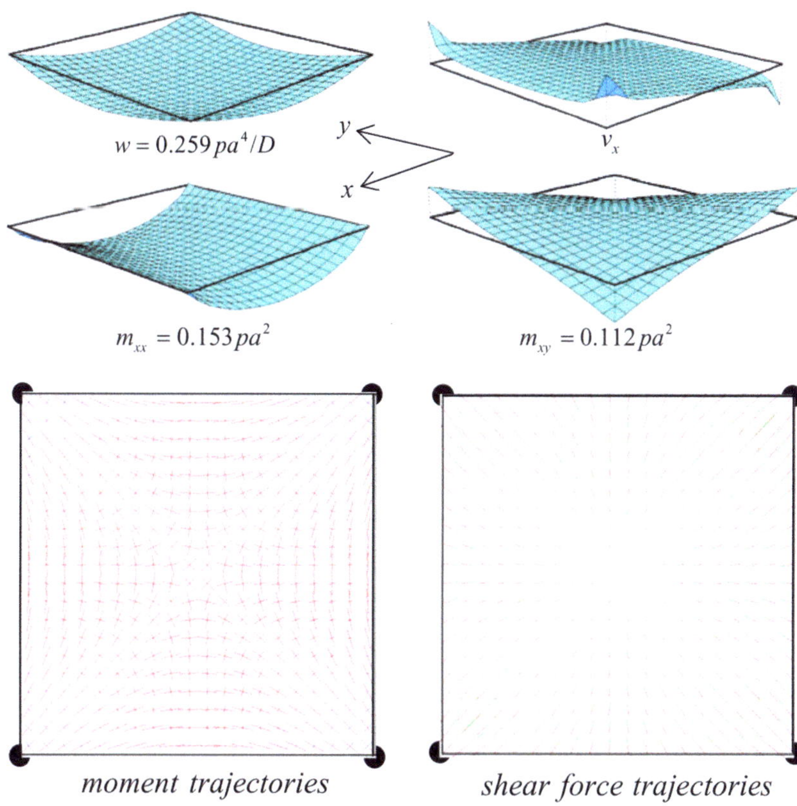

Figure 13.3 FE results. Square plate. Corner supports. Distributed load.

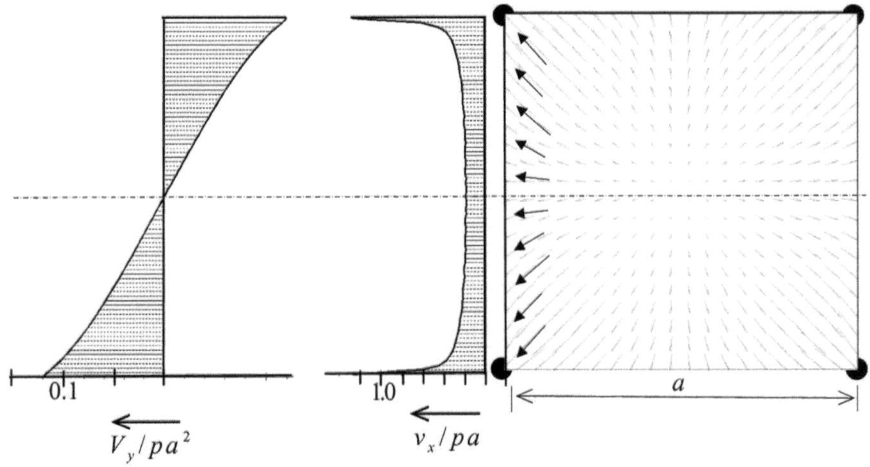

Figure 13.4 Load flow in shower analogy.

The engineer's feeling is that the load will flow in the direction of the corners, and high transverse shear forces occur in a quarter of a circle around the ball support. The structural engineer expects that all trajectories will be directed to the corner in a different way from the simply-supported case. However, reality is different. Figure 13.4 is very instructive about the transfer of loads to the supports. The shear trajectories are hardly different from the trajectories for the simply-supported structure in Figure 13.1. The load p flows as shear force v_x to the depicted edge. The value of v_x is almost constant along the major part of the edge. A load flow occurs along the free edge in the direction of the support, which is the concentrated force V_y in the y-direction. This force has the value of m_{xy} at the free edge. Starting from the middle of the edge, the concentrated shear force increases from zero to its maximum at the ball support. V_y is the integral of the shear force v_x. The shower analogy of Section 4.3 fully applies. The square plate is the m-hill, the free edges are gullies, and the ball support is a drain pipe. The distributed p-load is the shower. The water flows in the direction of the deepest slope to the gully, and from there through the gully to the drain pipe at the corner. The flow in the gully represents the concentrated shear force V_y.

The support reaction in the ball support is twice the maximum value of the concentrated shear force V_y. Therefore the expected maximum value of this shear force is $pa^2/8$. The FE analysis value for the reported mesh fineness 20×20 is $0.117pa^2$ instead of $0.125pa^2$. Continuous refinement to 40×40,

80×80 and 120×120 leads to coefficients 0.117, 0.120 and 0.121. The convergence to the exact value 0.125 is slow, but appears to be obtained. Therefore, all the distributed load p is transferred to the corner supports by the concentrated edge shear force. The distributed shear force v_y parallel to the edge is not zero at the edge and increases from zero at mid-span of the edge to its maximum at the ball support, but it remains finite as the element mesh is refined. The same holds for the principle shear force along the diagonal of the plate.

In reality corner supports will have some size. Let us assume that the support covers an area of a quarter of a circle with a radius of the order of the plate thickness t. Then the boundary between support and plate has a length of the order of $\pi t/2$, say $2t$. In Section 3.6 we learned that the concentrated edge shear force attenuates over a length of the order t. This means that the two concentrated forces which arrive at a corner do not spread nicely over boundary of length $2t$, but rather remain concentrated at the two ends of the boundary.

13.2.2 Point Load

The results for the point load are assembled in Figure 13.5. After the discussion of the point load for the simply-supported edge and the distributed load for the ball support, no further comment is needed. In the plate centre the correspondence with a circular plate is seen again. Figure 13.5 confirms the shower analogy Distributed shear forces v_x transfer the point load to the free edge and a concentrated shear force V_y (equal to m_{xy}) carries the distributed shear forces v_x to the ball supports.

13.3 Edge Beams

In Section 5.3 we touched on the subject of flexurally rigid, torsionally weak beams as a way to simulate simply-supported edges. It was found that for zero Poisson's ratio the bending moment M in the edge beam is 50% higher than expected on the basis of the load which flows to the edge. In Section 5.3 this study was done on the basis of a two-way sine load. In the discrete model of Section 9.4 we touched on the same subject for a homogeneously distributed load p. In the present section we again use a homogeneous distribution and check whether similar results are produced in a FE analysis.

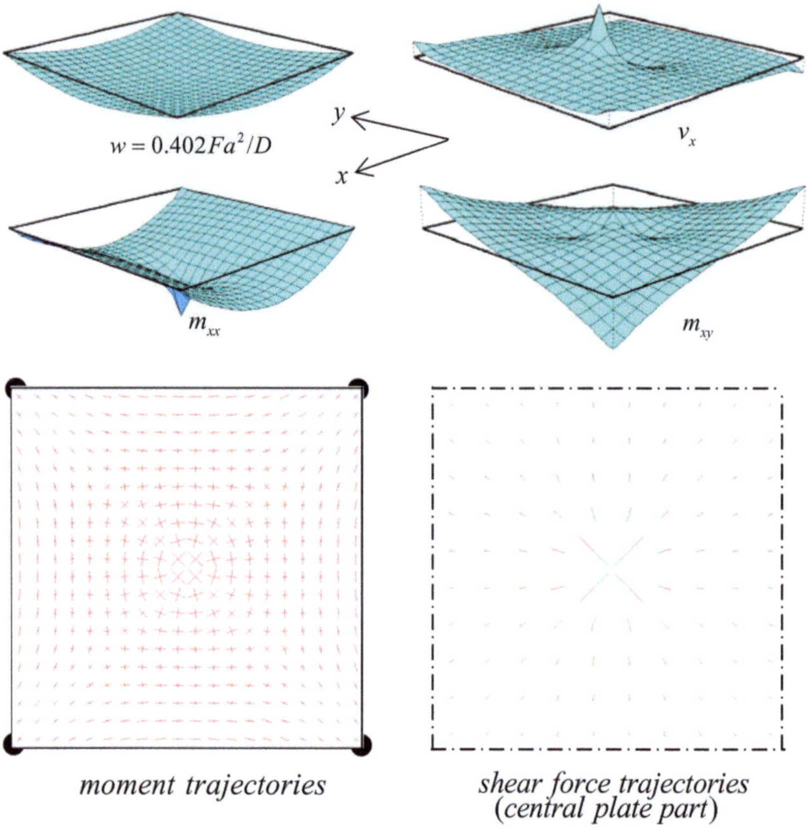

Figure 13.5 FE results. Square plate. Corner supports. Central point load.

13.3.1 Rigid Beams

In this section the four edge beams have infinitely large flexural rigidity, but zero torsion rigidity. The flexural stiffness in the FE analysis is chosen a thousand times larger than the total plate stiffness aD. The beams are supported by balls at the corners. We do not repeat plots for moments and shear forces, for they are precisely the same as for the simply-supported plate in Figure 13.1. Here we are interested in the bending moment M at mid-span of the beam, and the maximum shear force V at the beam end. In Figure 13.6 we have plotted the shear force v_x and twisting moment m_{xy} in the plate along the edge beam, and the moment M and shear V of the beam itself. We can check what moment and shear force in the beam occur if we load it by the shear force v_x only. For this purpose we first adapt the values

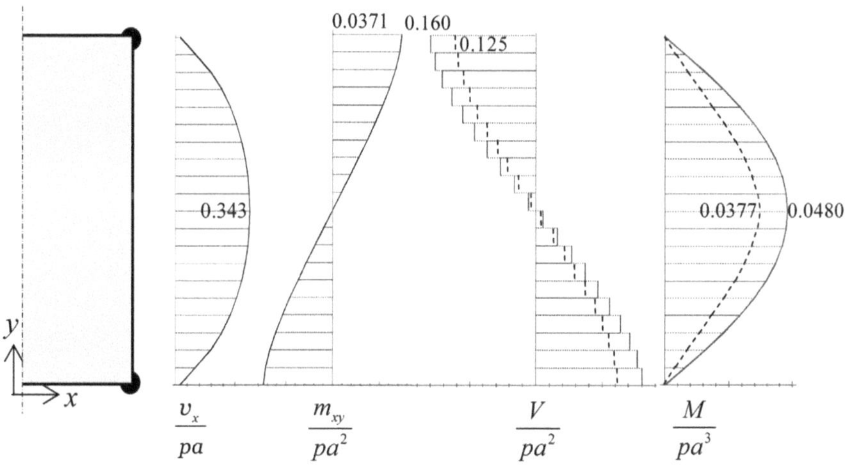

Figure 13.6 Check on excessive moment in rigid edge beam.

of v_x because shear forces are less accurate than moments; the reason is that shear forces are calculated in the centre line of elements. We know that the area of the v_x-diagram must be $0.250pa^2$. Assuming that the distribution shape is correct, we increase the maximum value of v_x shown in Figure 13.1 from $0.313pa$ to $0.343pa$. Numerical integration of the shear force over the half span of the beam then leads to $0.125pa^2$. Integrating twice leads to the bending moment $0.0377pa^3$ in the beam. The integration results are the dashed lines in the moment and shear diagram of the beam. We see that the FE analysis confirms the theory that the beam moment is larger than would be expected on the basis of the distributed load v_x only. The difference is caused by the twisting moments, which are an additional load on the edge beam. The area of the twisting moment diagram over half the span is $0.0106pa^3$. Adding this to $0.0377pa^3$ we obtain $0.0483pa^3$ which is very close to the FE result $0.0480pa^3$. For the two-way sine load and Poisson's ratio 0.2, the beam moment is 1.4 times the expected value. In the present case of homogeneously distributed load, a factor $0.0480/0.0377 = 1.27$ applies.

In Section 5.3.4 we expressed the expectation that the twisting moment will not influence the shear force in the beam, based on the assumption that the twisting moment can be carried to the web of the edge beam. This is not confirmed by the analysis. The analysis leads to a beam shear force which is a factor $0.160/0.125 = 1.28$ too large, practically the same amplification factor as for the moment. Apparently, a concentrated force of opposite sign

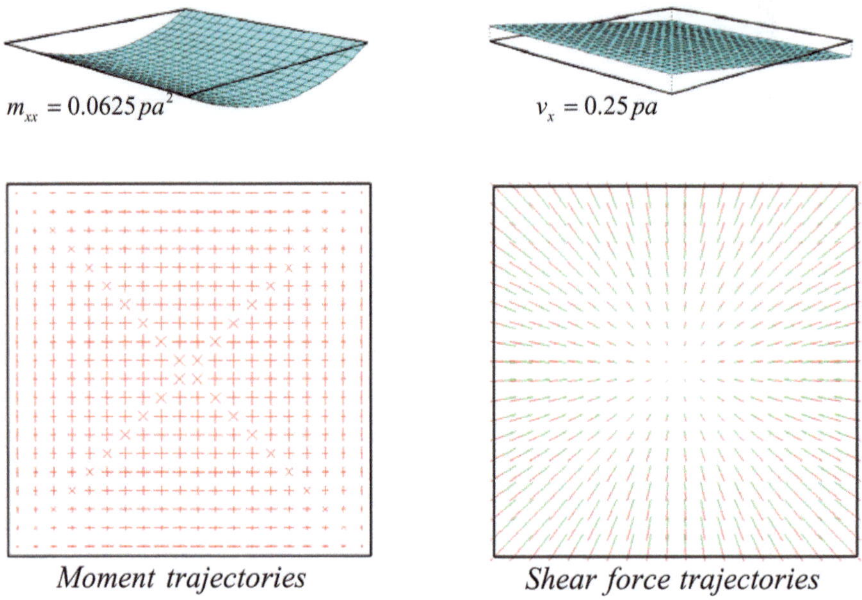

Figure 13.7 FE results. Square plate. Twistless case.

still remains in the plate edge, which balances the overestimated shear force in the beam. The reason is the use of Kirchhoff theory. Here the thick plate theory of Mindlin performs better.

13.3.2 Flexible Beams

For flexural edge beams leading to zero twisting moments we must choose the flexural rigidity $aD/2$, where D is the plate rigidity and a the plate span, corresponding with the discussion in Section 5.4. We expect that half the distributed load p is transferred in the x-direction and half in the y-direction. The bending moment m_{xx} is independent of y, and m_{yy} is independent of x. Similar considerations hold for the shear forces. The total load pa^2 is homogeneously distributed over the circumference $4pa$. The support reaction is $0.25pa$. The maximum moment and shear force in the plate are $0.0625pa^2$ and $0.25pa$, respectively. FE results for a 20×20 mesh are shown in Figure 13.7. The bending moment m_{xx} and shear force v_x indeed are clearly constant in the y-direction. The moment trajectories are different from Figure 13.1 for rigid edge beams. The direction is parallel to x- and y-direction all over the plate. The moments m_{xx} and m_{yy} have become principal moments because the twisting moment is zero. At the plate diagonals, the bending moments are equal and Mohr's circle becomes a point. Then the direction of the trajectories is indeterminate. The FE program computes them parallel and normal to the diagonal. The shear trajectories have also changed compared to the case of rigid edge beams. Similar to what we have seen for corner supports in Figure 13.3, they are no longer normal to the edge. For symmetry reasons the moment m_{yy} and shear force v_y need not be shown. The twisting moment is zero at each position. This plot is also skipped.

We now consider the shear trajectories more closely. In a twist-less slab they are straight lines, originating from the plate centre. The explanation is straight forward. Consider a set of x- and y-axes with origin in the plate centre. Then $v_x = \frac{1}{2}px$ and $v_y = \frac{1}{2}py$. According to Eq. (4.23) the trajectory direction is calculated from $\beta_o = \arctan(y/x)$. In a straight line, starting in the centre, y/x is constant, so a constant trajectory direction β is obtained along the straight line.

When we consider a length ds along the edge beam, the load p on a triangle with area $\frac{1}{2} \times a/2 \times ds$ flows to this edge part. The total load on this triangle is $\frac{1}{4}pa \times ds$, and this flows to an edge part of length ds. Therefore the shear force per unit length is $\frac{1}{4}pa$, which we had decided on earlier on other grounds.

According to classical beam theory the maximum moment in the edge beam is $M = \frac{1}{8} \times \frac{1}{4}pa \times a^2 = 0.03125pa^3$ and the maximum shear force $V = \frac{1}{4}pa \times a/2 = 0.125pa^2$. The value of the beam moment M is due to the load v_x only on the one edge and to v_y on the other edge. The FE analysis with the 20 × 20 mesh delivers $M = 0.03125pa^3$ and $V = 0.11875pa^2$ respectively. The bending moment is exact. The shear force is 5% less than the exact value. The difference is easily explained. The shear force in the FE analysis is constant over each beam element, so in fact holds true in the middle of the element. In reality the shear force increases linearly from the beam centre to the column. We have used 10 elements over the half beam length, therefore we are missing a half element size over 10 elements. This explains the 5% error. The shear force will approach the exact value $0.125pa^2$ for increasing mesh fineness.

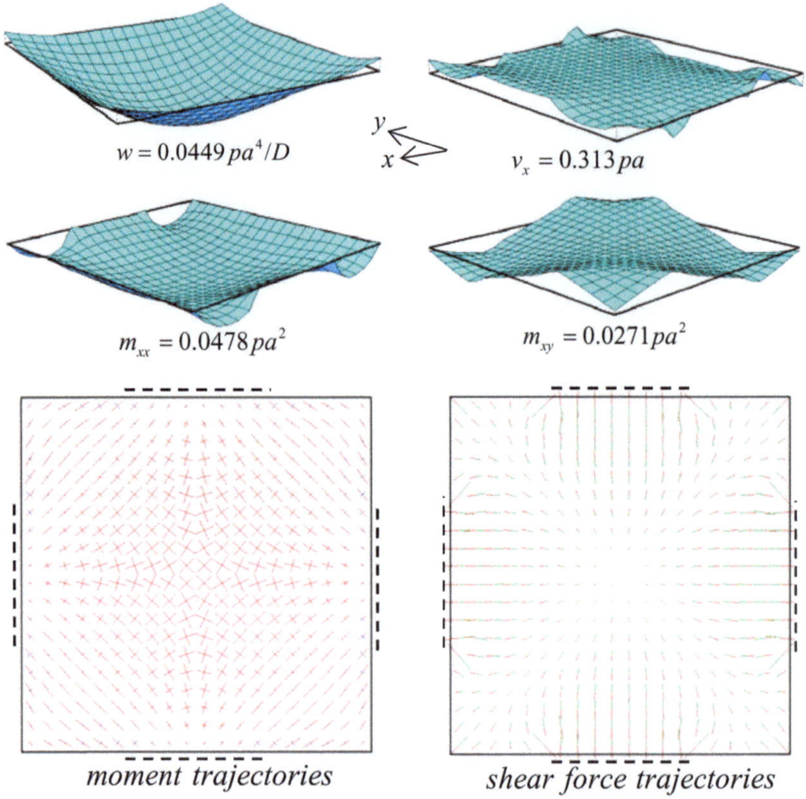

Figure 13.8 FE results. Square plate. Pressure only. Distributed load.

13.4 Pressure-Only Support

In Section 13.1 we considered a simply-supported plate subjected to a homogeneously distributed load. It was tacitly assumed that the support is able to transfer both pressure and tension reactions. And indeed the result showed that concentrated tensile reactions do occur in the corners. Physically this presupposes either a very good fixing to, say, a wall below the slab, or the presence of a wall on top of the slab edge, which provides the needed downward vertical reaction force in the corner. When neither the one nor the other is secured, we must reckon with another moment distribution. In Chapter 5 we noticed the high efficiency of the simply-supported plate, and that the diagonal beam action takes care for it to a large extent. This contribution will be reduced if no tensile reaction forces can occur. We must expect lower twisting moments at the cost of higher bending moments in the plate centre.

If we use a linear-elastic program we must do the analysis in an iterative way. In the first run, all edge nodes are fixed ($w = 0$). Then the computation is repeated with the nodes released where a tensile force occurs. This must be done until there are no tensile reactions. Figure 13.8 shows results of such an iterative analysis. If we compare the results with moments for the ideal simple support in Figure 13.1 the following conclusions hold. The plate corners lift from the supports. This occurs over about 30% of the edge length near each corner. Only 40% of the edge remains where we find compressive reactions. The maximum deflection increases 11%. The twisting moment m_{xy} at the plate edge is zero in the uplifted corner as it should be according to theory. Its maximum has shifted to a position at some distance from the corner. The maximum value has become 27% smaller. The bending moment m_{xx} in the plate centre has increased about 8%. Most probably not many structural engineers are aware of this phenomenon. It is always wise to make allowance for unexpected loadings and imperfect support conditions.

13.5 Message of the Chapter

- Whatever boundary conditions, we find a nearly axi-symmetric state in the neighbourhood of a point load.
- Twisting moments and shear forces are zero on lines of symmetry at symmetric loading. Bending moments are maximal on the lines of symmetry.
- For simply-supported and corner-supported plates, there are large twisting moments at the plate corners. Their signs are different, and so are the signs of the corner reactions. In the corner supported plate the reaction is compressive; in the simply-supported plate it is tensile.
- The two limit cases, simple supports and corner supports, can be simulated by edge beams with flexural rigidities, which are infinite or zero, respectively. An ideal twist-less case can be obtained by a proper in-between choice of the flexural rigidity of the edge beam.

- If tensile support reactions cannot occur, the twisting moments in the corner reduce substantially at the cost of higher bending moments and a larger deflection.

- FE codes should not only offer the output option of contour plots, but also three-dimensional presentations of deflections, moments and shear forces. The latter are more appealing to structural engineers, and concentrate attention to spots with large moments and shear forces.

Chapter 14
Handling Peak Moments

This chapter is devoted to the subject of local peaks in moment distributions. These occur on top of columns and at receding walls, and for more than one reason structural engineers do not know how to handle them. It is not clear how seriously such peaks must be taken, and how they can be smoothed. An additional problem is the dependency on the mesh fineness; the finer the mesh, the higher the peak. So, the engineer may think to be punished for being serious. This chapter intends to provide practical hints on choosing mesh fineness and designing reinforcement.

14.1 Peaks at Columns

We state the problem for the structure in Figure 14.1. A rectangular concrete slab of length 20 m, width 10 m and thickness 0.6 m with free edges is supported by four inner columns as shown in the figure. The plate is subjected to a homogeneously distributed load $p = 10$ kN/m^2. The material properties are $E = 3 \times 10^7$ kN/m^2 and $\upsilon = 0.2$. A set of axes x, y is chosen with x in the direction of the long edge and y in the short direction. In this section the column is introduced as point support in one single node of the mesh, acting as a ball. We start with square elements of size 0.5 m, which is about slab thickness. This mesh is shown in Figure 14.1.

Figure 14.2 is an assemblage of contour plots and 3D representations for this slab problem. We call the two left-hand columns the left support, and the two right-hand columns the right one. The plots for the displacement w and

J. Blaauwendraad, *Plates and FEM: Surprises and Pitfalls*, Solid Mechanics and Its Applications 171, DOI 10.1007/978-90-481-3596-7_14,

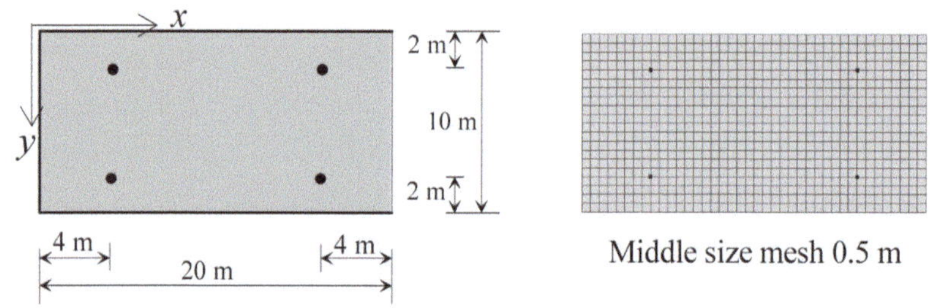

Figure 14.1 Rectangular slab on inward columns.

the bending moment m_{xx} show an almost cylindrical bending between the two supports. In the line over the supports the moment m_{xx} concentrates on top of the columns, which in another way is noted in Figure 14.3. A similar concentration holds for the bending moment m_{yy}. The plot for the twisting moment m_{xy} in Figure 14.2 supports the statement that the horizontal line and vertical line at the supports are more or less lines of symmetry. There the twisting moments are zero. In the four quadrants around the column we notice cloverleaf-type distributions.

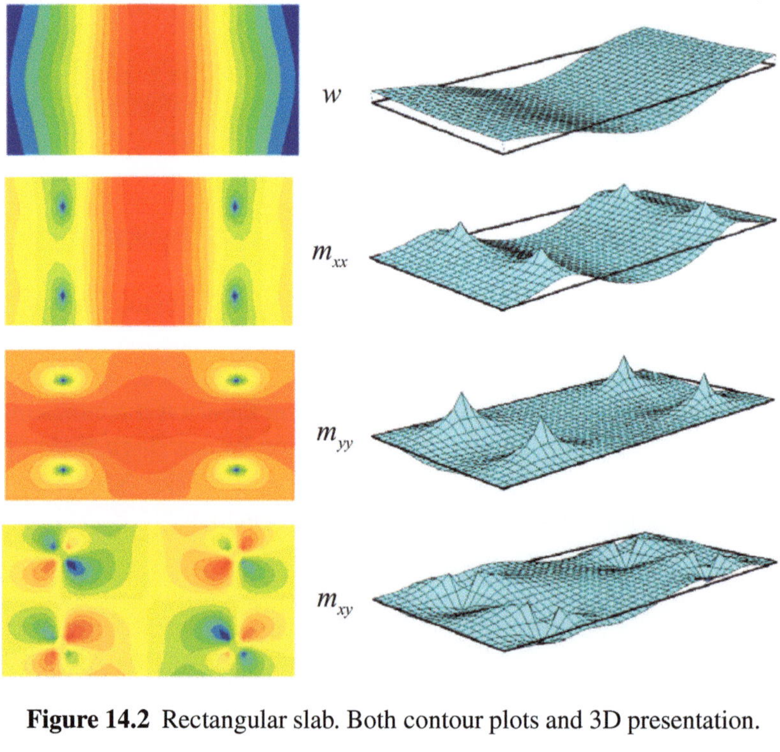

Figure 14.2 Rectangular slab. Both contour plots and 3D presentation.

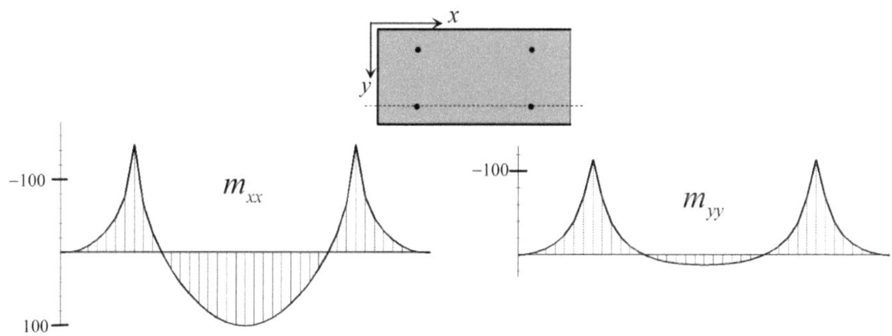

Figure 14.3 Moment diagrams in section over columns.

Next we examine the influence of the mesh fineness, using a commercially available FE code. Figure 14.4 shows the distribution of the moment m_{xx} in the section above the two right columns for three different meshes. Starting from the mesh size 0.5 m we doubled the mesh size to 1.0 m, and then halved it to 0.25 m. When we call the peak moment of the coarsest mesh 100%, then the peak in the medium coarse mesh is 128%, and in the finest 153%. This is not very helpful to the structural designer. But we have other information. We know the exact value of the integral of the moment m_{xx} over the section of the plate; it can be determined from the free body equilibrium of the plate part right of the section under consideration and the p-load on it. We can compare this exact moment with the value which we obtain when we integrate m_{xx} from the FE analysis over the section. If we call the exact moment 100%, then the finest mesh FE mesh delivers 99.2%, the middle 98.9% and the coarsest 97.6%. The largest error in the integrals is about 2%. While the difference in peak between the coarsest and finest mesh is 53%, the integral differs less than 2%! Another observation is that the large difference in value rapidly disappears in the neighbourhood of the column. The moment in the field between two columns is within 1% for all three mesh finenesses.

Lesson on moment peaks

We must not look at the peak value, but at the area of the moment diagram over the section. For this area, it is sufficient to consider a part of the section which extends left and right of the column.

To offer a helping hand, a plate length two times the column width at each side of the column will do, in total five times the column width as in Fig-

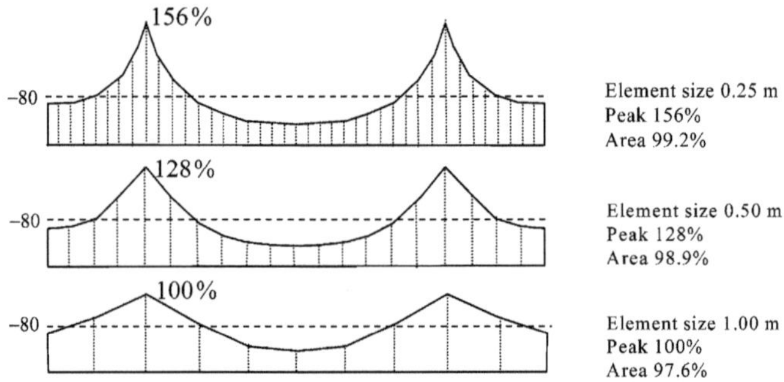

Figure 14.4 Moment distribution for different mesh fineness. The peak values differ much, the areas do not.

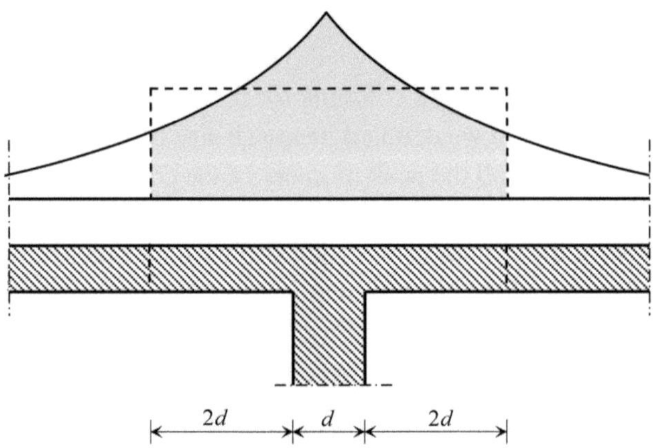

Figure 14.5 Smearing out of moment peak.

ure 14.5. The integral over this section part determines the reinforcement which is needed in this section part. The structural engineer may spread this total amount equally over the width of the section part or choose to spread part of it and concentrate the remaining part above the column. Whatever is chosen, the total amount is the same and is sound.

14.2 Column Reaction Distribution

The previous section has made clear that precise knowledge of the peak moment is not essential for dimensioning the reinforcement. Yet it may support our understanding if we know more about what is happening at a column. For that purpose we performed an exploratory analysis of the stress problem in the connection between the column and the plate. We simplified the problem to a two-dimensional stress state and performed a membrane-type of analysis. It is understood that the real stress state is axisymmetric, but precise accuracy is not considered necessary. The computed structure is depicted in the left top part of Figure 14.6. It consists of the parts of a horizontal beam and vertical column, cast in one course to a monolithic structure. The left and right edges of this beam are points of counter-flexion with zero moment. There is a downward shear force, which is the load of the structure under consideration. A parabolic load distribution over the thickness is supposed. In the horizontal section between column and beam the vertical stresses are plotted and the position of the resultant for each column half is determined. It appears that this resultant is at a distance two-third of the half column width from the centre line, as depicted in the top right part of the figure. Figure 14.7 shows the distribution over depth of the bending stress at different positions in the beam. Halfway the beam end and the column we see a linear distribution of bending stresses as in beam theory. In the cross-section above the column face the stress distribution has become nonlinear, particularly in the compression zone. In the column centre line the distribution has

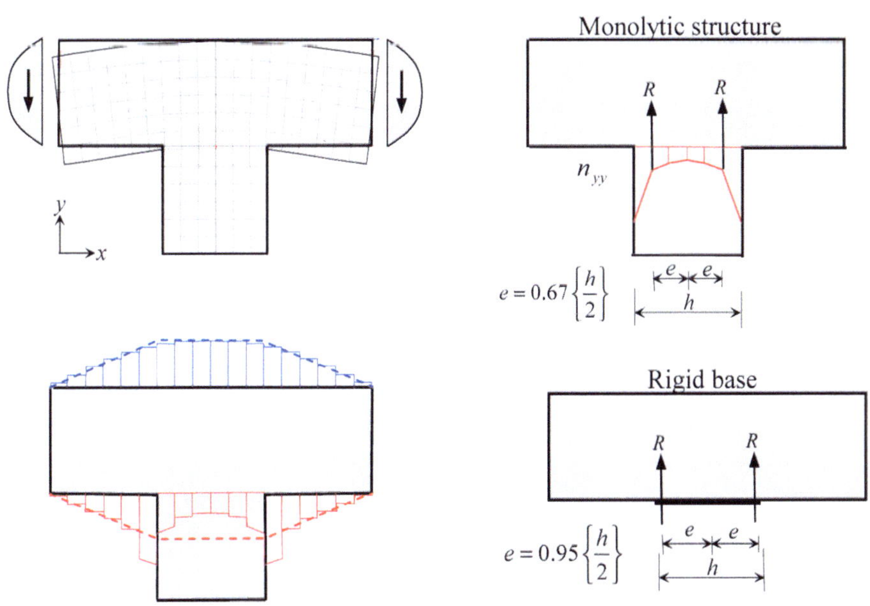

Figure 14.6 Resultant of column reactions.

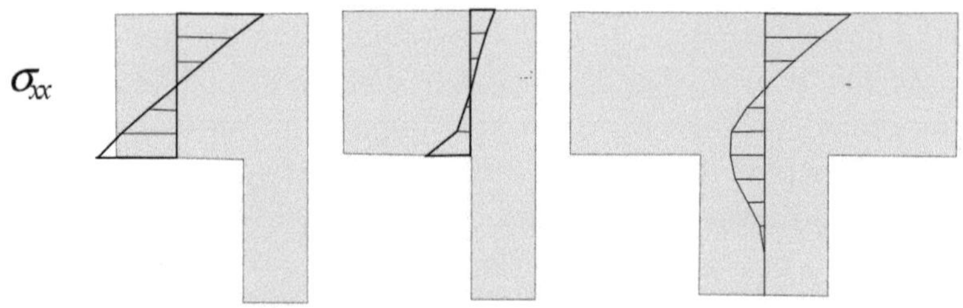

Figure 14.7 Bending stresses in monolytic structure.

even changed further, because now part of the column is taking part in the distribution.

The bottom left part of Figure 14.6 shows the distribution of the bending stresses in the top and bottom faces of the beam. The dashed line presents the stress which would apply according to classical beam theory when just centre-lines are used. Outside the column this classic theory applies practically over the full length. Above the column it does not; there it is disturbed.

Until now in this section we supposed the support of the plate was flexible. We change it into a rigid one in which only compression stresses can be transferred. Such a calculation needs some iteration. The position of the support reactions is shown in the bottom right part of Figure 14.6. The resultant of the vertical stresses over one half column width acts along a line which almost coincides with the face of the column. The distance to the centre line is 95% of the half column width.

The real column situations will lie in the bandwidth which we have covered with the two different cases. The first one is flexible and able to transfer both tensile and compression stresses, and the other is rigid and transfers compression stresses only. The resultant is at a distance to the column face in the order of 10% of the half width ($\frac{1}{2}d$) of the column. We expect that this distance would have been even smaller if we had carried out an axisymmetric analysis. Then the outer material of the column behaves more stiffly because more material is present further from the central axis. We conclude that the reaction is definitely not homogeneously distributed. Instead, it is justified to accept the rule that the reaction acts at the face of the column.

14.3 Application

The message of Section 14.2 will be applied to the structure of Figure 14.1. There the column was introduced as a point support acting on a single node of the mesh. Now we suppose that the column has an area of 0.4 × 0.4 m^2. This time we have taken an element size of 0.2 m in the analysis. The mesh is unnecessarily small, much smaller than the plate thickness, but this facilitates fitting the column in the mesh. Over the area of the column cross-section there are four elements and nine nodes, of which one coincides with the column centre.

From symmetry considerations we know that the column reaction is 500 kN. We perform three analyses. The first is a repetition of the analysis of Section 14.1, in which the support reaction of 500 kN is introduced as an upward point load in the centre node above the column. In the second analysis we apply the support reaction as a homogeneously distributed load over the column area. In the third analysis the support reaction is introduced as a line load over the perimeter of the column. In all three cases the vertical displacement of the node in the centre of the column is constrained in order to prevent rigid body motions. After completion of the FE analysis the support reaction will be zero, because we have introduced the support reaction as an upward load.

The results of the three analyses for the bending moment m_{xx} in the section over two columns are assembled in Figure 14.8. The maximum moments are −205, −150 and −125 kN respectively. Expressed in terms of the size of the support reaction $R = 500$ kN the results are in absolute values $R/2.3$, $R/3.3$ and $R/4.0$ respectively. The moment in the first analysis (point reaction) is more than 60% higher than in the third (edge load). The value $R/4.0$ for the edge load is in good agreement with the bandwidth expectation of $R/6$ to $R/4$, at which we arrived in Section 7.7 on the basis of the same supposition.

14.4 Cast-Connected Column

Usually columns are fixed to the slab so that moments can be transferred from column to slab. Modelling the column as a one-dimensional line element and fixing it to the slab in one single node will in general evoke high moment peaks in the slab. The same occurs if the column is replaced by a rotational spring in one single slab node. In these cases the conclusions of

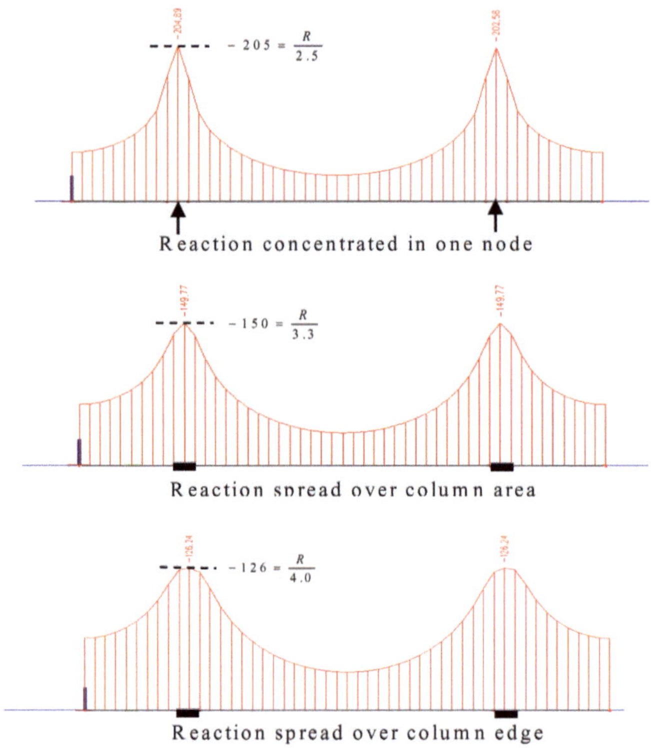

Figure 14.8 Moments in section for different column distribution assumptions.

the preceding section still apply. It is not the peak that is important, but the area of the moment diagram over a section.

There is an alternative way to look at this problem. Structural engineers wonder how to introduce the column in the analysis so that the structural response is closer to the real stress state at the top of the column. One approach is to use distributed springs instead of a lumped rotational spring, either homogeneously distributed over the column cross-section area or as a line distribution along the edge of the area. On the basis of the exploration in Section 14.2 we prefer the edge distribution.

Consider the rectangular cross-section of Figure 14.9 and suppose that this column fits in a mesh of elements. The dimensions of the column cross-section are h and b respectively, where $b < h$. We assume that elements are used with corner nodes only. A lumped spring of constant K is placed in each node. The units of the spring constant are kN/m. Structural engineers tend to choose the value of K such that the axial stiffness EA/l of the column is

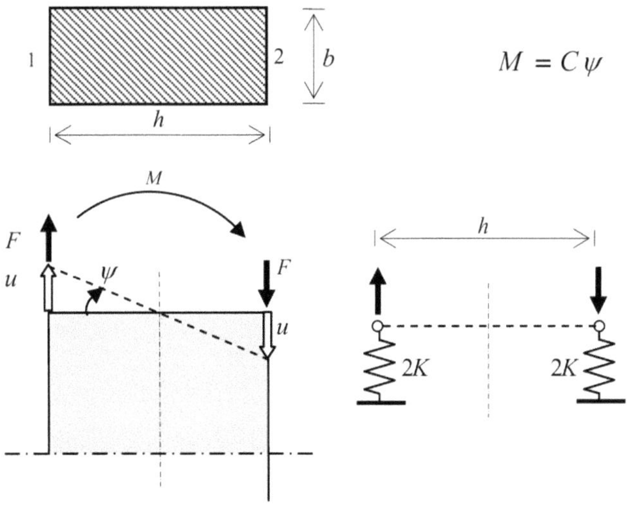

Figure 14.9 Springs on column edge for rotational stiffness.

spread over the four lumped springs, therefore $4K = EA/l$. We will examine how realistic this choice is with respect to the rotational stiffness.

We consider bending of the column about the short axis and apply a moment M, which causes a rotation ψ. We determine the stiffness relation $M = C\psi$, in which C is the rotational spring constant. At a rotation ψ, the two left nodes displace over a distant u, and the right nodes over the same distant in the opposite direction. The force F in two nodes is $F = 2Ku$, and the relation between the displacement and the rotation is $u = \frac{1}{2}h\psi$. Therefore we obtain $F = Kh\psi$. Because of the relation $M = hF$ we find $M = Kh^2\psi$. Therefore, the rotational spring constant is

$$C = Kh^2 \tag{14.1}$$

From the definition of K at the start of the derivation we know

$$K = \frac{1}{4}\frac{EA}{l} \tag{14.2}$$

Substituting this equation in Equation (14.1), and remembering that $I = bh^3/12$, we obtain

$$C = 3\frac{EI}{l} \tag{14.3}$$

This result will appeal to structural engineers: it is the rotational stiffness of a column which is pin-connected at the base, see Figure 14.10. If a column

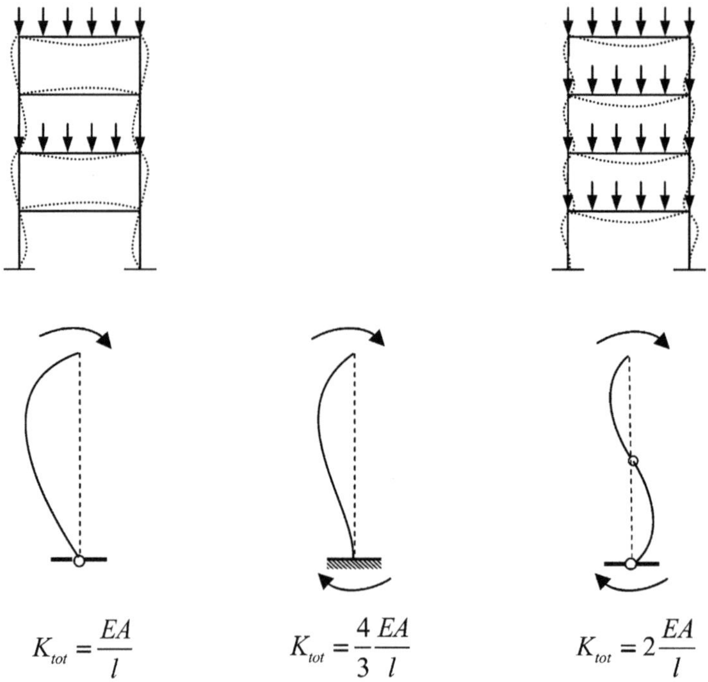

Figure 14.10 Spring constant for different column conditions.

must be simulated which is clamped at the base, the stiffness K must be multiplied by 4/3. If the point of counter-flexion is to act halfway up the column, we must choose a factor 2.

Once more it is stressed that this way of modelling columns requires a mesh, at least around the column, which may be finer than the plate thickness. This approach possibly satisfies the structural engineer, because the maximum moment has some physical meaning. But even in this approach it is wise to determine the integral of the moment distribution in a section in order to know what total reinforcement is required. The improved knowledge about the peak moment then suggests how the reinforcement should be spread over the section.

14.5 Dependence on Program

In the current section we discuss two items together. We meet another type of moment peak, and we address the influence of the choice of program. Figure 14.1 shows the structure. There is a reinforced concrete floor slab of an apartment building with a balcony. The thickness of the floor is 200 mm and of the balcony 120 mm. The material properties are $E = 3 \times 10^7$ kN/m^2 and $\nu = 0.2$. The floor is subjected to 1.0 kN/m^2 and the balcony to 0.8 kN/m^2. Two providers, I and II, of commercial software have been invited to submit a solution for the bending moments in this plate. We specified that the boundary line coinciding with the y-axis must be considered as clamped and the other wall supports as simply-supported. All three balcony edges are free.

14.5.1 Review of FEM Results

Results of provider I are depicted in Figure 14.11. We will comment on these results first and after that make a comparison with results of provider II. The moment diagram for m_{xx} is drawn for the section at mid-span of and parallel to the simply-supported edges. A negative moment m_{xx} = –8.35 kN occurs at the clamped edge, the moment changes to a positive sign in the middle of the floor, is negative again at the boundary between floor and balcony, and ends with a zero value at the free balcony edge. Along the same section also moment m_{yy} is shown. At the clamped edge it is also negative and has the value -1.66 kN. This value is very close to the product of Poisson's ratio $\nu = 0.2$ and the moment $m_{xx} = -8.35$ kN. This confirms the statement in

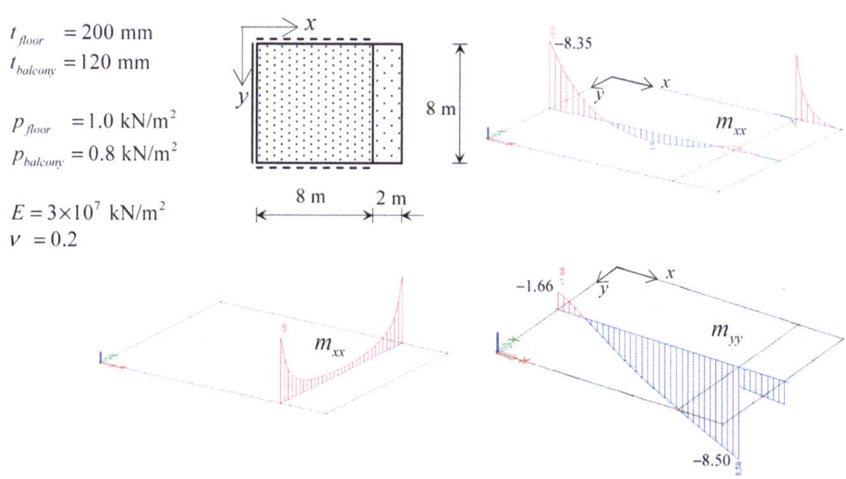

Figure 14.11 Apartment floor with balcony.

Section 4.4.1 that at a clamped edge $m_{yy} = \nu m_{xx}$, where m_{xx} is the moment normal to the edge and m_{yy} in the direction of the edge.

At the boundary between floor and balcony the bending moment m_{xx} is continuous, but the moment m_{yy} appears not to be. According to theory, this must be due to the different flexural rigidities of floor (D_{fl}) and balcony (D_{ba}). It is easily explained for zero Poisson's ratio, but holds generally. Let us write $m_{xx,fl}$ and $m_{xx,ba}$ for floor and balcony respectively. In the same way we write $m_{yy,fl}$ and $m_{yy,ba}$. At the boundary between floor and balcony we have

$$\begin{aligned} m_{xx,fl} &= D_{fl}\,\kappa_{xx,fl} \\ m_{xx,ba} &= D_{ba}\,\kappa_{xx,ba} \end{aligned} \tag{14.4}$$

$$\begin{aligned} m_{yy,fl} &= D_{fl}\,\kappa_{yy,fl} \\ m_{yy,ba} &= D_{ba}\,\kappa_{yy,ba} \end{aligned} \tag{14.5}$$

At the boundary continuity of the moments m_{xx} is required for reasons of equilibrium, and continuity of the curvatures κ_{yy} for reasons of compatibility. Therefore

$$m_{xx,fl} = m_{xx,ba}, \qquad \kappa_{yy,fl} = \kappa_{yy,ba} \tag{14.6}$$

Equality of $m_{xx,fl}$ and $m_{xx,ba}$ delivers, according to Eq. (14.4, different values of $\kappa_{xx,fl}$ and $\kappa_{xx,ba}$, because of the difference between D_{fl} and D_{ba}. This does not play a further role in our consideration. If $\kappa_{yy,fl}$ and $\kappa_{yy,ba}$ are equal, then Eq. (14.5) yields different values of $m_{yy,fl}$ and $m_{yy,ba}$. This explains the discontinuity in m_{yy} in the diagram of Figure 14.11.

14.5.2 Program Comparison

In the discussion of the results of provider I in Section 14.5.1 we did not show a graph for the distribution of moment m_{xx} along the boundary line between floor and balcony. This is the subject of the present section.

We show the diagram for m_{xx} at the boundary between floor and balcony in Figure 14.12 for both provider I and provider II. High concentrations are found at both line ends near the supporting walls. If we consider the free body equilibrium of the balcony we can compute the total moment due to the balcony load which must be transferred in the boundary. Apparently this moment is not evenly distributed along the line, but heavily concentrated at the ends. If we compare the results of the two companies we notice different

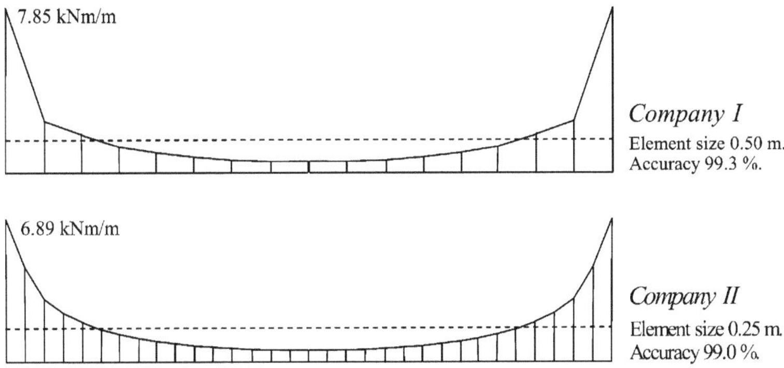

Figure 14.12 Comparison of two FE codes.

peak values, 7.85 and 6.89 kN respectively. The difference is 14%. However, the message, learned for columns, holds here as well. If we integrate the diagram over the length, we obtain moment sums of 99.3 and 99.0% of the exact value, respectively. Again, it is not the peak value that provides the best information, but the area of the moment diagram. Again, the moments at some distance of the peak are virtually the same in both cases. So it is sufficient to determine the diagram area over a short distance at the end of the line. This area needs to be examined for detailing reinforcement in the x-direction at the ends of thc boundary linc.

Graph output in sections

It is important that FE codes offer the output option for graphs about sections or part of sections.

We now draw attention to the peculiar fact that provider I with the highest peak used the coarsest mesh. In this FE analysis the element size is 0.5 m, while provider II with the lower peak chose a smaller size 0.25 m. This underlines the statement in Section 10.2 that unexpected results may be obtained. It is possible that a higher-order element in combination with a coarse mesh performs better than a fine mesh in combination with a lower-order element.

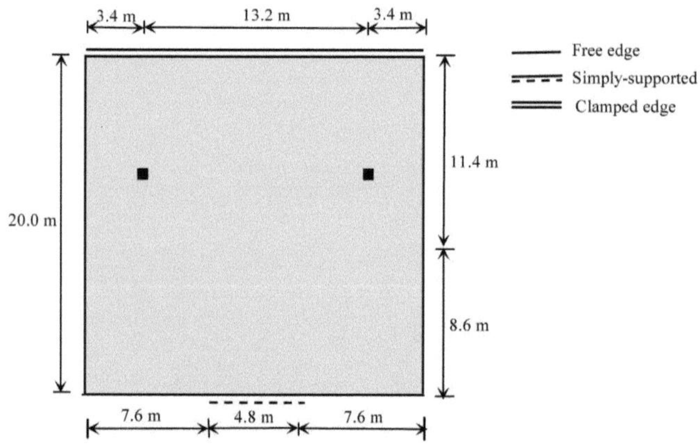

Figure 14.13 Slab for study about user dependency.

14.6 Dependence on User

In this section we show the influence of users on results. This is carried out on a structure with a receding wall. It is another example with a moment peak. The considered floor slab is drawn in Figure 14.13. It has two free edges, one clamped edge (double lines), one edge partly free and partly simply-supported, and two inward placed columns. In this example the rotational stiffness of the columns is specified. The area of the reinforced concrete slab is 20×20 m^2 and the thickness 400 mm. The columns have a cross-section 400×400 mm^2. The design load is 18 kN/m^2, which includes self-weight and variable load, each with accompanying load factors.

We invited structural engineers who attended a continuing education course to design the reinforcement of this slab. At the start we specified a number of of items:

- The concrete cover is 35 mm.
- The mass of hair pins and cutting losses is neglected.
- Concrete properties are $E = 3 \times 10^7$ kN/m^2, $v = 0.2$.
- Steel yield stress is 435 N/mm^2.
- If possible, Kirchhoff theory is used.
- Ignore crack widths and deflections; only ULS is considered.
- Codes of practice requirements for minimum reinforcement are disregarded.
- The lever arm in the cross-section is 0.9 times the effective plate thickness.

The structural engineers were asked to submit

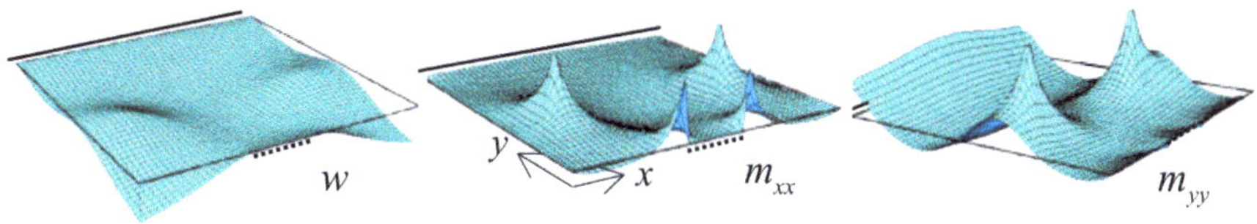

Figure 14.14 Deflection and moments for slab of Figure 14.13.

- The peak values of the moments at the column support.
- The total mass of reinforcement in kilograms.
- A subdivision of the total amount of reinforcement needed for the four meshes, two in the top layers and two in the bottom layers.

The invitation to structural engineers was repeated for five subsequent courses in place-country-region Belgium and The Netherlands. We show the result of one of these five courses, which is representative for all. Eighteen people in the course participated, and they used six different commercial FE codes. Two of them just provided moment values and did not submit an amount of reinforcement. We will first show some FE results. They come from an analysis with small elements such that the column support matches with 2×2 elements. Figure 14.14 shows a 3D plot of the deflection and bending moments m_{xx} and m_{yy}. Figure 14.15 shows the distribution of m_{xx} and m_{yy} over two important sections. The values of m_{xy} are small and therefore omitted. The peak values of the bending moment in Figure 14.15 are in the bandwidth 400 to 450 kNm/m: the peak moment is approximately M = 425 kNm/m. The column support reaction R is 1880 kN. The ratio of R to

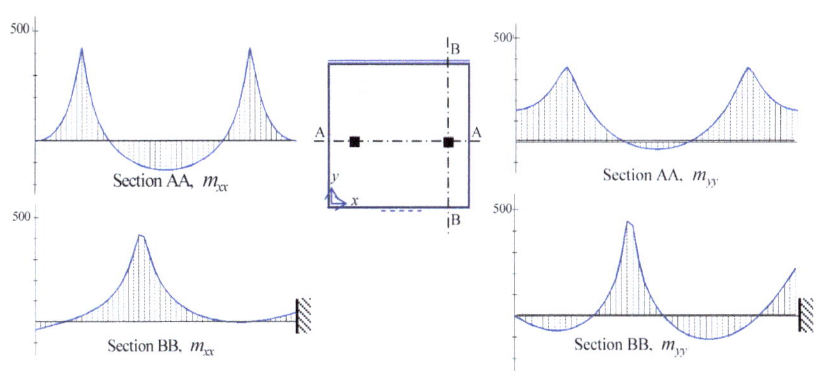

Figure 14.15 Moment diagrams in sections.

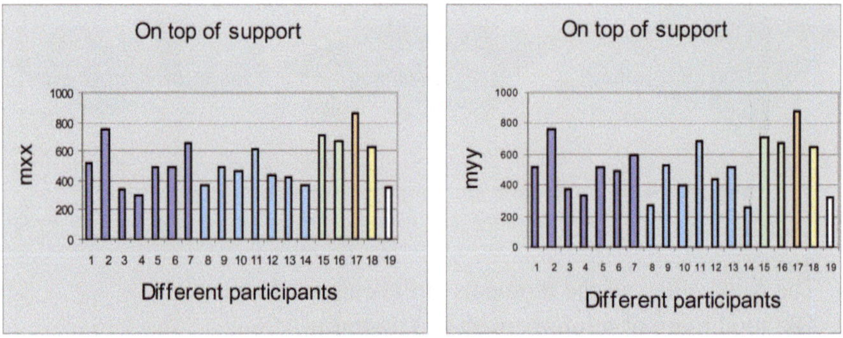

Figure 14.16 Moments at column by different participants.

m is 4.5, close to the expected ratio 5 which we expect in a grid of columns, as discussed in Section 7.7 and depicted in Figure 7.13.

We now will discuss the submissions of the participants. Figure 14.16 is a bar chart for the moments m_{xx} and m_{yy} at the column support. The 18 participants obtained moment peaks differing by more than a factor of two. Bars with the same colour have been obtained with the same FE code.

Lesson

Using the same program does not necessarily lead to similar results; the user can make a difference.

In Section 14.1 we explained that the difference in peak value should not be a big problem in the design of reinforcement; we work with areas of the moment diagram rather than with peak values. Then no large differences in reinforcement amount are to be expected. However, reality is different. In Figure 14.17 both the total amount is shown and the subdivision over the two orthogonal reinforcement layers in both top and bottom of the slab. The differences between the participants are substantial.

As stated, the experiment has been repeated several times with other groups of experienced structural designers, always with a similar outcome. Do engineers think 'iron is cheaper than brains'? Many engineers place reinforcement where it is not necessary for strength reasons.

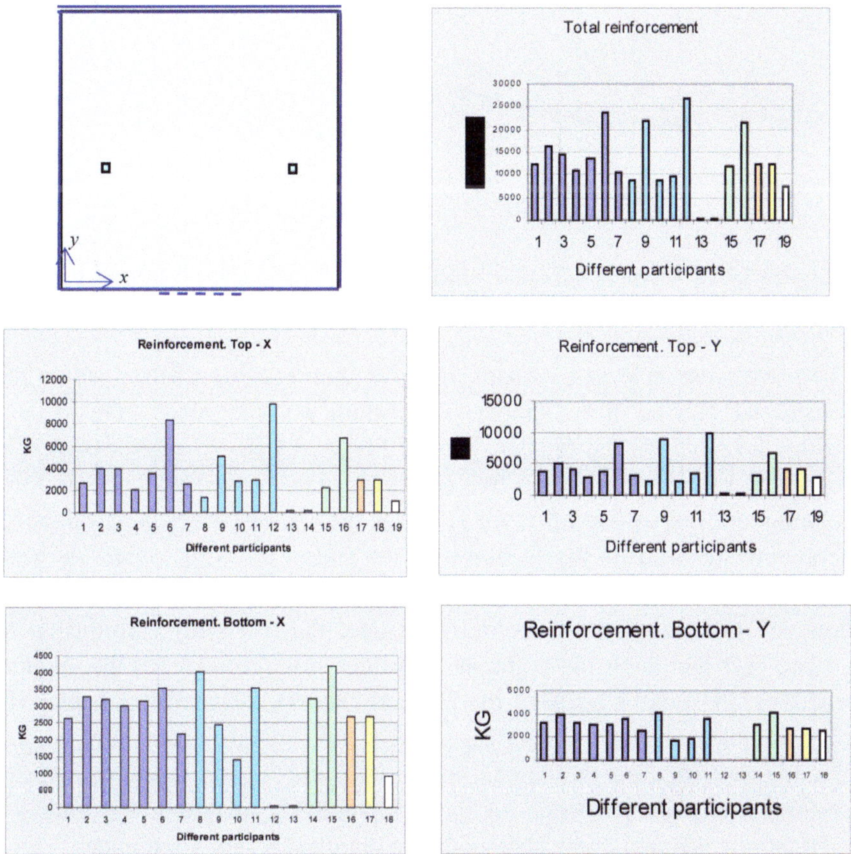

Figure 14.17 Reinforcement demand by different participants.

Lesson

Different structural engineers come to very different amounts of reinforcement for the same structure and load.

14.7 Impact of Support Flexibility and Concrete Cracking

In Sections 14.1 and 14.5.2 we considered the moment integral over a section in the plate and compared it to the exact moment due to the load.

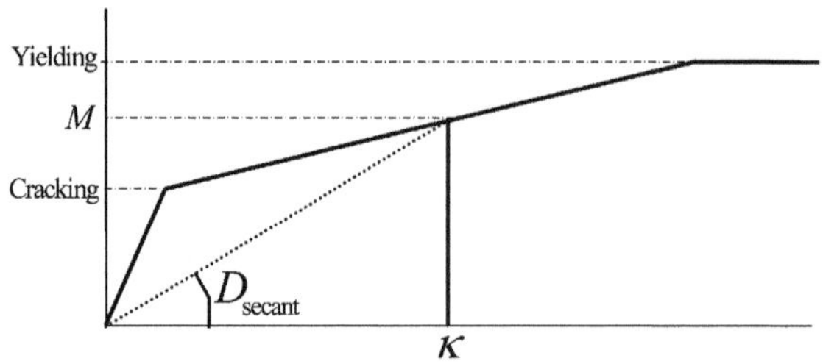

Figure 14.18 Moment-curvature diagram of reinforced concrete cross-section.

There we were able to calculate the total moment due to the load in the considered section from simple equilibrium considerations. This is not always possible. In the top left part of Figure 14.18 we show a structure supported by both a wall and two columns, such that the wall end and the columns are not in one line. We consider the FE result for m_{yy} (direction parallel to the wall) in the section over the end of the wall. Again we want to compare the area of this moment diagram to the exact value of the total moment. For that purpose we must consider the free body equilibrium of a plate part including the columns. We must now account for the support reactions; these are a result of the FE analysis and, as stated in Chapter 10, very reliable. They are in perfect equilibrium with the loads. So, we are able to calculate the total moment due to load and support reactions and can perform a check on the area of the computed m_{xx}-diagram. We omit the calculations and just report that again we obtained a very good match.

We proceed with the example to show the impact of introducing flexibility of the wall support and accounting for cracking. For that reason we again consider the distribution of the moment m_{yy} in the section over the end of the wall and perform three calculations. In the first calculation all nodes in the line where the wall occurs are rigid supports. There the displacement w is zero. In the second analysis we replace the rigid supports by flexible ones by inserting springs in the z-direction. The spring constant is EA/l where E is Young's modulus of the wall concrete, A is the cross-section of the wall part which is assigned to a node, and l the height of one storey. In the third analysis we return to the rigid supports and reduce the plate rigidity D in order to account for cracking. Due to cracking, the rigidity easily halves at large moments compared to D for uncracked concrete. Figure 14.18 illustrates this statement. Large negative moments occur on the tops of the wall and columns. Therefore we introduce a reduction of 50% for the two rows of elements left and right of the wall and the four elements around

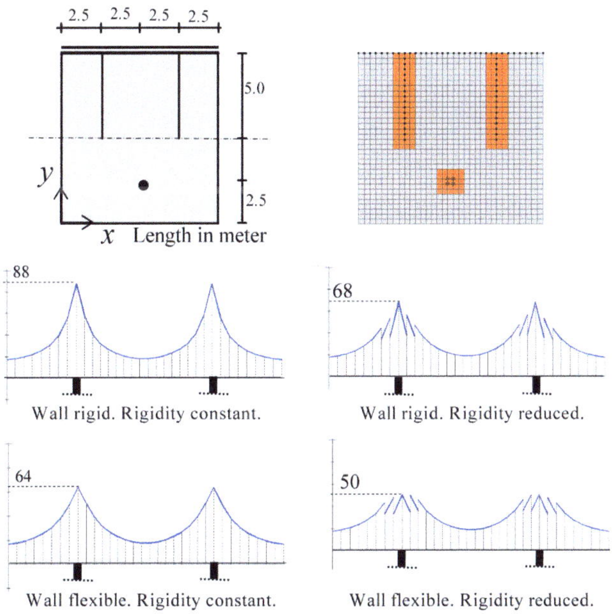

Figure 14.19 Study about impact of flexible wall and reduced rigidity.

the columns. In the fourth analysis we finally combine wall spring supports with reduced plate rigidity.

Figure 14.19 shows the results for the four analyses. We notice considerable differences in the peak values of the moment m_{xx}. For rigid wall supports and uncracked plate rigidity the maximum moment value is 88 kN. Inserting only wall flexibility makes the value reduce to 64 kN, a reduction of 28%. Reducing only the plate rigidity results in 68 kN, a reduction of 23%. The combination of springs and cracking leads to a moment 50 kN. Now the reduction has increased to 43%. We conclude that adapting plate rigidity decreases moments in the order of 20%, and the combination of rigidity decrease and wall flexibility in the order of 40%.

Use of flexible supports

This small investigation teaches us that it is worthwhile to introduce realistic conditions. The correct value of the spring constant is not a critical issue. Introducing flexibility makes the difference. After that, halving or doubling the spring constant no longer has much effect. Similarly it is worth accounting for cracking by reducing the plate rigidity.

The moment reduces significantly, and (not shown in this example) deflections will increase substantially. Of course, the choice made here is subjective and has just the intention to show the effect. It makes clear that realistic moment-curvature diagrams must be consulted for fixing rigidities when FE codes make predictions about deflections in the serviceability state.

Commercially available FE codes are starting to offer checking facilities on deflections. The programs must be able to handle orthotropic plate properties as discussed in Section 3.5. In this orthotropic case m_{xy} and m_{yx} are equal so $m_{av} = m_{xy}$. Also D_{xy} and D_{yx}, defined in Eq. (3.46), are equal so $D_{\text{av}} = D_{xy}$. Therefore we must know the rigidities in the constitutive equations

$$\begin{Bmatrix} m_{xx} \\ m_{yy} \\ m_{xy} \end{Bmatrix} = \begin{bmatrix} D_{xx} & D_{\nu} & 0 \\ D_{\nu} & D_{yy} & 0 \\ 0 & 0 & D_{xy} \end{bmatrix} \begin{Bmatrix} \kappa_{xx} \\ \kappa_{yy} \\ \rho_{xy} \end{Bmatrix} \tag{14.7}$$

A first analysis is done with constant plate rigidity D to determine the moments and design the reinforcement. Then the moment-curvature diagrams for x- and y-direction are calculated for each element. From the first analysis we know the bending moments m_{xx} and m_{yy}, and from the moment-curvature diagrams we derive the curvatures κ_{xx} and κ_{yy}. From this new rigidities D_{xx} and D_{yy} are found. It is not clear which off-diagonal rigidity D_{ν} must be used. A practical value is

$$D_{\nu} = \nu\sqrt{D_{xx} D_{yy}} \tag{14.8}$$

which leads to the correct term in the isotropic case. The value of Poisson's ratio ν can be made dependent on the size of the moments. For an uncracked cross-section the value 0.2 is fine, and as the reinforcement starts to yield the value should be set to zero. Between the cracking moment and yield moment a linear interpolation might be chosen.

It remains to choose a value for the torsion rigidity D_{xy} which holds for the value m_{xy} from the first analysis. With reference to Section 11.6.2 and Chapter 16 we may assume that this twisting moment raises reinforcement forces which are conforming with bending moments $m_{xx} = m_{xy}$ and $m_{yy} =$

m_{xy}. At these bending moment values we read rigidities D_{xx} and D_{yy} from the moment-curvature diagrams. Then we may apply the formula

$$D_{xy} = \frac{1}{2}(1-\nu)\sqrt{D_{xx}D_{yy}} \tag{14.9}$$

A physical ground for this approach does not exist other than the statement that the correct rigidity is obtained for the limiting case of isotropy.

The second linear-elastic FE analysis with adapted rigidities will yield moment values different from the first analysis. Therefore the procedure of rigidity adaptation must be repeated and a further analysis is necessary. This is done until the results do not change anymore in a practical sense. Therefore, the final moment values and deflections are the result of an iterative procedure.

After completing the iteration procedure we can also compute crack widths. From the moments we know the steel strains in the reinforcement bars and can pronounce an expectation on the crack widths, following the codes of practice rules.

14.7.1 Application of Finite Element Program

We show an application with the commercial FE code *Diamonds* (Buildsoft, Belgium). This program chose its cracking formula from the Eurocode 2 (EN 1992-1-1). The curvature α in the cracked state is calculated by the formula

$$\alpha = \zeta\alpha_{\|} + (1-\zeta)\alpha_{|} \tag{14.10}$$

where $\alpha_{|}$ is the curvature for the uncracked section and $\alpha_{\|}$ the curvature for the fully cracked section (the section consisting of concrete compression zone and reinforcement). The distribution coefficient ζ is

$$\zeta = 1 - \beta(\sigma_{\mathrm{sr}}/\sigma_s)^2 \tag{14.11}$$

Here β is a coefficient accounting for the influence of the duration of the loading or of repeated loading on the average strain (1.0 for single short-term loading; 0.5 for sustained loads or many cycles of repeated loading); σ_s is the stress in the reinforcement calculated on the basis of a cracked section; σ_{sr} is that same stress but under the loading conditions causing first cracking. The curvature α is used to derive the reduced rigidity in the element stiffness matrix.

According to Eurocode 2 (EN 1992-1-1) the crack width is

$$w_k = s_{r,\max}(\varepsilon_{sm} - \varepsilon_{cm}) \tag{14.12}$$

where $s_{r,\max}$ is the maximum crack spacing in the principal stress direction. It is calculated from

$$\frac{1}{s_{r,\max}} = \frac{\cos\theta}{s_{r,\max,x}} + \frac{\sin\theta}{s_{r,\max,y}} \tag{14.13}$$

Here θ is the angle between the axes of the principal stress and the direction of the reinforcement, and $s_{r,\max,x}$ and $s_{r,\max,y}$ are the crack widths in the reinforcement directions; they are defined by

$$s_{r,\max} = k_3\, c + k_1 k_2 k_4\, \phi / \rho_{p,\text{eff}} \tag{14.14}$$

where

k_3 = 3.4
C = cover of longitudinal reinforcement
k_1 = 0,8 (high bond bars)
k_2 = $(\varepsilon_1 + \varepsilon_2)/2\, \varepsilon_1$
ε_1 and ε_2 are (tensile) strains at the boundaries of the cracked section;
ε_1 is the greater stress
k_4 = 0.425
ϕ = bar diameter
$\rho_{p,\text{eff}}$ = $A_s/A_{c,\text{eff}}$

The factor $\varepsilon_{sm} - \varepsilon_{cm}$ in Equation (14.12) is the difference between the mean strain in the reinforcement and the mean strain in the concrete between cracks under the relevant combination of loads

$$\varepsilon_{sm} - \varepsilon_{cm} = \frac{\sigma_s - k_t \frac{f_{ct,\text{eff}}}{\rho_{p,\text{eff}}}\left(1 + \alpha_e \rho_{p,\text{eff}}\right)}{E_s} \geq 0.6 \frac{\sigma_s}{E_s} \tag{14.15}$$

where

σ_s = stress in the tension reinforcement assuming a cracked section
α_e = ratio E_s/E_{cm}
k_t = 0.4 for long term loading
$f_{ct,\text{eff}}$ = tensile strength of the concrete

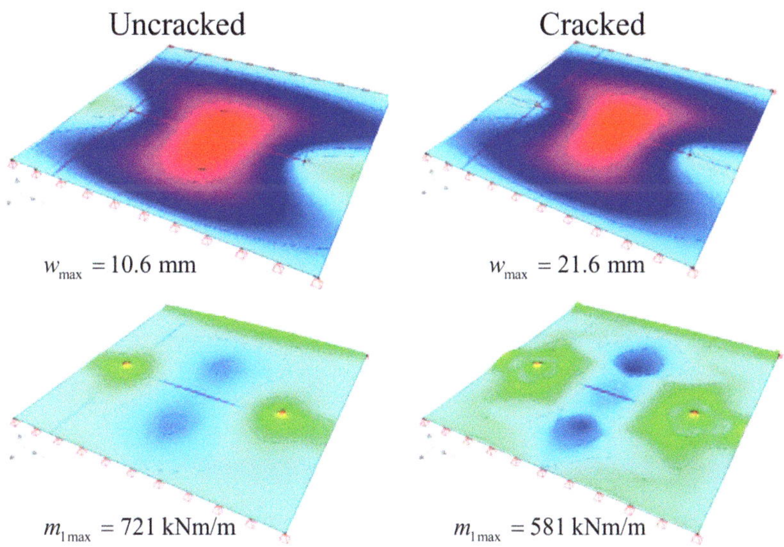

Figure 14.20 Results of FE code for deflections and moments.

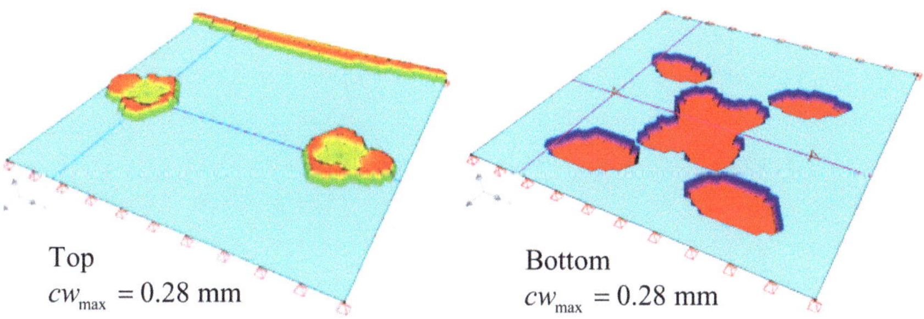

Figure 14.21 Results of FE code for crack widths.

Normally the bar diameter ϕ is not known in the design process. In this case a fictive diameter is used defined as $\sqrt{(2A_s/3\pi)}$; (six bars lead to A_s).

Figure 14.20 shows results of an analysis for a rectangular slab, supported at two parallel edges and two inner columns, both for ULS and SLS. The ultimate limit state was considered for the moments (safety), and the serviceability limit state for the deflections (normal practice). The left three plots in the figure are for the un-cracked slab; the right plots are for the cracked slab. Note that the deflection increases from 10.6 to 21.6 mm due to cracking. The maximum value of the principal moment m_1 for ULS decreases from 721 to 581 kNm/m; this is the order of 20%, mentioned earlier. Notice that the peak moment localizes in the uncracked state, and smears out in the cracked state. Figure 14.21 shows crack widths for SLS in the upper and lower face.

14.8 Message of the Chapter

- We need to focus on the area of the bending moment diagram and not on the peak bending moments. That area, representing the total moment to be transferred by the section, is in equilibrium with the load and support reactions.
- It is possible to pass a judgement on the real peak value. The column reaction can be best distributed over the perimeter of the cross-section of the column. The value of the bending moment is, in conformance with Chapter 7, in the bandwidth of one-sixth to one-fourth of the column reaction force.
- Cast-connected columns, capable of loading the slab by a lumped moment, can be modelled by springs on the edge of the column cross-section.
- FE analysis results are dependent on the chosen FE code and on the user. Different programs for the same problem lead to different solutions. Higher-order elements may produce a lower peak than lower-order elements with a finer mesh.
- Design of reinforcement depends strongly on the individual structural engineer.
- The introduction of springs at hard support spots and accounting for severe cracking at moment peaks makes moments spread out over larger widths and reduces maximum values.
- FE codes should offer the output option of section graphs for deflections, moments and shear forces. It should also be possible to specify a special part of a section. The FE code should output integrals over the (part of the) section.

Chapter 15
Sense and Nonsense of Mindlin

In Chapters 3 and 4 we presented the theories of Mindlin (more properly Mindlin and Reissner) and Kirchhoff, without explaining which theory must be used in a particular practical case. Commercially provided FEM software usually offers both options and even may have chosen one of them as the default option. The goal of this chapter is to help users make a proper choice.

15.1 Result Dependence on Analyst and Program

We start with the comparison of plate bending results obtained by different programs and analysts. Four providers of commercial software accepted the invitation to participate in the computation of a plate structure. They were asked to perform a linear-elastic computation. The plate, shown in the left top part of Figure 15.1, has the shape of a carpenter's square. The two long edges have a length of 4 m and the short edges 2 m. The modulus of elasticity is 40×10^6 kN/m^2 and Poisson's ratio is 0.2. The plate is subjected to a distributed load of 1 kN/m^2. We choose a set of orthogonal axes x, y along the long edges. The z-axis is normal to the plate. The corner is supported by a ball support which prevents the displacement w and permits free rotations ψ_x and ψ_y. The two short edges of length 2 m are simply-supported. All other edges are free.

J. Blaauwendraad, *Plates and FEM: Surprises and Pitfalls*, Solid Mechanics and Its Applications 171, DOI 10.1007/978-90-481-3596-7_15,

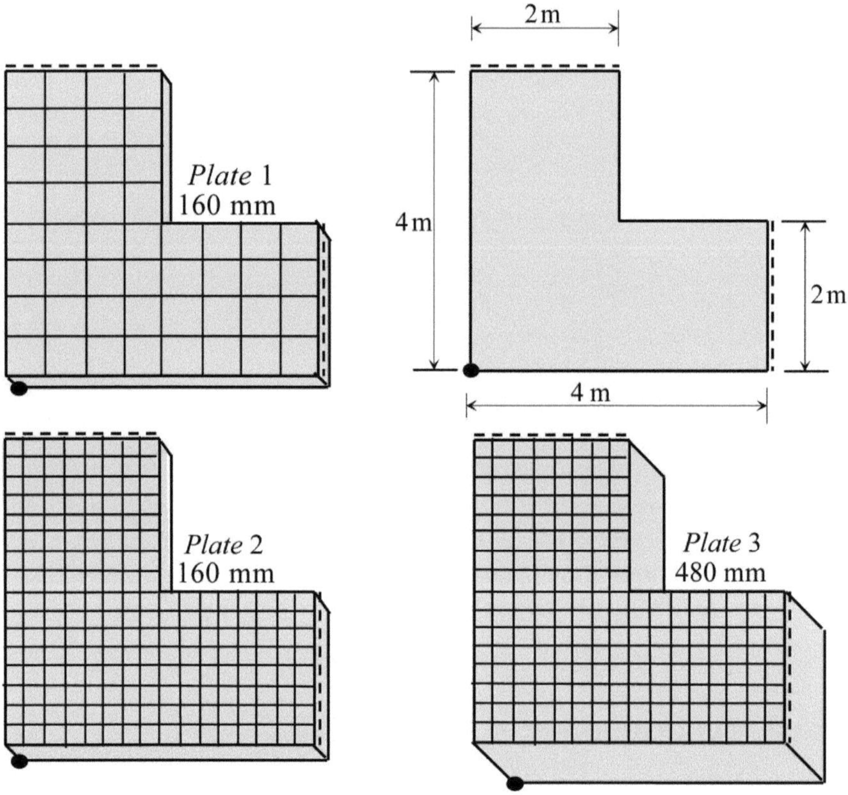

Figure 15.1 Three plates. Mesh fineness and thickness varied.

15.1.1 Invitation

The software houses were invited to perform three calculations with square elements for three versions of the plate.

- Plate 1. Mesh 8 × 8 elements (spacing 500 mm). Thickness 160 mm.
- Plate 2. Mesh 16 × 16 elements (spacing 250 mm). Thickness 160 mm.
- Plate 3. Mesh 16 × 16 elements (spacing 250 mm). Thickness 480 mm.

These different plates are shown in Figure 15.1. The required output should include

- a plot of the shear force v_x in section $x = 0.5$ m.
- a plot of the moment m_{xx} in the section $x = 2.0$ m.
- a plot of the twisting moment m_{xy} along the edge $x = 0$.

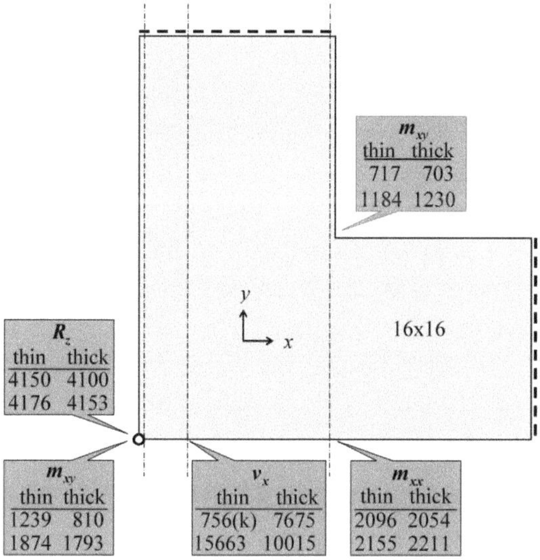

Figure 15.2 Big scatter in submitted results for twisting moment and shear force. Units in N and m.

The ultimate goal was to investigate which plate theory should be used: Kirchhoff or Mindlin.

A span of 4,000 mm and a thickness of 160 mm imply a span-depth ratio of 25. Without any doubt, this is a thin plate. This holds for the first two analyses; any differences are due to the mesh fineness. The element size 500 mm is about three times the thickness, and the element size of 250 mm about one and a half times. In both analyses the element size is not smaller than the plate thickness. If related to the longest span of 4,000 mm, the element size 500 mm means that only eight elements occur between the ball support and the simply-supported edge; this must be considered a coarse mesh, at least in the neighbourhood of the ball support. The element size 250 mm means the application of 16 elements, which sounds acceptable for the case.

The second and third analyses have the mesh fineness in common. Here the difference is in thickness. In the third analysis the span-thickness ratio is slightly greater than eight; to the perception of structural engineers, this plate is thick.

In conclusion, the thin plate 2 may be considered representative for the choice of structural engineers in practice. Variant plate 1 is intended to show the effect of mesh change and variant plate 3 the effect of slenderness.

15.1.2 Submitted Results

Three software providers performed the analysis primarily with Mindlin theory and one chose Kirchhoff theory. Figure 15.2 depicts the scatter of results. Each time, we report the smallest and largest submitted value, and do it for both the thin plate 2 and the thick plate 3. The support reaction R at the ball support is insensitive to the plate thickness. Differences are less than 1%. The differences in the bending moment m_{xx} are not significant too. However, large differences appear in the values of the twisting moment m_{xy}, both at the re-entrant corner and at the ball support, and also in the shear force v_x. Except in one case, the lowest and largest values come from a Mindlin analysis. The exception is marked with (k) of Kirchhoff.

We computed the same three plates with the program *Kola*, switching on the Reissner theory (as the program calls Mindlin theory). Figure 15.3 is the result for the moment m_{xx} in section $x = 2.0$ m. This moment is insensitive to mesh fineness or slenderness, and to the applied theory. Values vary within 2%. Note that the bending moment near the re-entrant corner is not correctly zero at the free edge. Only at a couple of elements distant from the corner does the moment become zero. This is in agreement with the expectation, as explained in Section 11.4.

Figure 15.4 is the result for the twisting moments at the edge $x = 0$, and Figure 15.5 for the shear force in the section $x = 0.5$ m. Now large differences are exposed in a convincing way. The values of the twisting moment differ by a factor of 1.3. The shear forces at the edge differ by a factor of 2.0. The output submitted by the participating providers is very similar to the results of the program *Kola*.

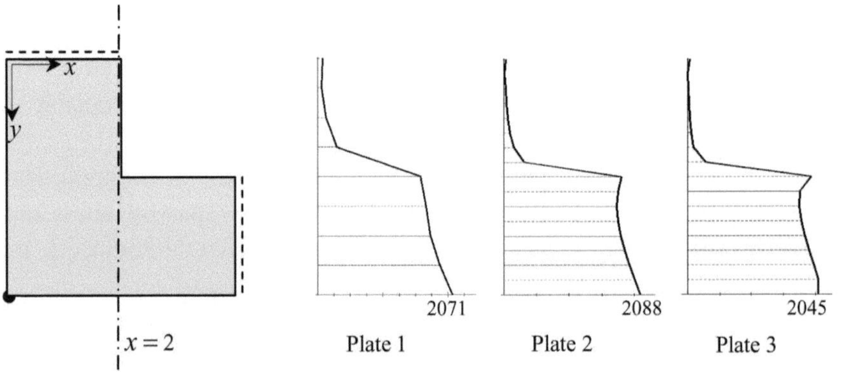

Figure 15.3 Bending moments m_{xx} (Nm/m) in section $x = 2$ m for Mindlin theory.

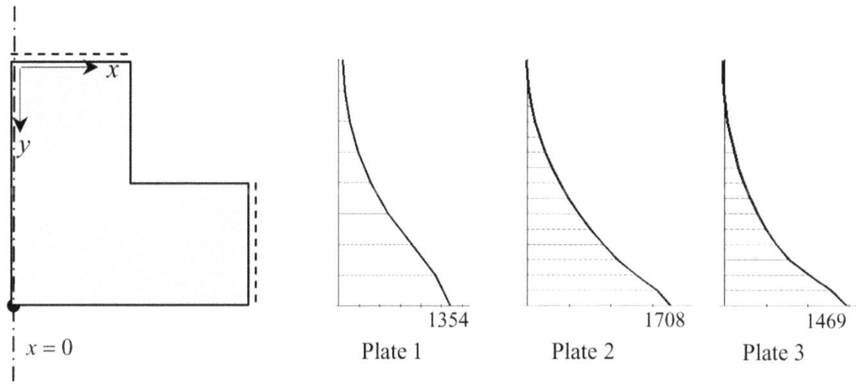

Figure 15.4 Twisting moments m_{xy} (Nm/m) at edge $x = 0$ for Mindlin theory.

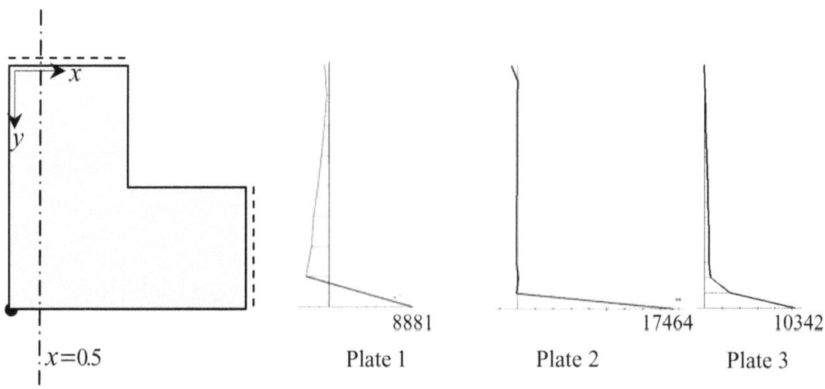

Figure 15.5 Shear force v_x (N/m) in section $x = 0.5$ m for Mindlin theory.

15.2 Explanation of the Differences

In order to understand the large differences in the twisting moment and shear force we must recall the starting points of the Kirchhoff and Mindlin theories. Classically, plates have been analyzed by the Kirchhoff theory for thin plates. Only since the broad availability of FE codes has Mindlin theory come on the scene. The user of commercial software is expected to make a choice from the two theories, but often one of them is a default option without the user being aware of this. Kirchhoff theory holds for plates in which the deformation by shear forces can be neglected, which is the case for a sufficiently large span-thickness ratio l/t. The slenderness $l/t > 10$ is sufficient, and most slabs will satisfy $l/t \geq 20$.

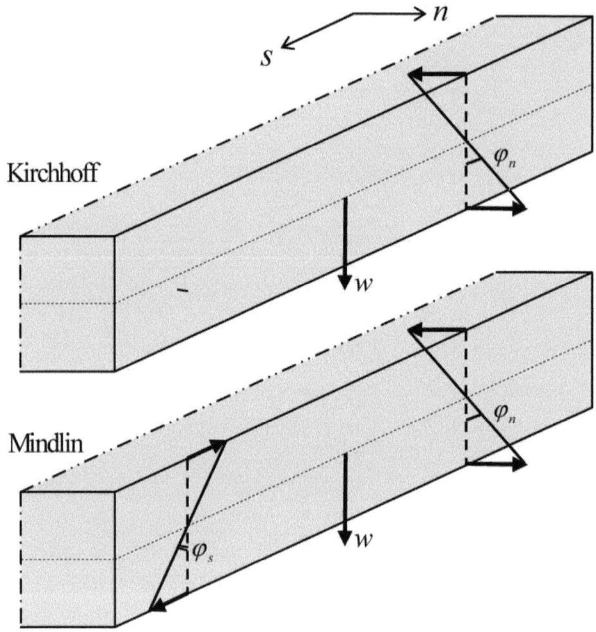

Figure 15.6 Different boundary conditions for Kirchhoff and Mindlin.

We must investigate which circumstances justify or even command the choice of Mindlin theory. We have presented his theory in Section 3.2 to which we refer. For the purpose of explaining differences in computational results it is helpful to repeat the discussion about boundary conditions at a free edge. For convenience we choose axes n normal to and s parallel to the edge. In Mindlin theory there are three independent degrees of freedom at the edge, the displacement w, the rotations φ_n normal to the edge and φ_s in the plane of the edge. This is visualized in Figure 15.6. Note that we return tot the rotation φ as used in Chapter 3 for the derivation of the differential equations. In general there are three edge load components: a distributed force f in the direction of w, a distributed torque t_n in the direction of φ_n and a distributed torque t_s in the direction of φ_s. These edge loads are one-to-one equal to the shear force v_n, the bending moment m_{nn} and the twisting moment m_{ns}, respectively. Usually t_n and t_s are zero, therefore both the bending moment m_{nn} and the twisting moment m_{ns} are zero. Eq. (15.1) summarizes this

$$\begin{Bmatrix} w \\ \varphi_n \\ \varphi_s \end{Bmatrix} \rightarrow \begin{Bmatrix} f \\ t_n \\ t_s \end{Bmatrix} = \begin{Bmatrix} v_n \\ m_{nn} \\ m_{ns} \end{Bmatrix} \rightarrow \begin{matrix} v_n = f \\ m_{nn} = 0 \\ m_{ns} = 0 \end{matrix} \tag{15.1}$$

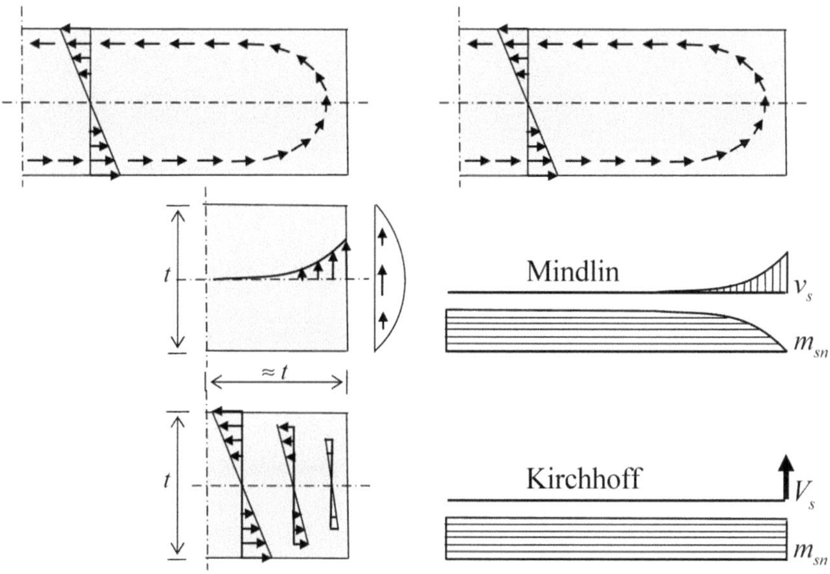

Figure 15.7 Close look at stress state near free edge.

In Kirchhoff theory there are only two degrees of freedom at the free edge: the displacement w and the rotation φ_n normal to the edge. The rotation φ_s in the plane of the edge is a slave of the displacement w because of the relation $\varphi_s = \partial w/\partial s$. Now only two edge loads can be applied, f in the direction of w, and t_n in the direction of φ_n. Yet, in general all three plate quantities v_n, m_{nn} and m_{ns} can occur at the edge and may be non-zero. In Section 4.4 we have seen which relations exist between these three quantities and the two edge loads. We repeat them here.

$$\begin{Bmatrix} w \\ \varphi_n \end{Bmatrix} \rightarrow \begin{Bmatrix} f \\ t_n \end{Bmatrix} = \begin{Bmatrix} v_n + \dfrac{\partial m_{sn}}{\partial s} \\ m_{nn} \end{Bmatrix} \rightarrow \begin{matrix} v_n \neq f;\; m_{ns} \neq 0 \\ m_{nn} = 0 \end{matrix} \tag{15.2}$$

A zero edge load t_n will lead to a zero bending moment m_{nn}, however a zero edge load f does not in general lead to a zero shear force v_n and zero twisting moment m_{ns}. At an unloaded free edge ($f = 0, t = 0$) there will be both a twisting moment and a shear force. This phenomenon is closely related with the concentrated edge shear force V_s in sections normal to the edge, as we found in Section 4.4. This concentrated shear force in Kirchhoff theory does not appear in Mindlin theory. That theory is able to compute distributed shear forces v_s in a narrow edge zone. In order to understand what makes the difference, we repeat in Figure 15.7 the shear stress flow due to a twisting moment in a section normal to the free edge. At a sufficient distance from the edge, the twisting moment causes a linear distribution of shear stresses over the depth of the plate, with zero value in the mid plane. Close to the edge, the shear flow must turn around within the section, because the edge face is stress free. This happens over a plate part with a length of about plate thickness t. We want to describe the stress state in this small end part of the plate in terms of plate quantities v_s and m_{ns}. At the mid plane of the plate the shear stresses have a vertical direction, and a distribution which increases from zero to a maximum value at the edge. The distribution is not linear. At the edge, the shear stress is vertical over the full thickness of the plate, with a distribution which is (close to) parabolic, becoming zero at the top and bottom of the edge. Outside the mid plane and at some distance from the edge we can decompose the shear stress into vertical and horizontal components. The integral of the vertical components delivers a vertical shear force v_s, which is zero at a distance from the edge of about one thickness, and becomes maximum at the edge. The integral of all horizontal components leads to a twisting moment m_{ns}, which decreases in the opposite direction and becomes zero at the free edge. Here the principal difference between Mindlin theory and Kirchhoff theory becomes apparent. Mindlin is able to describe the discussed distribution of the shear force and twisting moment and Kirchhoff is not. In Mindlin theory we can handle the boundary condition $m_{ns} = 0$, whereas we cannot in Kirchhoff theory. Instead, Kirchhoff determines the integral of all the local vertical stress components and concentrates them into one shear force V_s located at the very edge. At the same time, Kirchhoff is not able to have the twisting moment diminish to zero, and instead keeps it constant up to the edge, see Figure 15.7.

Once more we want to stress the fact that these differences happen in a plate length of about one plate thickness. In this domain a big gradient occurs in the n-direction. To cover this in a FE analysis needs a big number of elements over a short distance.

15.3 Supporting Side Study

What we have explained on the basis of the theory can be supported by a case study. We computed a simply-supported square plate by both Kirchhoff and Mindlin, each for a thin and a thick plate. The length l of the edges is 9 m. The thickness t is 200 mm for the thin plate, and 2,250 mm for the thick plate. The span-thickness ratios are 45 and 4 respectively. The first is

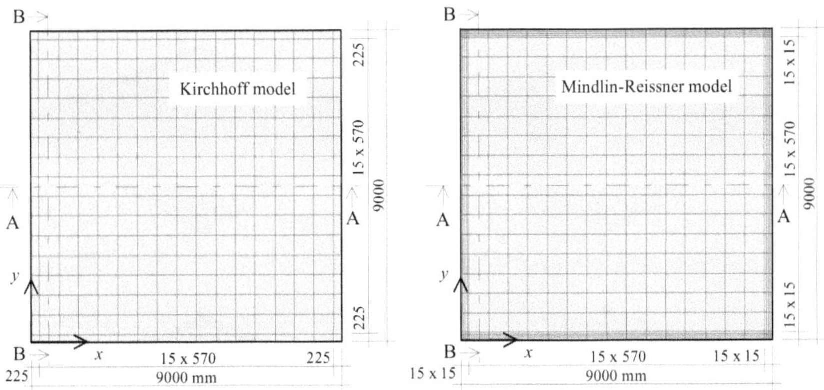

Figure 15.8 Mesh for thin plate analyses.

clearly a thin plate, and the second a thick plate. In all calculations the same distributed load $p = 3$ kN/m^2 is applied. In the plots to follow, Kirchhoff is depicted on the left and Mindlin on the right. The chosen meshes are displayed in Figure 15.8. An element size of 570 mm is used over the main area of the plate outside the edge zone. For the thin plate this is almost three times the plate thickness and for the thick plate it is about a quarter of the plate thickness. In the edge zone one element of 225 mm width is used in the Kirchhoff analysis. A smaller element size has no effect because the localized shear force in the edge zone anyhow is replaced by a concentrated shear force at the edge. This is different for the Mindlin analysis; there this 225 mm edge zone has been divided into 15 very small elements of 15 mm width each. Usually we would avoid a large aspect ratio, but we can use it for the purpose of this study, as no large gradients are expected in the direction parallel to the edges.

15.3.1 Thin Plate Results

We start with the analysis for the thin plate. Figure 15.9 shows the displacement and bending moment. The difference between Kirchhoff and Mindlin is of the order of 2%, both for displacement and moment. A difference occurs in the corners, where Mindlin leads to an isolated peak, which is absent for Kirchhoff. The twisting moments and the shear forces are given in Figure 15.10. The twisting moment in the Kirchhoff-analysis is non-zero at the edge, whereas the Mindlin analysis manages to make the twisting moment

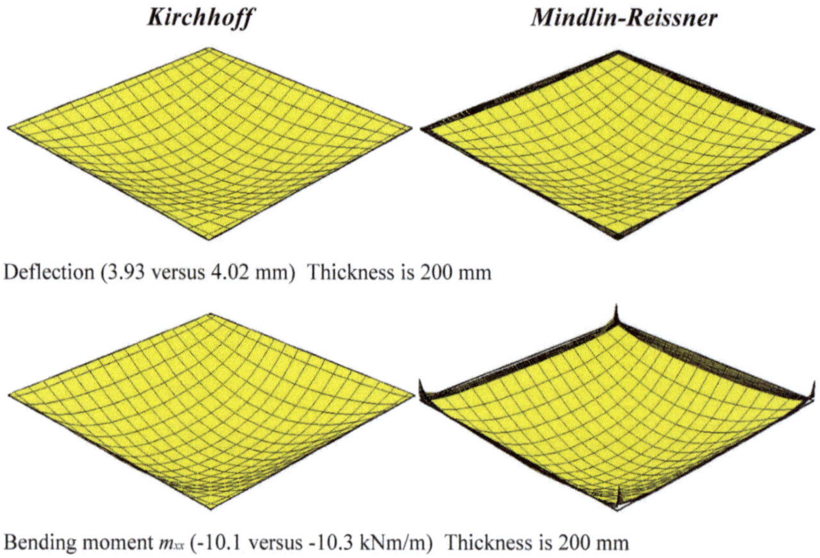

Figure 15.9 Deflection and bending moment in thin plate.

practically zero. However, this must be obtained in a very small edge zone. The maximum value occurs very close to the corner and is only 3% smaller than in the Kirchhoff analysis, where the maximum value is found exactly in the corner. Remembering that the Mindlin analysis requires an impractically fine mesh to provide the correct solution, our conclusion is justified that Kirchhoff in combination with a practical mesh delivers a good result. The shear force distributions are at first glance very different. However, we must bear in mind that the Mindlin analysis is supposed to reproduce the local shear force distribution at the edge due to the returning twisting moment shear flow, and that these local shear stresses very much dominate the plot.

We have repeated the Mindlin analysis for the practical mesh (in fact still a fine mesh) of the Kirchhoff analysis. The result is presented in Figure 15.11. Now the maximum appears at a distance of about one tenth of the span from the edge near the corner and is 9% smaller than the correct value. Mindlin in combination with a practical mesh does apparently not offer an advantage, but rather decreases the accuracy of the computation. As the saying goes, it is 'Neither fish, flesh nor good red herring'. We recommend using Kirchhoff; this theory gives good results for both moments and shear forces. The structural engineer must be aware that there is a local concentrated shear force V_s along the edge, and must be able accounting for it.

Kirchhoff ***Mindlin - Reissner***

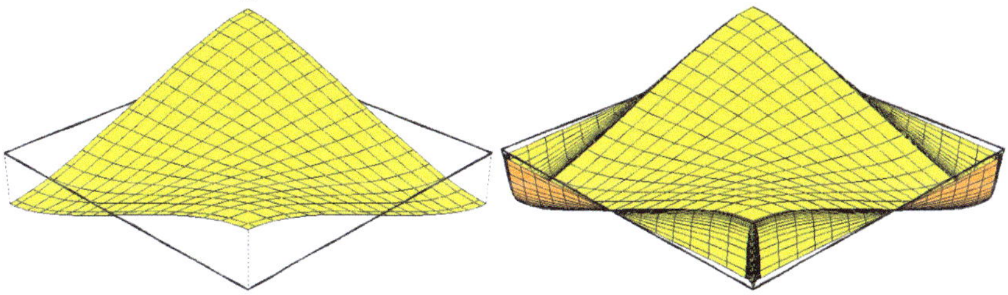

Torsion moments m_{xy} (9.69 versus 9.39 kNm/m). Thickness 200 mm

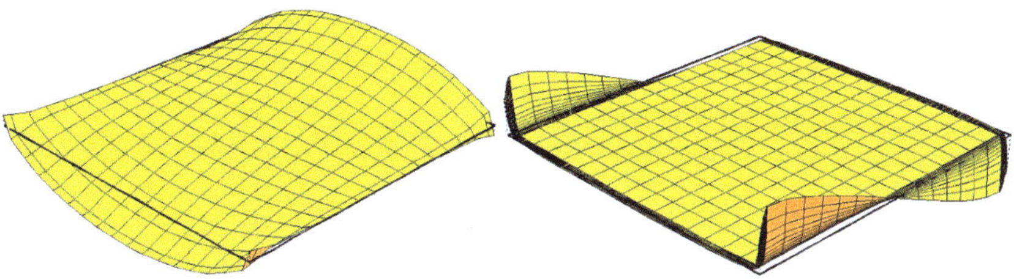

Shear force v_x (8.77 versus 11.53 kN/m). Thickness 200 mm

Figure 15.10 Twisting moment and shear force in thin plate.

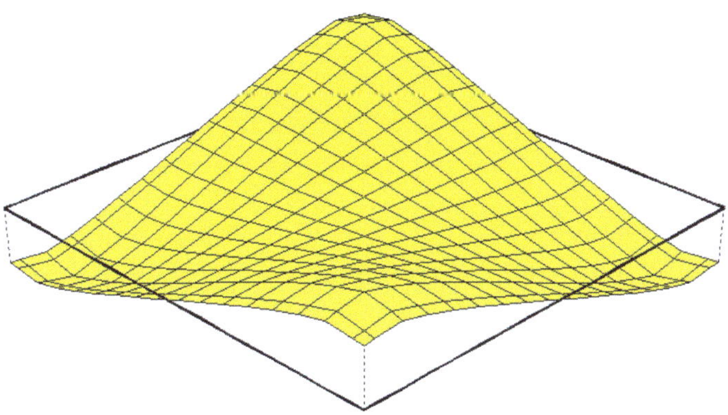

Figure 15.11 Mindlin analysis for thin plate with practical mesh. Neither fish, nor flesh, nor good red herring.

We stress that there is no point in choosing smaller elements than plate thickness. Edge zone effects are disturbed anyhow. At simple supports and clamped edges, there is always an edge disturbance with a three-dimensional stress state. The same holds at columns and at intermediate supports of continuous plates. Point loads actually apply over non-zero areas. Therefore,

Kirchhoff ***Mindlin-Reissner***

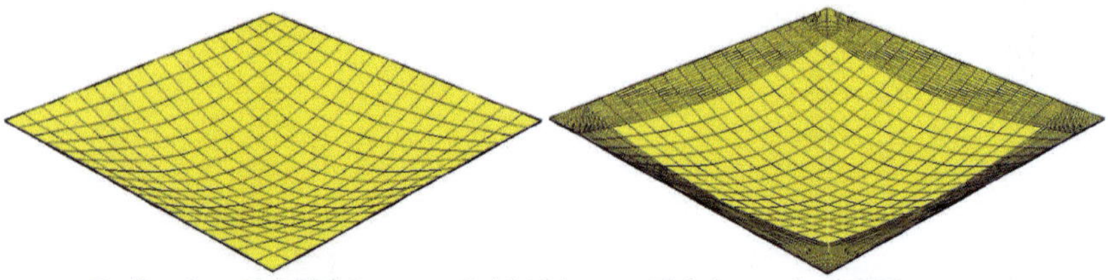

Deflection (0.00276 versus 0.00401 mm. Thickness is 2,250 mm

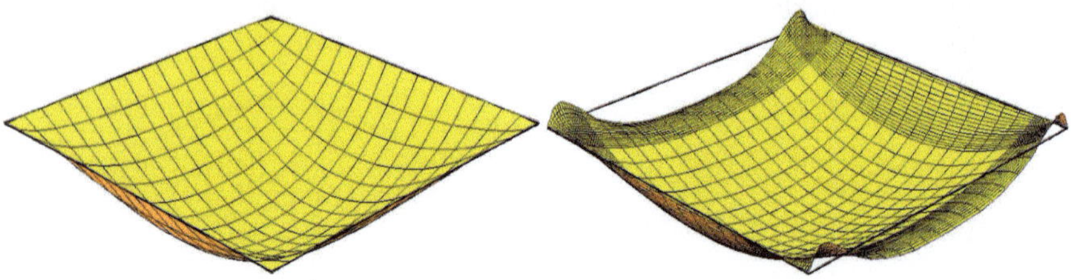

Bending moment m_{xx} (10.2 versus 11.9 kNm/m). Thickness is 2,250 mm

Figure 15.12 Mesh, deflection and bending moment for thick plate.

considering all these comments, we justify adopting the rule: never use an element size smaller than the order of magnitude of the plate thickness.

> ***Lesson***
>
> In a thin plate analysis we must use Kirchhoff. The Mindlin analysis requires a senseless fine mesh to produce practically the same results. Choosing Kirchhoff, we need never use element sizes smaller than the plate thickness.

15.3.2 Thick Plate Results

We now turn to the results for the thick plate with slenderness 4. A mesh of 17 × 17 elements is used in the Kirchhoff analysis. The mesh in the Mindlin analysis is 15 × 15 for the regular area outside the edge zones. In the edge zones again a fine mesh is used. Displacements and bending moments are presented in Figure 15.12. Now substantial differences are seen. The deflection in the Mindlin analysis is about 1.5 larger than in the Kirchhoff-analysis.

Kirchhoff ***Mindlin-Reissner***

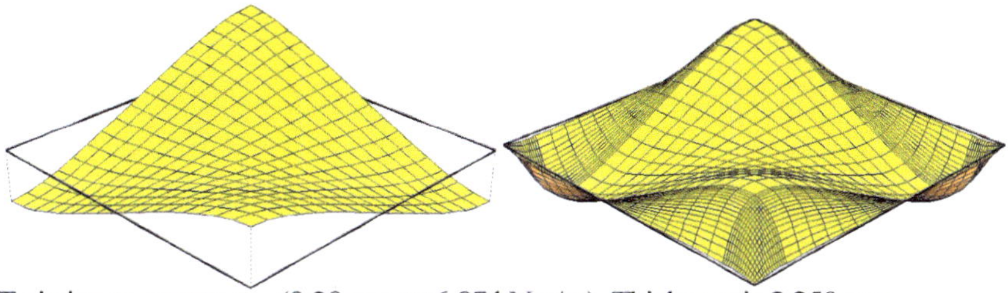

Twisting moments m_{xy} (9.28 versus 6.87 kNm/m). Thickness is 2,250 mm

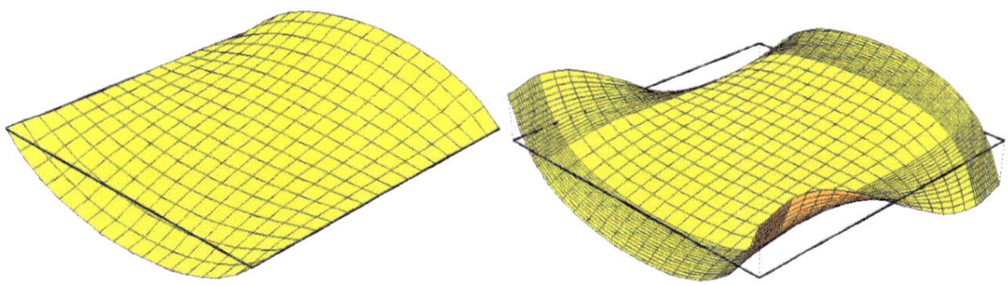

Shear force v_y (7.73 versus 12.1 kN/m). Thickness is 2,250 mm

Figure 15.13 Twisting moment and shear force in thick plate.

Moment values are 17% larger in the Mindlin analysis. A Kirchhoff analysis would seriously underestimate the bending moment. Figure 15.13 shows the results for the twisting moment and shear force. Now the edge zone in the Mindlin analysis is a noticeable part of the plate and the peak value of the twisting moment occurs at a point of the plate far away from the boundary. The value is about three quarter of the corner value in the Kirchhoff analysis. Apparently, a thick plate reduces the twisting moment at the cost of a higher bending moment. The shear force plot in the Mindlin analysis is more regular than for the thin plate and needs no further explication. The concentrated edge shear force V_s in a thin plate is nicely spread over a wide zone in a thick plate and is just part of the overall distribution of the shear force.

Lesson

For thick plates we must use Mindlin. The mesh must be refined in an edge zone of width equal to about the plate thickness. It is sufficient to take five elements over the edge zone.

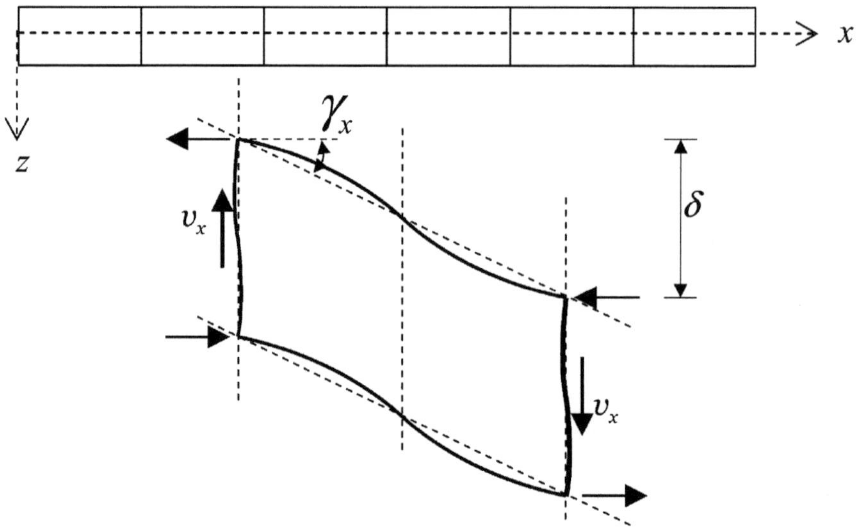

Figure 15.14 For slender multi-cell plates Mindlin theory must be applied.

Thin Orthotropic Plates

All conclusions in this chapter refer to plates of isotropic homogenous material. There is an exception, in which we always must analyze a thin plate with Mindlin theory. That is for a plate of relative small depth and small shear rigidity, like a multi-cell slab. The distortion of the cells in the x-direction, as shown in Figure 15.14, is interpreted as a shear deformation γ_x. For the shear rigidity of such plates, we refer to Section 21.3.

15.4 Comparison in Hindsight

The knowledge obtained in the preceding sections helps us understand why there was so much scatter in the submitted results for the three different plate analyses in Section 15.1. Return to Figure 15.4 for the twisting moment at the free edge. In a Mindlin analysis the twisting moment must become zero. In plate 1 the thickness is 160 mm and the element size 500 mm. There is no chance at all that a Mindlin analysis will lead to a reasonable result. At least five elements must be chosen in an edge zone of 160 mm. The value of the twisting moment m_{xy} at the ball support in a Kirchhoff analysis is half the support reaction, so 2,025 N. In the Mindlin analysis of plate 1 we should obtain $m_{xy} = 0$, however the moment becomes 1,448 N. It is less than 2,025 N but not zero. Again, 'neither one thing nor the other'. In plate 2 the element size is 250 mm, still about a factor 1.5 larger than thickness (160 mm), instead of a factor five smaller. Plate 3 has a three times larger thickness (480 mm). The edge zone in which we should refine the mesh is about twice the element size (250 mm). We still must refine the row of elements along the edge substantially in order to get reasonable results. As a matter of fact, we must conclude that plate 3 is hardly a thick plate; the edge zone of 480 mm is about one eighth of the span (4,000 mm), still a rather small zone, the more so if one remembers that the concentration of the edge shear force in fact occurs in the half of the disturbed edge zone. Mindlin has no chance to perform well with the applied mesh, and the result proves it. The corner value of the twisting moment, which should be zero for Mindlin and 2,025 N for Kirchhoff, is somewhere in between (1,212 N). Again the shear force is 'neither one thing nor the other'.

The story for the shear force in the section $x = 500$ mm can be short. For the explanation we refer to Figure 15.5. Near the free edge there is a concentrated shear force of the size of the twisting moment at that point. This value will be a little smaller than the corner value 2,025 N. In the Mindlin analysis this concentrated value is part of the smeared shear force v_x. The concentrated shear force acts in an edge zone of about 160 mm in plate 1 and 2 and in an edge zone of about 480 mm in plate 3. The Mindlin analysis cannot predict these localized shear forces with the applied large element sizes. Comparison of plate 1 to plate 2 shows immense element size dependence. The smoothest result is reached in plate 3, but even this is misleading: the computed maximum shear force has no physical meaning.

15.5 Message of the Chapter

Thin Plates

- Thin plates should preferably be calculated with Kirchhoff theory.
- If Mindlin theory is used for thin plates, this must be done at the cost of a very fine mesh, with results hardly different from Kirchhoff.

- Thin plates with Mindlin and a practical mesh are 'neither fish, flesh nor good red herring'.
- Application of Kirchhoff theory requires an element size not smaller than about plate thickness.
- If Kirchhoff theory is chosen and the FE-program offers the option of a graph for the shear force diagram across a section, also the concentrated edge shear force should be shown.
- If Kirchhoff theory is chosen and the FE-program is able to determine the resultant of shear forces and twisting moments (total force, total torque) over a section, also concentrated edge shear forces must be accounted for. Otherwise equilibrium is violated. This also holds at plate boundaries with edge beams.
- If Kirchhoff theory is chosen and edge beams are applied, the bending moment in the beam is correct, but the shear force must be obtained as the sum of the concentrated edge shear force V_{edge} and the beam shear force V_{beam}.

Thick Plates

- A thick homogeneous isotropic plate must be analyzed by Mindlin theory.
- An edge zone must be chosen of a width equal to about plate thickness, in which a sufficiently fine mesh is applied.
- Sufficiently fine is five or more elements over the edge zone.

Chapter 16
Reinforcement Design Using Linear Analysis

Contributed by Paulo B. Lourenço, Minho University, Guimarães, Portugal

Design of reinforced concrete structures can be described by the following consecutive steps:

1. Select the initial dimensions of all the structural elements using simple rules of thumb or experience. These dimensions should be able to satisfy the serviceability and ultimate limit states, and should fulfill the requirements for adequate site execution and any other applicable requirement (e.g. acoustic isolation, fire protection, etc.).
2. Perform a global structural analysis to calculate the internal forces (or stresses) due to the combination of loads defined in the codes. The method almost used exclusively today is the finite element method and the behaviour of the structure is assumed to be linear elastic at this stage.
3. Verify concrete initial dimensions and calculate the reinforcement capable of resisting the calculated internal forces. At this stage, the ultimate capacity of the individual cross-sections is considered; this which is typically associated with nonlinear constitutive laws.

The main advantage of the above process is that linear elastic finite element analysis is well established and is straightforward to apply. In addition, multiple load cases can be easily incorporated and reinforcement is placed in the locations where tensile stresses appear. These regions correspond to the initial crack locations, helping to control crack propagation.

Of course there are also some disadvantages in the process described: stress redistribution can be difficult to incorporate, providing more expensive reinforcement arrangement; no real information is obtained about the collapse load of the structure, even if a lower bound estimate is obtained when ductility is enforced, and no real information is provided on inelastic

J. Blaauwendraad, *Plates and FEM: Surprises and Pitfalls*, Solid Mechanics and Its Applications 171, DOI 10.1007/978-90-481-3596-7_16,

phenomena as crack width, crack spacing or maximum deflection, even if they can be estimated for beam-type structural elements. A consequence of the process is that detailing guidelines need to be used to ensure ductility and serviceability demands.

Only in very few selected cases of structures with unusual size, shape, or complexity, would a full nonlinear analysis of the previously designed structure be made for assessment, tracing out the entire behaviour through the uncracked, cracked, and ultimate stages. Such an analysis generally requires significant time for pre-processing, computation and post-processing, which is not compatible with cost and time demands. Also, as nonlinear analysis requires the definition of geometry and reinforcement, it should not be regarded as a design tool but, mostly, as an assessment tool.

The simulation of concrete walls, slabs, assemblages of walls and slabs and shells using finite element analysis is becoming a standard in structural analysis tools for building design, meaning that adequate methodologies for the design of these elements are necessary.

In this chapter, we discuss design methods for membrane states (walls), bending states (slabs) and combinations as may occur in spatial assemblages of plates and in shells. Hereafter they are referred to as shells. Design of reinforced concrete elements subjected to membrane states has been developed since 1960s by authors like Baumann, Braestrup and Nielsen, only to name of few. The book of Nielsen is a classical reference [19]. This process resulted in formulas for reinforcement design and check of concrete strength in the CEB-FIB Model Code 1990 for Concrete Structures [20].

Reinforcement design for slabs and shells has also received attention in the Model Code 1990. For that purpose a three layer sandwich model was introduced. Pioneers of this approach are Gupta and Marti. The introductory version of Eurocode 2 suggested a different method based on the normal yield criterion. That result is alternatively referred to as the Wood–Armer equations [21]. It applies to slabs only and not for shells. The later version EN 1992-1-1:2004 of Eurocode 2 [22] removed this method. The version EN 1992-2:2005 of Eurocode 2 again included a solution in its appendices, returning to the three layer sandwich model. This solution has received a place in Part 2 of Eurocode 2 on Bridges [23]. *fib* has published in 2008 a *Practitioners' guide to finite element modeling of reinforced concrete structures* [24]. This document also presents the three layer sandwich model. Readers interested in a more complete review of the historical development of the different methods are referred to [25, 26].

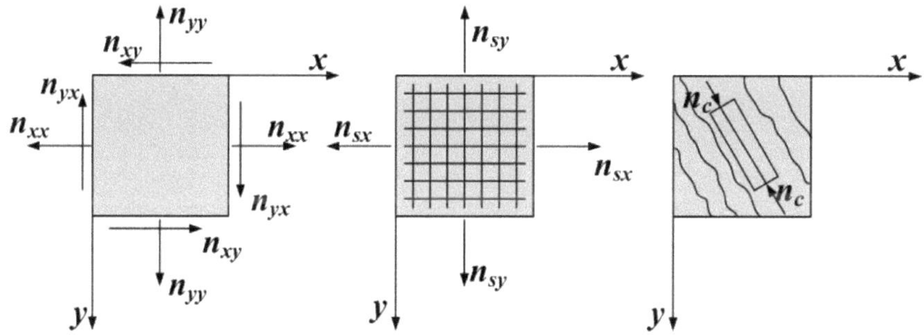

Figure 16.1 Model for membrane element: (a) applied forces; (b) reinforcement contribution; (c) concrete contribution.

Here we will refer to the three layer sandwich model of Eurocode 2 and *fib* practitioners' guide as *basic model*. We call it *basic* because the concept is very useful, but the working-out still requires improvement, because internal lever arms are only approximated and some terms are neglected in the equilibrium equations. After the presentation of the basic model, we introduce an *advanced method* which is consistent with all terms in the equilibrium conditions.

Both the basic and advanced model will first be discussed for cases of moderate transverse shear forces which can be carried by the concrete. After that we make the extension to slabs with larger transverse shear forces; these require transverse shear reinforcement, an extension we owe to Marti [27].

16.1 Design of Membrane States

Consider a membrane element with a thickness h, subjected to applied in-plane forces n_{xx}, n_{yy} and n_{xy}, as shown in Figure 16.1a. The reinforcement consists of two orthogonal sets of rebars parallel to the x, y-axes. a_{sx} and a_{sy} are the needed reinforcement areas per unit length in this co-ordinate system. They are calculated from forces n_{sx} and n_{sy} respectively. The purpose of this section is to find formulas for n_{sx} and n_{sy}.

The applied forces will be resisted by the reinforcement and concrete contributions. It is assumed that the concrete is subjected to uni-axial compression n_c parallel to the cracking orientation, at an angle θ with the y-axis. The two rebar sets in Figure 16.1b and the concrete struts of Figure 16.1c must together carry the applied loads of Figure 16.1a. For the sign convention of the applied n_{xx}, n_{yy} and n_{xy} loads we refer to Chapter 1. The forces n_{sx} and

n_{sy} are always positive or zero, and the membrane force n_c in the concrete is negative or zero.

In the chosen x, y-coordinate system, the shear resistance of the reinforcement is zero and the state of stress of the concrete is uni-axial. The first principal membrane force is zero and the compressive force n_c occurs in the second principal direction. The stress state in Figure 16.1a is equivalent to the combination of the states in Figures 16.1b and 16.1c when the following equilibrium conditions are satisfied:

$$\begin{aligned} n_{xx} &= n_{sx} + n_c \sin^2\theta \\ n_{yy} &= n_{sy} + n_c \cos^2\theta \\ n_{xy} &= -n_c \sin\theta\cos\theta \end{aligned} \tag{16.1}$$

The second principal stress σ_c must be smaller than the compressive strength f_c of concrete

$$n_c \geq -hf_c \tag{16.2}$$

The applied forces are in the left member of Eq. (16.1) and the internal forces are in the right member. It should be remembered that n_c is negative, so n_{sx} is not smaller than n_{xx} and n_{sy} is not smaller than n_{yy}. The cases of $\theta = 0$ and $\theta = \pi/2$ are trivial, meaning that the principal directions are aligned with axes x and y. If $\theta \neq 0$ and $\theta \neq \pi/2$, Eqs. (16.1) to (16.2) can be recast such that the steel and concrete forces are in the left member and the applied forces are in the right:

$$\begin{aligned} n_{sx} &= n_{xx} + n_{xy}\tan\theta \\ n_{sy} &= n_{yy} + n_{xy}\cot\theta \\ n_c &= -\frac{n_{xy}}{\sin\theta\cos\theta} \end{aligned} \tag{16.3}$$

The third expression in Eq. (16.3) indicates that n_{xy} and θ must have the same sign, so that n_c is negative, or in compression. The total amount of reinforcement can be obtained from the first two expressions in Eq. (16.3), and is

$$n_{sx} + n_{sy} = n_{xx} + n_{yy} + n_{xy}(\tan\theta + \cot\theta) \tag{16.4}$$

Note that the last term in this equation is always positive, as n_{xy} and θ have the same sign. Thus, the minimum amount of reinforcement corresponds to $\theta = \pm\pi/4$. For these values of θ, noting that the reinforcement must be always subjected to tension, i.e. $n_{sx} \geq 0$ and $n_{sy} \geq 0$, the first two expressions of Eq. (16.3) gives $n_{xx} \geq -|n_{xy}|$ and $n_{yy} \geq |-n_{xy}|$ respectively. Otherwise the θ value must be changed. Therefore four different cases of reinforcement have to be considered.

Case 1: Reinforcement in X- and Y-Direction Needed

For this case it holds

$$\begin{aligned} &n_{xx} \geq -|n_{xy}|, && n_{yy} \geq -|n_{xy}| \\ &n_{sx} = n_{xx} + |n_{xy}|, && n_{sy} = n_{yy} + |n_{xy}| \\ &\theta = \pm\frac{\pi}{4}, && n_c = -2|n_{xy}| \end{aligned} \tag{16.5}$$

Case 2: Only Reinforcement in Y-Direction Needed

For this case the following equations hold:

$$\begin{aligned} &n_{xx} < -|n_{xy}| \rightarrow n_{sx} = 0 \\ &\tan\theta = -\frac{n_{xx}}{n_{xy}} \\ &n_{sy} = n_{yy} - \frac{n_{xy}^2}{n_{xx}}, \quad n_{sy} \geq 0 \rightarrow n_{yy} \geq \frac{n_{xy}^2}{n_{xx}} \\ &n_c = n_{xx} + \frac{n_{xy}^2}{n_{xx}} \end{aligned} \tag{16.6}$$

Case 3: Only Reinforcement in X-Direction Needed

For this case the following equations hold:

$$\begin{aligned} &n_{yy} < -|n_{xy}| \rightarrow n_{sy} = 0 \\ &\tan\theta = -\frac{n_{xy}}{n_{yy}} \\ &n_{sx} = n_{xx} - \frac{n_{xy}^2}{n_{yy}}, \quad n_{sx} \geq 0 \rightarrow n_{xx} \geq \frac{n_{xy}^2}{n_{yy}} \\ &n_c = n_{yy} + \frac{n_{xy}^2}{n_{yy}} \end{aligned} \tag{16.7}$$

Case 4: No Reinforcement Needed in Any Direction

No cracking occurs and the stress state is biaxial compression. In the concrete two principal membrane forces n_{c1} and n_{c2} are present:

Case	n_{sx}	n_{sy}	n_c
$n_{xx} \geq -\lvert n_{xy}\rvert$ $n_{yy} \geq -\lvert n_{xy}\rvert$	$n_{xx} + \lvert n_{xy}\rvert$	$n_{yy} + \lvert n_{xy}\rvert$	$-2\lvert n_{xy}\rvert$
$n_{xx} < -\lvert n_{xy}\rvert$ $n_{yy} \geq \frac{n_{xy}^2}{n_{xx}}$	0	$n_{yy} - \frac{n_{xy}^2}{n_{xx}}$	$n_{xx} + \frac{n_{xy}^2}{n_{xx}}$
$n_{xx} \geq \frac{n_{xy}^2}{n_{yy}}$ $n_{yy} < -\lvert n_{xy}\rvert$	$n_{xx} - \frac{n_{xy}^2}{n_{yy}}$	0	$n_{yy} + \frac{n_{xy}^2}{n_{yy}}$
other	0	0	$\frac{n_{xx}+n_{yy}}{2} - \sqrt{\left(\frac{n_{xx}+n_{yy}}{2}\right)^2 + n_{xy}^2}$
$a_{sx} = \frac{n_{sx}}{f_{syd}}$		$a_{sy} = \frac{n_{sy}}{f_{syd}}$	$f_c = \frac{n_c}{h} \leq f_{cd}$

Figure 16.2 Overview of four cases for membrane reinforcement.

$$
\begin{cases} n_{xx} < -|n_{xy}| \\ n_{yy} < \dfrac{n_{xy}^2}{n_{xx}} \end{cases} \quad \text{or} \quad \begin{cases} n_{yy} < -|n_{xy}| \\ n_{xx} < \dfrac{n_{xy}^2}{n_{yy}} \end{cases}
$$

$$
n_{sx} = 0, \quad n_{sy} = 0 \tag{16.8}
$$

$$
n_{c1,c2} = \frac{n_{xx} + n_{yy}}{2} \pm \sqrt{\left(\frac{n_{xx} - n_{yy}}{2}\right)^2 + n_{xy}^2}
$$

Rebar Design and Check on Concrete Stress

The four cases are summarized in Figure 16.2. The formulas correspond to the optimum direction of concrete compression, i.e., the θ value leading to the minimum amount of reinforcement. The reinforcement design is obtained from

$$
a_{sx} = \frac{n_{sx}}{f_{syd}}, \quad a_{sy} = \frac{n_{sy}}{f_{syd}} \tag{16.9}
$$

where a_{sx} and a_{sy} are steel areas per unit length and f_{syd} is the design yield strength of the reinforcement. The concrete stress is given by

$$f_c = -n_c/h \tag{16.10}$$

which must be checked against the design compressive strength f_{cd}. For this strength we can apply the Model Code 1990 or the Practitioners guide of *fib*.

Model Code 1990

The Model Code 1990 recommends

$$\begin{array}{ll} \text{Case 1 to 3} & f_c \leq f_{cd2} \\ \text{Case 4} & f_c \leq K\ f_{cd1} \end{array} \tag{16.11}$$

where

$$\begin{aligned} f_{cd1} &= 0.85\left[1 - \frac{f_{ck}}{250}\right] f_{cd} \\ f_{cd2} &= 0.60\left[1 - \frac{f_{ck}}{250}\right] f_{cd} \\ K &= \frac{1 + 3.65\alpha}{(1+\alpha)^2}, \quad \alpha = \frac{\sigma_2}{\sigma_1} \end{aligned} \tag{16.12}$$

Here f_{cd} is the design strength of the concrete, f_{ck} is the characteristic strength of the concrete, and σ_1 and σ_2 are the two principal compressive stresses. These formulas are based on experimental studies on biaxial concrete behaviour of Kupfer.

fib Practitioners' Guide

The practitioners'guide of *fib* recommends following planned changes to the ACI code. The proposed formula for the concrete strength is

$$f_{cd} = 0.85\beta f_{ck}/\gamma_c \tag{16.13}$$

where the factor 0.85 accounts for the variation between the *in-situ* and cylindrical strengths, β accounts for influence of transverse tensile strain, f_{ck} is the characteristic compression strength, and γ_c is the partial safety factor. The formula for β is

$$\beta = \frac{1}{0.8 + 170\varepsilon_1} \tag{16.14}$$

Herein ε_1 is the major principal strain normal to the direction of the concrete struts. For this strain, the yield strain of the steel reinforcement might be chosen, so $\varepsilon_1 = f_{syd}/E$, where E is the Young modulus of steel.

Remark

At this point, it should be pointed out that the discontinuous variation of concrete compressive strength between Cases 3 and 4, or between Cases 2 and 4 does not seem acceptable. This gains special relevance as the expression for f_{cd2} in Eq. (16.12) corresponds practically to an absolute minimum of cracked reinforced concrete. However, this seems to be the price to pay for a simplified design approach.

16.2 Design of Slabs – Normal Moment Yield Criterion

As in an element in membrane state, dimensioning of slabs and shells from internal forces obtained in a finite element analysis is based on an equilibrium model at ultimate state. While careful consideration of the limited ductility of concrete is important in the dimensioning of membrane elements, such a concern is lower for slabs because such structures are typically under-reinforced. Failure is usually governed by yielding of reinforcement, with the exception of point loads, which may result in brittle punching failures in slabs and in shells without transverse reinforcement.

The stress resultants acting in a slab are the bending moments m_{xx} and m_{yy} and twisting moments m_{xy}. For the derivation of the design equations a set of orthogonal axes is chosen in directions x and y, giving moments per unit length m_{xx}, m_{yy} and m_{xy}, such that $m_{yy} > m_{xx}$. The normal moment yield criterion requires that at any point within the element under consideration the moment capacity m_n^* is greater than the applied normal moment m_n for all values of θ, the orientation of the plane measured in the clockwise direction from the x-axis. The resulting formulas for the design moments are presented in Figure 16.4. Reinforcement is provided in the x- and y-directions to resist design ultimate moments m_{xb}, m_{xt}, m_{yb} and m_{yt}. The subscripts b and t indicate bending moments giving tension in the slab bottom and slab top, respectively. The bottom is at the positive z-side of the slab middle plane, and

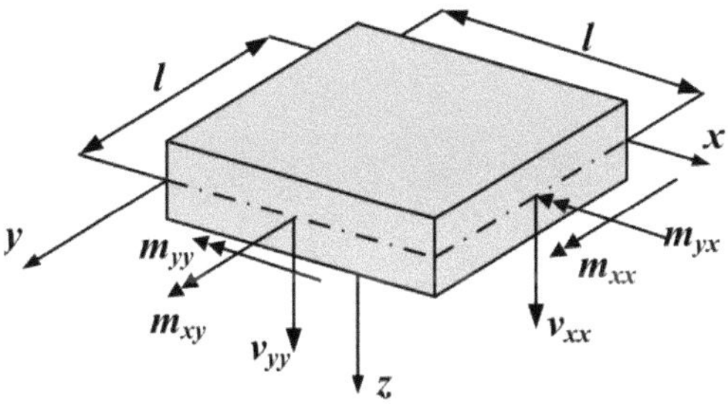

Figure 16.3 Stress resultants acting in a slab.

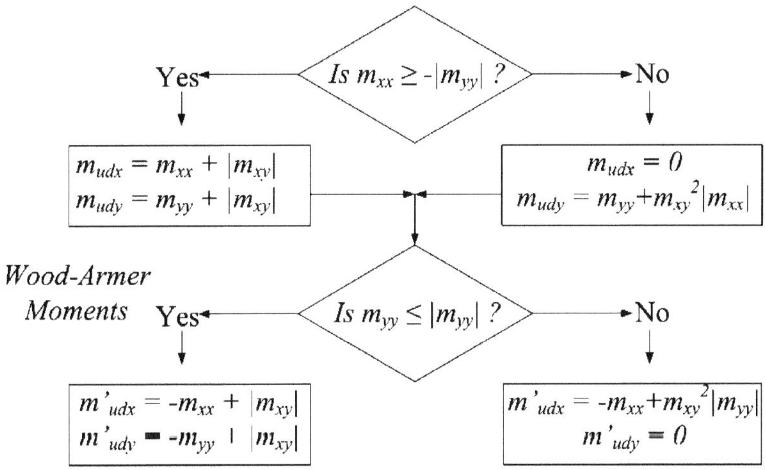

Figure 16.4 Reinforcement for slabs according to ENV 1992-1-1.

the top at the negative side. The shown equations are used in many software packages for slab reinforcement design. Often, only the top-left corner of the figure is used.

Evaluation

The use of the equations of Figure 16.4 is *discouraged* for a number of reasons. The equations are not able to take into consideration transverse shear forces, do not check for concrete crushing, and do not fulfill equilibrium requirements. We strongly recommend using the three layer sandwich model, which applies for slabs and shells. This is the subject of Sections 16.3 and 16.4.

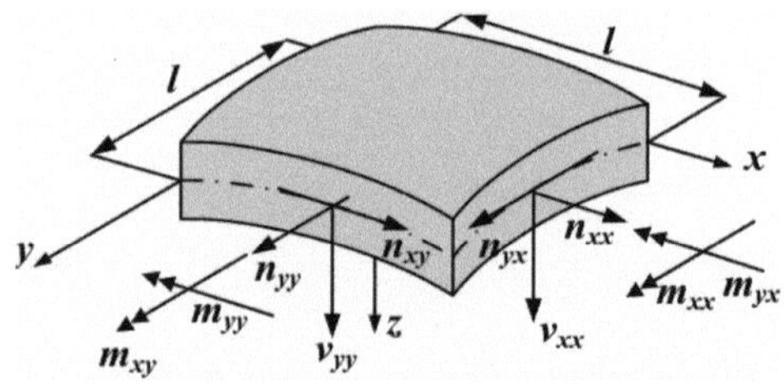

Figure 16.5 Stress resultants acting on a shell element.

16.3 Slab and Shell Elements – Basic Model

The problem to be discussed in this section is the design of a shell element of thickness h, subjected to combined membrane forces and bending moments and where the directions of the principal flexural and membrane forces do not, in general, coincide. A slab element is a special case of the stated problem. Figure 16.5 shows the applied forces and moments. These forces and moments have to be in equilibrium with the tensile forces in the reinforcement and the compressive forces in the concrete. We choose a set of x,y,z-axes as we did in Chapters 3 and 4, where z is pointing downward, see Figure 16.5. The reinforcement consists again of a mesh of orthogonal rebars parallel to the x,y-axes, now placed in a upper and lower layer. We refer to the upper or top layer by the subscript t (negative z-side) and to the lower or bottom layer by b (positive z-side).

The formulation of this problem is identical to that in Section 16.1 for membrane states. Again, the total resistance of the element is obtained by adding the concrete and reinforcement contributions. We model the shell element as a three layer sandwich, shown in Figure 16.6. The outer layers are covers of the sandwich and the inner layer is the core. The cover layers provide resistance to the in-plane effects of flexure and membrane loading, while the core provides a shear transfer between the covers. The thickness of the covers is a and the distance between the middle planes of the covers is d_v.

Dependent on its size, the transverse shear force has an impact on the amount of the reinforcement in the covers. Small values have no impact, large values do, To decide whether the shear forces are small we must consider the maximum shear force v_o as specified in Eq. (4.24).

$$v_o = \sqrt{v_x^2 + v_y^2} \tag{16.15}$$

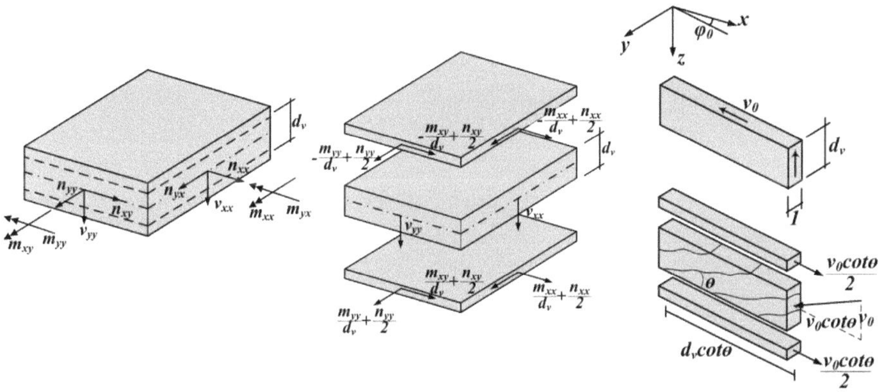

Figure 16.6 Sandwich model: (a) shell element; (b) layer forces; (c) transverse shear transfer in uncracked and cracked concrete, Marti (1991).

which acts in the direction of an angle β_o with the x-axis, defined by Eq. (4.23).

$$\tan\beta_o = \frac{v_y}{v_x} \tag{16.16}$$

This shear force is small if it is below the shear cracking resistance d_v, $\tau_{c,\text{red}}$, where $\tau_{c,\text{red}}$ is the nominal strength of the slab without transverse reinforcement. Then the core will remain uncracked. For the value $\tau_{c,\text{red}}$, we may apply ENV 1992-1-1, which provides

$$\tau_{c,\text{red}} = 0.25 f_{ctd}(1.6 - d_v)(1.2 + 40\rho_l) + 0.15\sigma_{cp} \tag{16.17}$$

Here, f_{ctd} is the design tensile strength, d_v is the internal lever arm in meters, ρ_l is the percentage of longitudinal reinforcement, and σ_{cp} is the in-plane normal compressive stress. If significant tensile membrane forces are applied to the element, $t_{c,\text{red}}$ should be taken to be zero. Provided that no significant tensile membrane forces exist, the expression in Eq. (16.17) can be simplified to a lower bound, neglecting the positive effect of the longitudinal reinforcement.

$$\tau_{c,\text{red}} = 0.30 f_{ctd}(1.6 - d_v) \tag{16.18}$$

16.3.1 Basic Model – No Cracking Due to Transverse Shear

We start with small shear forces. The core layer is supposed not to crack and to be able to carry transverse shear forces. Figure 16.6b depicts the sandwich model for this case. The need for reinforcement needs to be investigated only for the combination of membrane forces and bending and twisting moments. It is important to decide what thicknesses are assigned to the top and bottom layers. In the basic model these thicknesses are not explicitly considered. It is assumed that all reinforcement layers coincide in each outer sandwich layer, which also coincides with the resultants of the concrete compression forces. Therefore one lever arm d_v applies for both directions x and y. The membrane forces in the external layers are given by

$$\begin{aligned} n_{xxt} &= \frac{m_{xx}}{d_v} + \frac{n_{xx}}{2}, \quad n_{xxb} = -\frac{m_{xx}}{d_v} + \frac{n_{xx}}{2} \\ n_{yyt} &= \frac{m_{yy}}{d_v} + \frac{n_{yy}}{2}, \quad n_{yyb} = -\frac{m_{yy}}{d_v} + \frac{n_{yy}}{2} \\ n_{xyt} &= \frac{m_{xy}}{d_v} + \frac{n_{xy}}{2}, \quad n_{xyb} = -\frac{m_{xy}}{d_v} + \frac{n_{xy}}{2} \end{aligned} \tag{16.19}$$

Using the expressions provided above for the cover membrane elements, we can obtain the final expressions for the forces per unit width for the reinforcement design in shells (assuming that all four reinforcements are needed).

$$\begin{aligned} n_{sxt} &= \frac{m_{xx}}{d_v} + \frac{n_{xx}}{2} + \left| \frac{m_{xy}}{d_v} + \frac{n_{xy}}{2} \right| \\ n_{syt} &= \frac{m_{yy}}{d_v} + \frac{n_{yy}}{2} + \left| \frac{m_{xy}}{d_v} + \frac{n_{xy}}{2} \right| \\ n_{sxt} &= -\frac{m_{xx}}{d_v} + \frac{n_{xx}}{2} + \left| -\frac{m_{xy}}{d_v} + \frac{n_{xy}}{2} \right| \\ n_{syt} &= -\frac{m_{yy}}{d_v} + \frac{n_{yy}}{2} + \left| -\frac{m_{xy}}{d_v} + \frac{n_{xy}}{2} \right| \end{aligned} \tag{16.20}$$

From these forces we can derive reinforcement percentages:

$$\rho_{xt} = \frac{n_{sxt}}{hf_{syd}}, \quad \rho_{yt} = \frac{n_{syt}}{hf_{syd}}, \quad \rho_{xb} = \frac{n_{sxb}}{hf_{syd}}, \quad \rho_{yb} = \frac{n_{syb}}{hf_{syd}} \tag{16.21}$$

where f_{syd} is the design yield stress of steel.

16.3.2 Basic Model – Cracking Due to Transverse Shear

If the transverse shear forces are high enough to produce cracking of the sandwich core, additional reinforcement is required. Here we follow the approach as proposed by Marti [27]. An alternative proposal can be found in

Eurocode 2 [23]. The core is treated like the web of a girder of a flanged cross-section running in the β_o-direction of the maximal shear force. Figure 16.6c shows that concrete struts in the core come into being at an angle θ with the middle plane. To ensure equilibrium, additional membrane forces must occur in the upper and lower cover. Choosing $\theta = 45°$ leads to additional membrane forces in both covers of size v_o in the direction of the maximal shear force. The choice of 45° for the crack angle in the core conforms to the traditional Morsch truss for reinforced concrete beams. Decomposing the additional membrane force in the covers to membrane forces in the x- and y-direction leads to the following expressions:

$$\begin{aligned}
n_{xxt} &= \frac{m_{xx}}{d_v} + \frac{n_{xx}}{2} + \frac{v_x^2}{2v_o}, \quad & n_{xxb} &= -\frac{m_{xx}}{d_v} + \frac{n_{xx}}{2} + \frac{v_x^2}{2v_o} \\
n_{yyt} &= \frac{m_{yy}}{d_v} + \frac{n_{yy}}{2} + \frac{v_y^2}{2v_o}, \quad & n_{yyb} &= -\frac{m_{yy}}{d_v} + \frac{n_{yy}}{2} + \frac{v_y^2}{2v_o} \\
n_{xyt} &= \frac{m_{xy}}{d_v} + \frac{n_{xy}}{2} + \frac{v_x v_y}{2v_o}, \quad & n_{xyb} &= -\frac{m_{xy}}{d_v} + \frac{n_{xy}}{2} + \frac{v_x v_y}{2v_o}
\end{aligned} \tag{16.22}$$

Using these expressions for membrane elements, we find the final expressions for the design of reinforcement for shells (assuming that all four reinforcements are needed):

$$\begin{aligned}
n_{sxt} &= \frac{m_{xx}}{d_v} + \frac{n_{xx}}{2} + \frac{v_x^2}{2v_o} + \left| \frac{m_{xy}}{d_v} + \frac{n_{xy}}{2} + \frac{v_x v_y}{2v_o} \right| \\
n_{syt} &= \frac{m_{yy}}{d_v} + \frac{n_{yy}}{2} + \frac{v_y^2}{2v_o} + \left| \frac{m_{xy}}{d_v} + \frac{n_{xy}}{2} + \frac{v_x v_y}{2v_o} \right| \\
n_{sxt} &= -\frac{m_{xx}}{d_v} + \frac{n_{xx}}{2} + \frac{v_x^2}{2v_o} + \left| -\frac{m_{xy}}{d_v} + \frac{n_{xy}}{2} + \frac{v_x v_y}{2v_o} \right| \\
n_{syt} &= -\frac{m_{yy}}{d_v} + \frac{n_{yy}}{2} + \frac{v_y^2}{2v_o} + \left| -\frac{m_{xy}}{d_v} + \frac{n_{xy}}{2} + \frac{v_x v_y}{2v_o} \right|
\end{aligned} \tag{16.23}$$

From these forces, we derive reinforcement percentages by using the formulas in (16.21). Transverse reinforcement is needed with a percentage ρ_z given by

$$\rho_z = \frac{v_o}{d_v f_{syd}} \tag{16.24}$$

In practical problems we recommend increasing the slab or shell thickness so that transverse reinforcement is avoided.

16.3.3 Evaluation

The basic sandwich model is simple to apply, but definitely is an approximation to reality. We mention the following:

- It is assumed that the core does not contribute to transferring membrane forces. Compatibility requirements show that this cannot be correct.

It is assumed that both reinforcement layers in an outer layer are positioned in the middle plane of the cover; this is physically impossible, and that this middle plane coincides with the resultant of the respective compression force in the concrete, which is not normally the case.

The angle β_0 for the cracks in the core due to the transverse shear force has been tacitly assumed to be $\pm 45°$.

Even if global equilibrium is satisfied, several terms associated with different lever arms of all forces are not considered; this means that deviance from reality tends to increase for higher reinforcement percentages or large twisting moments. In these cases, the thickness of the concrete layer tends to increase and the basic model is unsafe.

In Section 16.4 we present an advanced sandwich model. It also starts from the supposition of a three layer sandwich model and division of force transfer, such that the covers carry the membrane forces, bending and twisting moments and the core carries the shear forces. For the rest the shortcomings of the basic model are fully repaired. We assign its own plane to each reinforcement layer, permit the thickness of covers to be adequately calculated and rigorously consider the missing terms. So a consistent set of hypotheses lays the foundation of the advanced model.

16.4 Formulation of the Advanced Three-layer Model

In the consistent model, the internal lever d_v is not assumed *a priori* and it is not equal in all directions, being calculated using an iterative process. Four different cases must be analyzed and treated separately: (a) reinforcement needed in both outer layers; (b) reinforcement needed only in the bottom layer; (c) reinforcement needed only in the top layer; (d) no need for reinforcement. The complete formulation of the problem, the software code and validation can be found in [25,26]. The described phenomena are simple but the resulting equations are reasonably complex, leading to an indeterminate system of nonlinear equations.

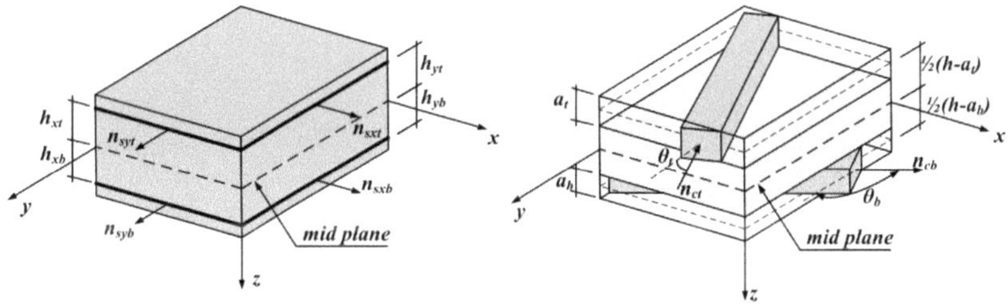

Figure 16.7 Definition of advanced sandwich model.

The geometry of the advanced model is shown in Figure 16.7. We introduce different distances h_{xt}, h_{yt}, h_{xb} and h_{yb} for the four reinforcement layers to the middle plane of the slab. The thicknesses of the outer layers are a_t and a_b, respectively. The core between these layers has thickness h_c. As for the basic model, we define resisting reinforcement forces n_{sxt}, n_{syt}, n_{sxb} and n_{syb}. The two forces for the x-direction are summed to n_{sx} and for the y-direction to n_{sy}. Correspondingly resisting reinforcement moments m_{sx} and m_{sy} are defined. For the concrete top and bottom layer we introduce resisting forces n_{ct} and n_{cb}, respectively, and resisting concrete moments m_{ct} and m_{cb}. Here, subscripts s and c indicate steel and concrete, respectively, and subscripts t and b indicate again top and bottom external layer, respectively.

Case 1: Reinforcement in Both Outer Layers

If reinforcement is needed in the outer layers, the resisting forces and moments for the reinforcement in the x- and y-directions are given by

$$\begin{aligned} n_{sx} &= n_{sxt} + n_{sxb} \\ n_{sy} &= n_{syt} + n_{syb} \end{aligned} \tag{16.25}$$

$$\begin{aligned} m_{sx} &= -n_{sxt} h_{xt} + n_{sxt} h_{xb} \\ m_{sy} &= -n_{syt} h_{yt} + n_{syt} h_{yb} \end{aligned} \tag{16.26}$$

and for the concrete by

$$\begin{aligned} n_{ct} &= -a_t f_c \\ n_{cb} &= -a_b f_c \end{aligned} \tag{16.27}$$

$$
\begin{aligned}
m_{ct} &= -\frac{1}{2}\,(h - a_t)\, n_{ct} \\
m_{cb} &= \frac{1}{2}\,(h - a_b)\, n_{cb}
\end{aligned}
\tag{16.28}
$$

Equations (16.25)–(16.27) provide the internal forces and moments. Equilibrium with the applied set of forces and moments leads to

$$
\begin{aligned}
n_{xx} &= n_{sx} + n_{ct} \sin^2\theta_t + n_{cb} \sin^2\theta_b \\
n_{yy} &= n_{sy} + n_{ct} \cos^2\theta_t + n_{cb} \cos^2\theta_b \\
n_{xy} &= -n_{ct} \sin\theta_t \cos\theta_t - n_{cb} \sin\theta_b \cos\theta_b
\end{aligned}
\tag{16.29}
$$

$$
\begin{aligned}
m_{xx} &= m_{sx} + m_{ct} \sin^2\theta_t + m_{cb} \sin^2\theta_b \\
m_{yy} &= m_{sy} + m_{ct} \cos^2\theta_t + m_{cb} \cos^2\theta_b \\
m_{xy} &= -m_{ct} \sin\theta_t \cos\theta_t - m_{cb} \sin\theta_b \cos\theta_b
\end{aligned}
\tag{16.30}
$$

Equation (16.29) correspond to the membrane forces, while Eq. (16.32) correspond to bending equations. If $\theta_t \neq 0, \pi/2$ and $\theta_b \neq 0, \pi/2$, the previous equations can be recast as

$$
\begin{aligned}
-n_{ct} &= \frac{(h - a_t)n_{xy} - 2m_{xy}}{h_c \sin 2\theta_t} \\
-n_{cb} &= \frac{(h - a_b)n_{xy} + 2m_{xy}}{h_c \sin 2\theta_b}
\end{aligned}
\tag{16.31}
$$

Reinforcement will be given upon solving Eqs. (16.25) to (16.30). The objective is to calculate the forces in the reinforcement n_{sxt}, n_{syt}, n_{sxb} and n_{syb}. The other unknowns are a_t, a_b, θ_t and θ_b. Therefore the system of six equations contains eight unknowns. This means that the values of θ_t and θ_b should be chosen so that the total amount of reinforcement is minimized. The values of $\theta_t = \theta_b = \pi/4$ and $a_t = a_b = 0.2h$ can be assumed as an initial guess. Setting the values of θ to $\pi/4$ minimizes the total reinforcement in membrane elements. Setting $a = 0.2h$ is usual for beam sections. The values are then adjusted by an iterative procedure until equilibrium is fulfilled. The reader is referred to [25,26] for a full description of the iterative method.

Compressive crushing is checked by enforcing that $a_t + a_b \leq h$ and tensile reinforcement is calculated assuming yielding of the reinforcement.

Case 2: Reinforcement in Bottom Layer Only

For biaxial compression in the top layer, reinforcement in the top layer is not needed. We indicate the concrete top layer membrane forces by n_{cxxt}, n_{cyyt} and n_{cxyt}. The forces and moments that the reinforcement resists in the x, y-directions are given by

$$\begin{aligned} n_{sx} &= n_{sxb} \\ n_{sy} &= n_{syb} \\ m_{sx} &= n_{sxb}h_{xb} \\ m_{sy} &= n_{syb}h_{yb} \end{aligned} \tag{16.32}$$

and by the concrete bottom layer are

$$\begin{aligned} n_{cb} &= -a_b f_c \\ m_{cb} &= \frac{1}{2}(h - a_b)n_{cb} \end{aligned} \tag{16.33}$$

Equilibrium with the applied set of forces and moments yields

$$\begin{aligned} n_{xx} &= n_{sx} + n_{cxt} + n_{cb}\sin^2\theta_b \\ n_{yy} &= n_{sy} + n_{cyt} + n_{cb}\cos^2\theta_b \\ n_{xy} &= n_{cxyt} - n_{cb}\sin\theta_b\cos\theta_b \end{aligned} \tag{16.34}$$

$$\begin{aligned} m_{xx} &= m_{sx} + m_{cxt} + m_{cb}\sin^2\theta_b \\ m_{yy} &= m_{sy} + m_{cyt} + m_{cb}\cos^2\theta_b \\ m_{xy} &= m_{cxyt} - m_{cb}\sin\theta_b\cos\theta_b \end{aligned} \tag{16.35}$$

with

$$\begin{aligned} m_{cxt} &= -\frac{1}{2}(h - a_t)\,n_{cxt} \\ m_{cyt} &= -\frac{1}{2}(h - a_t)\,n_{cyt} \\ m_{cxyt} &= -\frac{1}{2}(h - a_t)\,n_{cxyt} \end{aligned} \tag{16.36}$$

In the current case there are still eight unknowns. However, one extra equation must be added to the six equations of equilibrium, representing the biaxial state of stress in the concrete top layer

$$n_{ct} = -a_t f_c = \frac{n_{cxt} + n_{cyt}}{2} - \sqrt{\left(\frac{n_{cxt} + n_{cyt}}{2}\right)^2 + n_{cxyt}^2} \tag{16.37}$$

Here, f_c has a higher value than the uni-axial compressive strength of cylinders due to biaxial confinement. Nevertheless there are eight unknowns and seven equations, meaning that θ_b should be chosen so that the total amount of reinforcement is minimized.

Case 3: Reinforcement in Top Layer Only

The case of biaxial compression in the bottom layer is identical to the case of biaxial compression in the top layer, with a rotation of indices. Therefore establishing the equilibrium equations requires no additional explanation.

Case 4: No Reinforcement at All

Finally, in the case of biaxial compression in top and bottom layers, there is no need of reinforcement and the solution is unique. Assuming that the concrete top layer membrane forces are $n_{cxt,ncyt}$ and n_{cxyt} respectively in the x, y-direction and as shear force, and the concrete bottom layer membrane forces are n_{cxb}, n_{cyb} and n_{cxyb} with a similar meaning, the equilibrium equations might be written as

$$\begin{aligned} n_{xx} &= n_{cxt} + n_{cxb} \\ n_{yy} &= n_{cyt} + n_{cyb} \\ n_{xy} &= n_{cxyt} + n_{cxyb} \end{aligned} \tag{16.38}$$

$$\begin{aligned} m_{xx} &= m_{cxt} + m_{cxb} \\ m_{yy} &= m_{cyt} + m_{cyb} \\ m_{xy} &= m_{cxyt} + m_{cxyb} \end{aligned} \tag{16.39}$$

with

$$\begin{aligned} m_{cxt} &= -\frac{1}{2}(h - a_t)n_{cxt} \quad & m_{cxb} &= \frac{1}{2}(h - a_b)n_{cxb} \\ m_{cyt} &= -\frac{1}{2}(h - a_t)n_{cyt} \quad & m_{cyb} &= \frac{1}{2}(h - a_b)n_{cyb} \\ m_{cxyt} &= -\frac{1}{2}(h - a_t)n_{cxyt} \quad & m_{cxyt} &= \frac{1}{2}(h - a_b)n_{cxyt} \end{aligned} \tag{16.40}$$

The principal concrete compression forces in each layer may be calculated according to

$$
\begin{aligned}
n_{c,t} &= \frac{n_{cxt}+n_{cyt}}{2} \pm \sqrt{\left(\frac{n_{cxt}-n_{cyt}}{2}\right)^2 + n_{cxyt}^2} \\
n_{c,b} &= \frac{n_{cxb}+n_{cyb}}{2} \pm \sqrt{\left(\frac{n_{cxb}-n_{cyb}}{2}\right)^2 + n_{cxyb}^2}
\end{aligned}
\tag{16.41}
$$

and the layer thickness may be calculated according to the MC90 as

$$
a_t = -\frac{n_{ct,\max}}{Kf_{cd1}}, \quad a_b = -\frac{n_{cb,\max}}{Kf_{cd1}} \tag{16.42}
$$

As shown above, there are eight unknowns and eight equations (the six equilibrium equations and two equations to check the maximum compressive stress in the layers), meaning that the problem is determinate.

16.5 Applications on Element Level

In this section we illustrate the use of the basic and advanced model for two elements. The first one is subjected to a combination of a membrane force and bending moment. The second is a slab element subjected to a twisting moment.

16.5.1 Element with Membrane Force and Bending Moment

An element is subjected to an applied set of a bending moment and membrane shear force given by $m_{xx} = 235$ kNm/m and $n_{xy} = 1{,}806$ kN/m. The material properties of concrete and steel are $f_c = 41.8$ MPa and $f_{sy} = 492$ MPa. The location of the reinforcement is given by $h_{xt} = h_{xb} = 0.122$ m and $h_{yt} = h_{yb} = 0.100$ m. This element is chosen because an experimental result of Kirsher and Collins is available for the purpose of comparison [28]. The bending moment and membrane shear force mentioned above are the ultimate loads in the test. We will start from this load and determine how much reinforcement is needed.

The top row in Table 16.1 shows that in total 111.4 cm^2/m reinforcement is applied in the element in the test. Not all the reinforcement yielded at

Table 16.1 Reinforcement for membrane shear + bending moment.

	Reinforcement areas (cm^2/m)				
Method	x upper	y upper	x lower	y lower	Total
Experiment	41.8	13.9	41.8	13.9	111.4
Nonlinear	0.0	14.1	37.6	16.9	68.6
Basic method	0.0	15.7	39.9	18.4	74.0
Advanced method	0.0	16.6	36.8	17.9	71.3

failure. The second row in the table is the prediction of the needed reinforcement on the basis of a nonlinear analysis by an iterative computer program with optimization [29]. This provided a minimum amount of reinforcement equal to 68.6 cm^2/m. The third row presents the results of the basic sandwich model and the fourth row of the advanced model. So we use the two rows 'Experiment' and 'Nonlinear' are benchmark references for the basic and advanced method.

Basic Model

For the basic model an average distance of layer centres to the middle plane of the element of 0.111 m is chosen. Therefore $d_v = 0.222$ m. On the basis of Eq. (16.19) we find the following

$$n_{xxb} = 235/0.222 = 1{,}059 \text{ kN/m},$$

$$n_{xxt} = -235/0.222 = -1{,}059 \text{ kN/m},$$

$$n_{yyb} = 0,$$

$$n_{yyt} = 0,$$

$$n_{xyb} = 1{,}806/2 = 903 \text{ kN/m},$$

$$n_{xyt} = 1{,}806/2 = 903 \text{ kN/m}.$$

Using the expressions for membrane elements it is possible to obtain, for the top layer (x reinforcement not needed)

$$n_{sxt} = 0,$$

$$n_{syt} = n_{yyt} - \frac{n_{xyt}^2}{n_{xxt}} = 0 - \frac{903^2}{-1{,}059} = 770 \text{ kN/m}$$

$$n_{ct} = -1{,}059 + \frac{903^2}{-1{,}059} = -1{,}829 \text{ kN/m}$$

$$a_{sxt} = 0$$

$$a_{syt} = \frac{n_{syt}}{f_{sy}} = \frac{770}{492 \times 10^3} \times 10^4 = 15.7 \text{ cm}^2/\text{m}$$

$$a_t = -\frac{n_{ct}}{f_c^{\text{eff}}} - \frac{-1{,}829}{0.6 \times 41.8 \times 10^3} = 0.073 \text{ m}$$

Note that the value of the effective compressive strength was here assumed as $0.6 f_c$. Similarly, for the bottom layer

$$n_{sxb} = n_{xxb} + \left|n_{xyb}\right| = 1{,}059 + 903 = 1{,}962 \text{ kN/m}$$

$$n_{syb} = n_{xyb} + \left|n_{xyb}\right| = 0 + 903 = 903 \text{ kN/m}$$

$$n_{cb} = -2\left|n_{xyb}\right| = -2 \times 903 = 1{,}806 \text{ kN/m}$$

$$a_{sxb} = \frac{n_{sxb}}{f_{sy}} = \frac{1{,}962}{492 \times 10^3} \times 10^4 = 39.9 \text{ cm}^2/\text{m}$$

$$a_{sxt} = \frac{n_{sxt}}{f_{sy}} = \frac{903}{492 \times 10^3} \times 10^4 = 18.4 \text{ cm}^2/\text{m}$$

$$a_b = -\frac{-1{,}806}{0.6 \times 41.8 \times 10^3} = 0.072 \text{ m}$$

Advanced Model

The advanced sandwich model requires five iterations, provides the thickness of the layers equal to 0.072 m and 0.075 m for the top and bottom respectively, and the reinforcement results as given in the third row of Table 16.1. It can be seen that the results are almost the same as the basic sandwich model. If the results of the nonlinear analysis are assumed as reference values, the basic sandwich model provides +8% and the advanced sandwich model provides +4% of the total reinforcement.

16.5.2 Slab Element with Twisting Moment

A slab element is subjected to pure torsion by an applied twisting moment. The value of the twisting moment is one time chosen $m_{xy} = 42.5$ kNm/m

Table 16.2 Reinforcement for small twisting moment.

	Reinforcement areas (cm^2/m)				
Method	x upper	y upper	x lower	y lower	Total
Experiment	5.0	5.0	5.0	5.0	20.0
Normal moment	5.0	5.3	5.0	5.3	20.6
Basic model	5.6	5.6	5.6	5.6	23.4
Advanced model	5.0	5.0	5.0	5.0	20.0

and one time $m_{xy} = 101.5$ kNm/m. These values are chosen because results of a test. Marti, Leesti and Khalifa obtained them for a lightly reinforced (0.25%) and a severely reinforced (1.0%) element, respectively [30]. We will use these moments as load and determine the needed reinforcement. The material properties are $f_c = 44.4$ MPa, and $f_{sy} = 479$ MPa for the light reinforcement and $f_{sy} = 412$ MPa for the severe reinforcement. The location of the reinforcement is given by $h_{xt} = h_{xb} = 0.073$ m and $h_{yt} = h_{yb} = 0.084$ m, for the light one, and $h_{xt} = h_{xb} = 0.066$ m and $h_{yt} = h_{yb} = 0.082$ m, for the severe one.

Table 16.2 regards the small twisting moment. The line 'Experiment' lists the reinforcement existing in the tested element, which is the same in x- and y-direction, and in the top and bottom layer (5 cm^2/m). The second line in the table regards the normal moment method. This violates equilibrium as different reinforcements are calculated for each direction. This is in agreement with the formulation, as different lever arms are found for each reinforcement direction, but equilibrium requires the forces in all reinforcements to be the same. The basic and advanced sandwich models fulfill equilibrium correctly. The normal moment method provides a reasonable (conservative) value of reinforcement for the small twisting moment. The reinforcement is in x-direction about 27% less than the required value and 19% less in y-direction. The basic model is very safe for the small moment. The prediction by the advanced sandwich model is exact for the small moment.

Table 16.3 shows the result for the large twisting moment. In the experiment 20 mm^2/m was used for all four reinforcements. The normal moment method leads to different amounts in x- and y-direction, underestimates the reinforcement in x-direction by 27% and the total by 24%. The basic sandwich model leads to equal amounts, however 19% too small. The advanced sandwich model performs very satisfactory with only 3% difference with the amounts in the test.

Table 16.3 Reinforcement for large twisting moment.

	Reinforcement areas (cm^2/m)				
Method	x upper	y upper	x lower	y lower	Total
Experiment	20.0	20.0	20.0	20.0	80.0
Normal moment	14.6	16.3	14.6	16.3	61.8
Basic model	16.2	16.2	16.2	16.2	64.8
Advanced model	19.4	19.4	19.4	19.4	77.6

The reason for the bad predictions is that the location of the resultant for the forces in the concrete are incorrectly calculated and the interaction between the different forces in reinforcement and concrete are neglected. Therefore, the equations on basis of the normal moment yield criterion and the basic sandwich model should be used with much precaution, or not used at all. Obtained results from these models must be distrusted if high reinforcement ratios are obtained.

16.6 Applications on Structural Level

16.6.1 Deep Beam

We study the one-span deep beam with the geometry and loading illustrated in Figure 16.8(a), borrowed from the special report 'Toward a Consistent Design of Structural Concrete' of the *PCI Journal* [31]. The 0.4 m thick beam carries the axial load of two columns, with 0.5 by 0.4 m^2 cross-section. The span exhibits an opening that modifies the normal deep beam behaviour. The loads specified in Figure 16.8(a) are design loads and the materials employed are concrete class C25/30 ($f_{cd} = 16.67$ MPa) and the steel S500 ($f_{syd} = 435$ MPa).

We discretized the deep beam into eight-noded quadrilateral elements with the mesh schematically illustrated in Figure 16.8(b). We performed the linear-elastic FEM analysis of the deep beam subjected to the design loads, and obtained the in-plane forces n_{xx}, n_{yy} and n_{xy} in each element sampling point (2×2 in-plane integration). Defining the horizontal (x) and the vertical (y) directions as the two orthogonal reinforcement directions, we apply the design giving the amount of steel necessary to resist the calculated forces at each sampling point. Figure 16.9 illustrates the reinforcement needs in

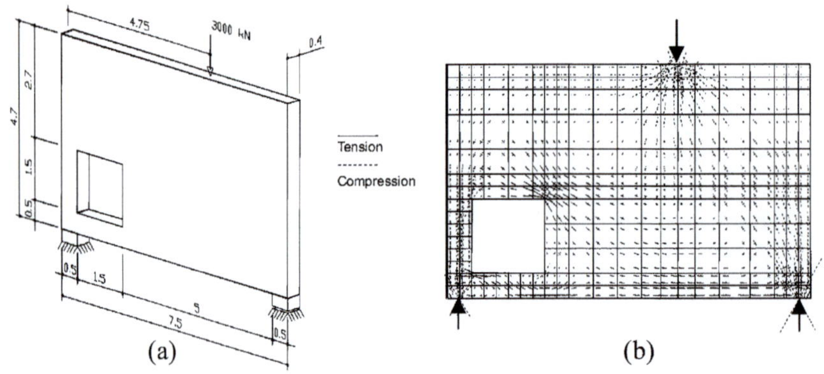

Figure 16.8 Deep beam: (a) geometry and loading (structure from *PCI Journal*, Vol. 32, No. 3); (b) mesh discretization, tension and compression forces and trajectories.

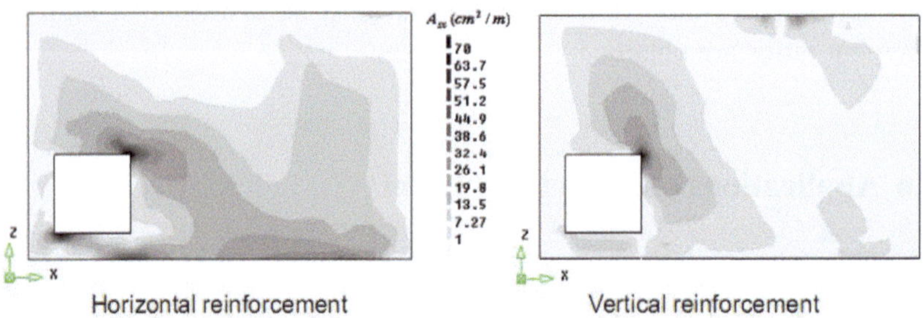

Figure 16.9 Deep Beam: reinforcement needs in the x- and y-directions.

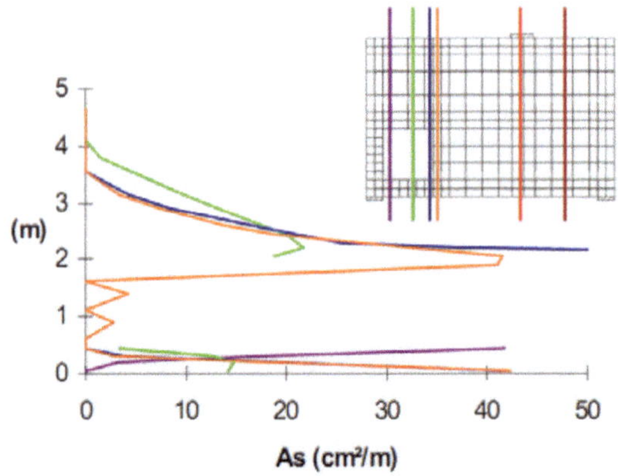

Figure 16.10 Deep beam. Diagrams of horizontal reinforcement at the critical sections.

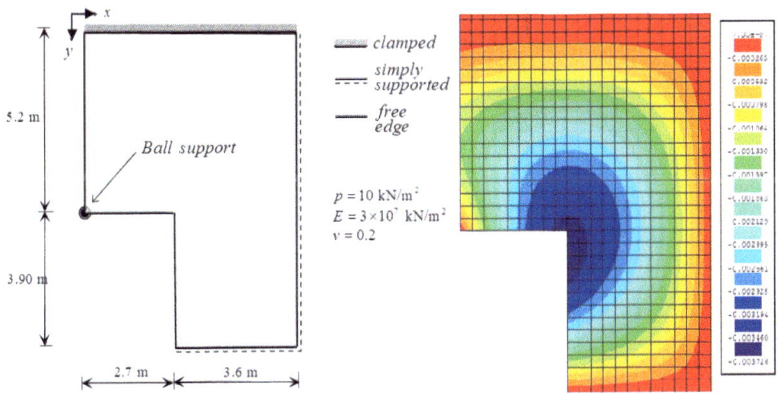

Figure 16.11 Slab: (a) geometry and loading; (b) vertical displacement (m).

the x- and y-directions, respectively. We can easily define the amount and distribution of the reinforcement to be employed by tracing out diagrams of reinforcement at the critical sections as illustrated in Figure 16.10 for the vertical and horizontal reinforcement.

16.6.2 Slab

The L slab, studied in Chapter 12, with the geometry and loading presented in Figure 16.11(a) is now analyzed. This slab has the support conditions

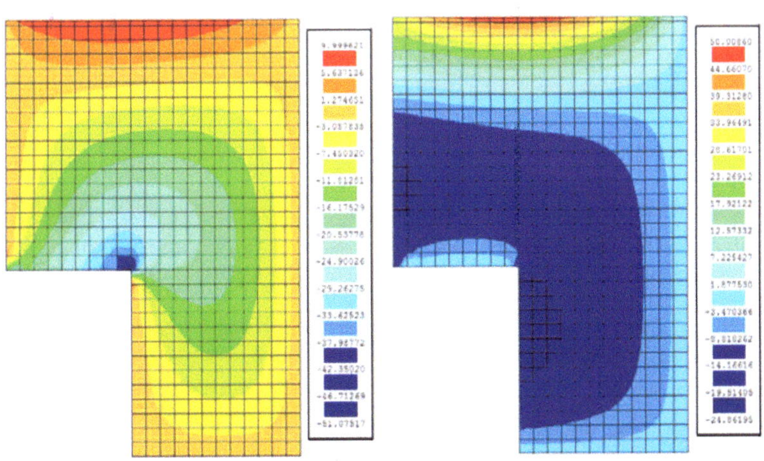

Figure 16.12 Bending moments in the x- and y-directions (kNm/m), respectively.

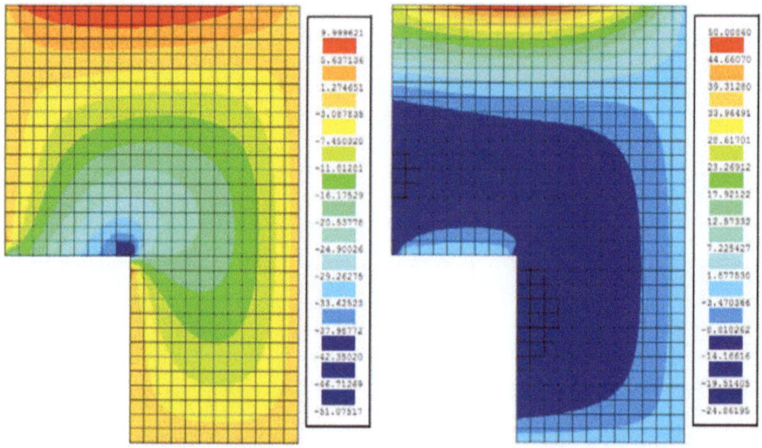

Figure 16.13 Twisting moment (kNm/m) and shear force in x-direction (kN/m).

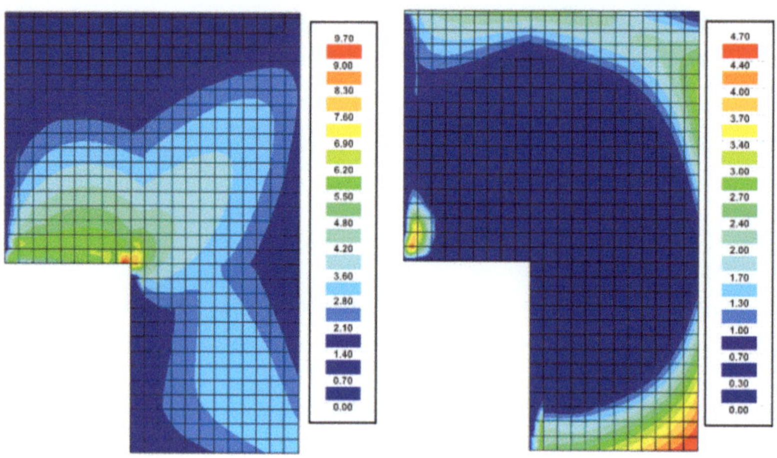

Figure 16.14 Bottom and top reinforcement needs in the x-direction, respectively (cm^2/m).

presented in the Figure 16.11(a) and has a thickness of 0.24 m. The load presented is the design load and the materials are concrete class C25/30 ($f_{cd} = 16.67$ MPa) and the steel S500 ($f_{syd} = 435$ MPa).

We discretize this slab into eight-noded quadrilateral elements with the mesh presented in Figure 16.11(b). We performed the linear-elastic FEM analysis of the slab, and obtained m_{xx}, m_{yy}, m_{xy}, v_x and v_y for all the 2×2 integration points. We can see in Figure 16.12 the bending moments in the x- and y-directions. We illustrate the twisting moment and the shear force in the

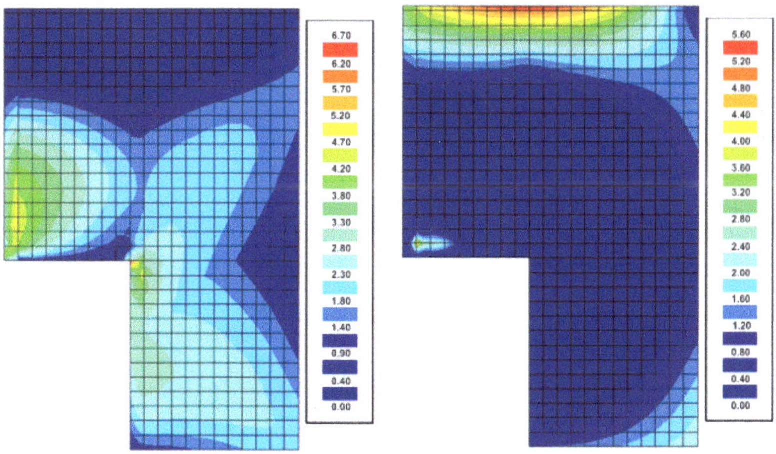

Figure 16.15 Bottom and top reinforcement needs in the y-direction, respectively (cm^2/m).

x-direction in Figure 16.13. Defining the x- and the y-directions as the two orthogonal reinforcement directions, we applied the design model giving the amount of steel necessary to resist the calculated forces at each integration point. Figure 16.14 shows the required reinforcement.

16.7 Message of the Chapter

- Reinforcement in a membrane state (wall) can be designed including the effect of shear forces. Four different cases must be considered, ranging from the need to apply reinforcement in two directions to no reinforcement at all.

- Design of reinforcement in a slab on the basis of the normal moment yield criterion leads to simple, easy to apply formulas. However, we cannot check on concrete crushing, and equilibrium is not satisfied.

- The three layer sandwich model for slabs and shell elements leads to design of reinforcement, a check on concrete crushing, and includes a reinforcement design method for transverse shear forces.

- Two variants of the sandwich model are known, basic and advanced. The basic model is easy to apply. However, it is an unsafe approximation when large twisting moments occur or high percentages of reinforcement are found. The advanced model consistently accounts for the real geometry of the element and yields adequate results.

Chapter 17
Special Slab Systems

17.1 Wide-Slab Floor

A wide-slab floor consists of a prefabricated and a cast-*in-situ* component. The prefabricated component is a thin floor unit of 2.4 m width. This is the dark shaded part in Figure 17.1. The thickness is between 50 and 70 mm. These wide elements are placed side by side without connection and are supported by scaffolding. They are the formwork for the second component, concrete which will be cast in situ on top of these floor units. This is the light shaded part in Figure 17.1. The total slab thickness is between 150 and 250 mm. On top of the prefabricated floor units there is a truss-shaped reinforcement (hat-shaped cross-section). Jointly the floor unit and reinforcement truss act as an integrated girder to provide sufficient stiffness during transport. Furthermore, the truss reinforcement assures sufficient interlock with the concrete above the floor units. The units are either reinforced by mild steel (passive reinforcement) or pre-tensioned.

We first describe the type with passive reinforcement. Apart from the truss-shaped reinforcement for transport purposes and bond, there are two layers of reinforcement in the floor unit, one to function as longitudinal reinforcement after completion of the slab, and one for flexural strength and stiffness in

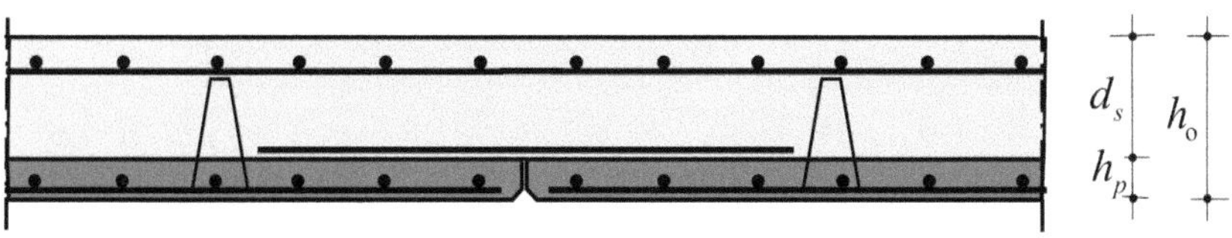

Figure 17.1 Slab with reinforced floor units.

J. Blaauwendraad, *Plates and FEM: Surprises and Pitfalls*, Solid Mechanics and Its Applications 171, DOI 10.1007/978-90-481-3596-7_17,

the lateral direction. After the floor units have been put in place, additional coupling reinforcement is placed above the seams in order to assure lateral flexural stiffness. This reinforcement must deliver the same flexural strength as the slab between the seams, so this reinforcement area is larger than in the floor units, because the lever arm in the effective depth is shorter. Where top reinforcement is needed in the slab, it is placed on top of the hat-shaped trusses.

The prefabricated floor units can also be placed as continuations of each other. Then there are not only longitudinal seams, but also lateral. Again coupling reinforcement is applied, now in the longitudinal direction. Normally the workmanship of these joints receives little attention.

In pre-tensioned floor units, the pre-stressing wires are put in the longitudinal direction. Dependent on national codes of practice, passive reinforcement in lateral direction may be required. In practice, this will be grid reinforcement, both longitudinal and lateral. If needed, longitudinal bar reinforcement can be added. Such floor units demand passive coupling reinforcement above the seams. The thickness of pre-tensioned floor units is of the order of 70 mm. If no lateral reinforcement in the floor unit is applied, then lateral reinforcement is placed above the floor unit, see Figure 17.2. Need for separate coupling reinforcement does not arise in this case.

In FE-based analyses, questions arise as to whether or not to account for the seams, and if so, in what way. This chapter intends to be a helping hand. It will appear that flexural rigidities in longitudinal and lateral direction can differ so much that orthotropic flexural properties may be considered. This chapter is intended for analyses with common FE codes which do not automatically adapt rigidities in an iterative way.

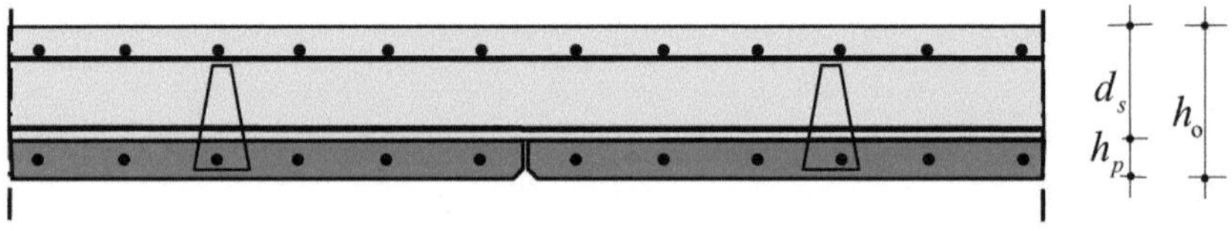

Figure 17.2 Slab with pre-tensioned floor units.

17.2 Reinforced Floor Unit

We suppose that the floor units are supported during construction such that no bending moments come into being in the curing phase of concrete. After hardening of the cast-*in-situ* concrete and removal of the provisional supports, the combination of floor units and cast-*in-situ* concrete act as one slab. Hereafter, we assume an x-axis in the longitudinal direction of the floor unit, and a lateral y-axis. Further we expect that the structural engineer has access to a FE code, which accommodates both isotropic and orthotropic calculations based on Kirchhoff theory. As explained in Chapter 4, the flexural rigidity relation for isotropic plates is

$$\begin{Bmatrix} m_{xx} \\ m_{yy} \\ m_{xy} \end{Bmatrix} = \frac{Et^3}{12\left(1-\nu^2\right)} \begin{bmatrix} 1 & \nu & 0 \\ \nu & 1 & 0 \\ 0 & 0 & \frac{1}{2}(1-\nu) \end{bmatrix} \begin{Bmatrix} \kappa_{xx} \\ \kappa_{yy} \\ \rho_{xy} \end{Bmatrix} \tag{17.1}$$

and the plate response is governed by the bi-harmonic differential equation

$$D\left(\frac{\partial^4}{\partial x^4} + 2\frac{\partial^4}{\partial x^2 \partial y^2} + \frac{\partial^4}{\partial y^4}\right) w = p \tag{17.2}$$

Here D is the flexural rigidity of the plate and p the distributed load. For orthotropic plates a generalized constitutive relationship between moment and curvature applies

$$\begin{Bmatrix} m_{xx} \\ m_{yy} \\ m_{xy} \end{Bmatrix} = \begin{bmatrix} D_{xx} & D_\nu & 0 \\ D_\nu & D_{yy} & 0 \\ 0 & 0 & D_{xy} \end{bmatrix} \begin{Bmatrix} \kappa_{xx} \\ \kappa_{yy} \\ \rho_{xy} \end{Bmatrix} \tag{17.3}$$

as discussed before in Section 14.7. Now the differential equation becomes

$$\left(H_{xx}\frac{\partial^4}{\partial x^4} + 2H_{xy}\frac{\partial^4}{\partial x^2 \partial y^2} + H_{yy}\frac{\partial^4}{\partial y^4}\right) w = p \tag{17.4}$$

The derivation of this equation proceeds along the same lines as for isotropic plates. Then the kinematic relations (17.2), the constitutive relations (17.3), and the equilibrium relations (17.5) apply, and substitution from the one in the other is done. The only difference is the replacement of the constitutive relations (17.3) by the new relations (17.3). We will find that H_{xx} coincides with D_{xx} and H_{yy} with D_{yy}, and the middle term becomes $H_{xy} = 2D_{xy}+D_\nu$. Standard FE codes with the option of orthotropic slab analyses require the input data D_{xx}, D_{yy}, D_ν and D_{xy}. Some codes treat orthotropic properties with

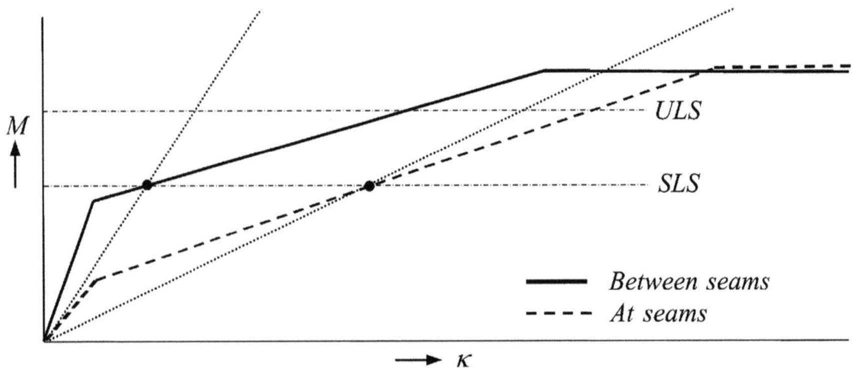

Figure 17.3 Moment-curvature diagrams for slabs with reinforced floor units.

constant thickness in combination with orthotropic material properties. Others may keep the material isotropic and introduce orthotropic cross-section second moments of area.

As stated, the question is whether or not we must account for the seams, and if so, how. It makes sense to consider the influence of seams on the moment-curvature diagram of the slab, see Figure 14.3. Between the seams, the depth of the slab is h_o, and at the seam d_s. For a thickness h_o of about 150 mm, thickness d_s is about 100 mm. A difference of factor 1.5 in thickness, leads to a difference of factor 3.4 in flexural rigidity in a linear-elastic analysis; then moments must be small enough to have a stress state which refers to the uncracked branch in the moment-curvature diagram. For strength reasons the lateral yield moment at the seam must be equal to the yield moment in the slab, so the reinforcement ratio of the coupling reinforcement will be larger than in the slab. Therefore, the third branch in the moment curvature (a horizontal one) is at the same moment value for the slab at the seam and slab between seams. The second branch in the moment-curvature diagram holds when cracking has developed. The moment at which the cracking branch starts is dependent on the square of the depths, therefore the cracking branch starts at a lower moment at the seams. The slope of this branch is highly dependent on the reinforcement ratio and the effective depth, however in a reversed way. Therefore, the branch stiffness is not expected to depend very much on their product, and the branches in the moment-curvature diagram may be more or less parallel for the slab at the seam and the normal slab between seams. Everything considered, we arrive at moment-curvatures as shown in Figure 17.3. In the Serviceability Limit State the difference between the curvatures at the seam and in between

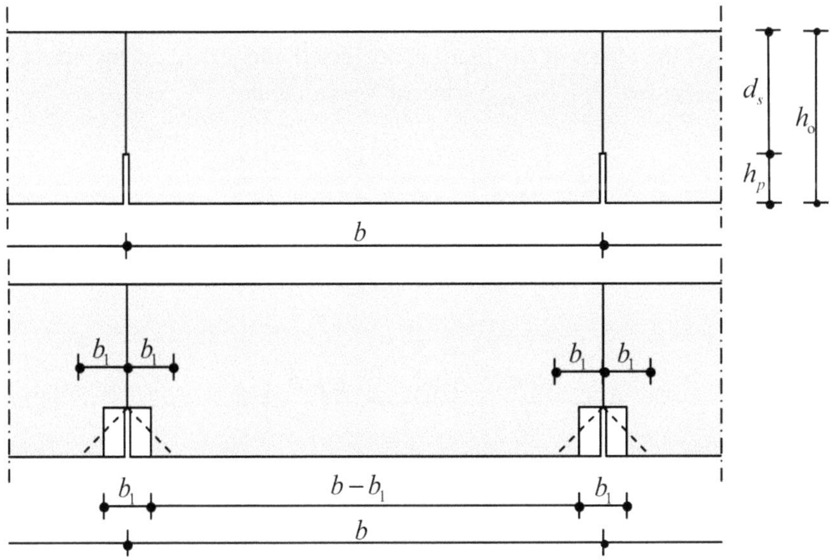

Figure 17.4 Geometry of seam in sensitivity analysis.

seams can easily surpass a factor two or three. The secant stiffness at the seam therefore is much smaller than for the normal slab. We will make use of this hereafter. For the Ultimate Limit State the difference is smaller, much closer to one.

17.2.1 Serviceability Limit State

We consider the slab near the seam, as depicted in Figure 17.4. Drawing inclined lines with angles of 45 degrees from the top of the seam downward, we expect that the slab material below these two inclined lines is almost stress-less. As an approximation, we assume the slab to consist of zones of constant depth, one of width b_1 at the seams, and one of width $b - b_1$ between seams. The depths are d_s and h_o, respectively. Note that the figure is not drawn to scale.

In order to determine the stiffness reduction due to the seams, we submit a slab part of width b to a constant bending moment m in the lateral direction. We determine the change of rotation between the two slab ends for the case with seam and the case without. The ratio is a measure for the stiffness reduction. The lateral flexural rigidity of the slab part with depth h_o is D, and for the slab part with depth d_s is D_1. The change of rotation over the width b is called $\Delta\varphi$ if the effect of the seam is neglected and $\Delta\varphi_{\text{seam}}$ if the effect of the seam is included. For $\Delta\varphi_{\text{seam}}$ and $\Delta\varphi$ we calculate

$$\begin{aligned}\Delta\varphi_{\text{seam}} &= \frac{\frac{1}{2}b_1 m}{D_1} + \frac{(b-b_1)m}{D} + \frac{\frac{1}{2}b_1 m}{D_1} \\ &= \frac{12(1-\nu^2)}{E}\left(\frac{E}{E_1}\frac{b_1}{d_s^3} + \frac{b-b_1}{h_o^3}\right)m\end{aligned} \tag{17.5}$$

$$\Delta\varphi = \frac{bm}{D} = \frac{12(1-\nu^2)}{E}\left(\frac{b}{h_o^3}\right)m \tag{17.6}$$

Here E is Young's modulus for uncracked concrete, and E_1 a reduced Young's modulus due to cracking in the seam zone. The ratio of the two rotation changes is

$$\frac{\Delta\varphi_{\text{seam}}}{\Delta\varphi} = \frac{\frac{E}{E_1}\frac{b_1}{d_s^3} + \frac{b-b_1}{h_o^3}}{\frac{b}{h_o^3}} = 1 + \left\{\frac{E}{E_1}\left(\frac{h_o}{d_s}\right)^3 - 1\right\}\frac{b_1}{b} \tag{17.7}$$

Realistic data are

b = 2,400 mm
b_1 = 50 mm
h_o = 150 mm
d_s = 100 mm
E/E_1 = 2 (estimate for average over width b_1)

On this basis we calculate

$$\frac{\Delta\varphi_{\text{seam}}}{\Delta\varphi} = 1 + \left\{2 \times \left(\frac{150}{100}\right)^3 - 1\right\}\frac{50}{2{,}400} = 1.12$$

The ratio of the flexural rigidities is the reciprocal

$$\frac{D_{\text{seam}}}{D} = 1.12^{-1} = 0.89$$

We conclude that due to the seam the deformation increases 12%, and the stiffness decreases 11%. In a similar way we may make an estimate for the torsional rigidity. We again expect 11% stiffness reduction. These amounts of reduction do not really justify accounting for the seam by making an orthotropic computation. Other suppositions in the mechanics model for slabs are of the same order of magnitude. The modulus of elasticity of concrete is just a guess, and differences in the position of the middle surface of the slab between seams and at the seam are not taken into account, not to mention more. The final conclusion is that we neglect the seam in the calculation of the moments and shear forces. We account for it only when designing reinforcement on the basis of the computed stress forces and moments.

17.2.2 Ultimate Limit State

In the Ultimate Limit State cracking will be fully developed. The lateral bottom reinforcement in the floor unit will become stressed; however this requires sufficient bond length. This length now determines the size of b_1. Codes of practice will give indications for the bond length. Often they indicate a length in the order of 20 to 30 times the rebar diameter. It must be reminded that such rules are intended for strength safety and not for stiffness. For strength a substantial safety margin is taken into account, whereas we now want to know reality. We rather should work with a bond length of about 10 times the diameter ϕ. Therefore

$$b_1 \approx 10\,\phi \tag{17.8}$$

For a diameter 16 mm, we obtain $b_1 = 160$ mm, the order of the slab thickness. Because we consider the ultimate limit state, we may assume that the slab will be also cracked in the thicker part remote from the seams. The ratio E/E_1 will be closer to 1. This leads to

$$\frac{\Delta\varphi_{\text{seam}}}{\Delta\varphi} = 1 + \left\{ \left(\frac{150}{100}\right)^3 - 1 \right\} \times \frac{160}{2{,}400} = 1.16$$

The reciprocal value 0.86 is the ratio of the stiffnesses. So, the lateral stiffness decreases by 14%, again not really a reason to introduce orthotropic rigidities.

We will examine the stiffness loss in an alternate manner. We have performed a FE analysis for a square simply-supported slab with one seam as shown in Figure 17.5. In this slab we have used two floor units. The edges of the slab are 5 m long. The slab thickness is 200 mm. The seam is modelled by a row of elements of width 0.2 m. These elements are considered as

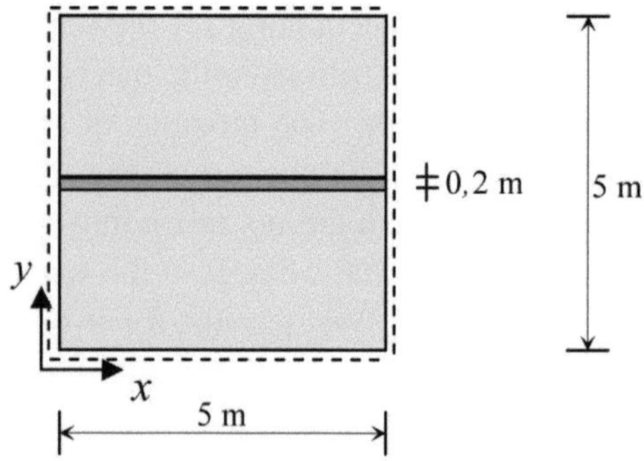

Figure 17.5 Computation of seam influence on bending moments.

isotropic elements of thickness 140 mm. A homogeneously distributed load is applied. Compared to a seamless slab the deflection increases 7%, and the moment in the lateral direction (y-direction) decreases 11%. This is reasonably in agreement with the above estimate of stiffness reduction of 14%. The increase of the deflection is half this decrease, which is understood, because the stiffness in the x-direction hardly changes. The decrease of the moment (11%) in the lateral direction is not exactly equal to the stiffness reduction (14%), however the order of magnitude is the same. This FE analysis supports the opinion that accounting for the stiffness change does not lead to noticeably different moments.

17.3 Pre-stressed Floor Unit without Lateral Reinforcement

We investigate in this Section slabs with pre-stressed floor units in combination with passive lateral reinforcement which is not in the floor unit. The passive reinforcement of the slab is a layer above the floor units.

17.3.1 Serviceability Limit State

No cracking will occur in the pre-stressed direction in the serviceability limit state (SLS), represented by the bold straight line in the moment-curvature diagram of Figure 17.6. For the passive lateral reinforcement another diagram applies, representing cracking and yielding, the dashed line in the figure. For SLS, the curvature κ at a moment M of the same size will differ substan-

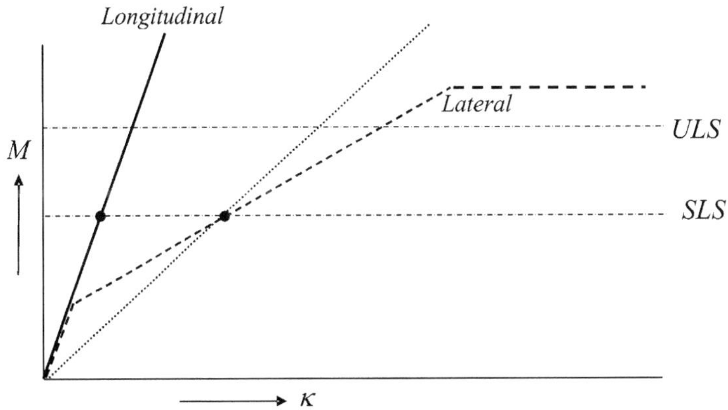

Figure 17.6 Moment-curvature diagrams for slab with pre-tensioned floor units. No lateral reinforcement in unit.

tially for the pre-stressed direction and the lateral direction. A factor two may easily occur in the secant stiffness (dotted line in the figure) and including orthotropic analysis starts making sense.

17.3.2 Ultimate Limit State

We may assume that cracking in the lateral direction has fully developed in the ultimate limit state (ULS). The difference in stiffnesses for the longitudinal direction and lateral direction will have become much larger. Introduction of orthotropic material properties is even more justified. There can be a factor of 5 difference, or even larger, between the bending stiffnesses in the x- and the y-direction. We conclude that the difference between the flexural rigidity D_{xx} in pre-stress direction and D_{yy} in lateral direction is due to different thickness and different Young's modulus. The ratio of D_{xx} and D_{yy} is

$$\frac{D_{yy}}{D_{xx}} = \frac{E_1}{E}\left(\frac{d_S}{h_o}\right)^3 \tag{17.9}$$

Here E_1 is the secant Young's modulus in the cracked lateral direction, and E in the pre-stressed longitudinal direction. Choosing $h_o = 150$ mm, $d_s =$ 100 mm and (cautiously supposed) $E/E_1 = 2.0$ we obtain

$$\frac{D_{yy}}{D_{xx}} = 0.5 \times \left(\frac{100}{150}\right)^3 = 0.15$$

Indeed, we have every reason to perform an orthotropic analysis. In fact we can conclude that the slab predominantly carries load in one direction. This supports engineering practice to consider this type of floor systems as one-way slabs.

Example

An example may clarify the impact of the difference in rigidities. We consider a square slab, simply-supported at all edges. The edges are 5 m long, the flexural rigidity in the pre-stress direction is 0.1×10^7 kNm, Poisson's ratio is 0.2 and the load 10 kN/m^2. An analysis as isotropic plate leads to the following maxima:

$$
\begin{aligned}
w &= 0.0253 \text{ mm} \\
m_{xx} &= 11.0 \text{ kNm/m} \\
m_{yy} &= 11.0 \text{ kNm/m} \\
m_{xy} &= 9.2 \text{ kNm/m}
\end{aligned}
$$

If we reduce the lateral flexural rigidity and torsion rigidity to 15% of the isotropic value the maximum values become

$$
\begin{aligned}
w &= 0.0691 \text{ mm} \\
m_{xx} &= 27.0 \text{ kNm/m} \\
m_{yy} &= 4.2 \text{ kNm/m} \\
m_{xy} &= 3.9 \text{ kNm/m}
\end{aligned}
$$

The deflection and moment in the pre-stress direction have increased by a factor two and a half, and the lateral moment and twisting moment are more than halved. The ratio of longitudinal and lateral moment is more than six. The effect of orthotropy is clearly very large. For this orthotropic plate the reinforcement scheme will be very different from the scheme which would apply for isotropic properties.

In zones of negative bending moment at inner supports of continuous slabs the pre-stressing is not active, because will be in the compression zone of the cross-section. This situation is depicted in Figure 17.7. There is no specific reason for the introduction of orthotropy. However, note that the isotropic flexural rigidity is far less than the flexural rigidity in the pre-stress direction in the domain of positive moments.

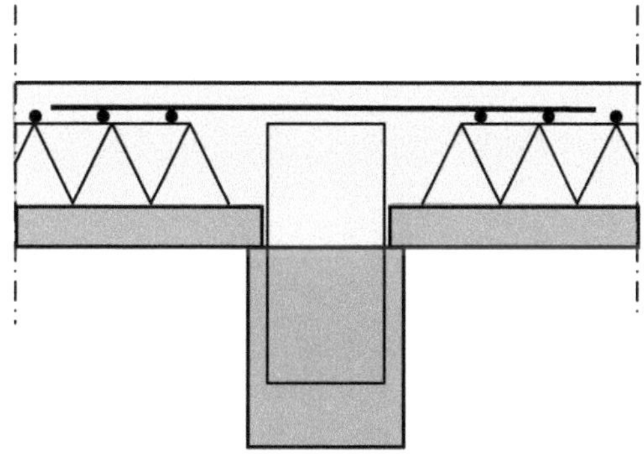

Figure 17.7 Continuous slab on beam support.

17.4 Pre-stressed Floor Unit with Lateral Reinforcement

In this section the passive lateral reinforcement is in the floor unit. Coupling reinforcement must be applied above the seams.

17.4.1 Serviceability Limit State

For the serviceability limit state we expect little cracking in the lateral direction, see also Figure 17.8. We have no reason to introduce orthotropic

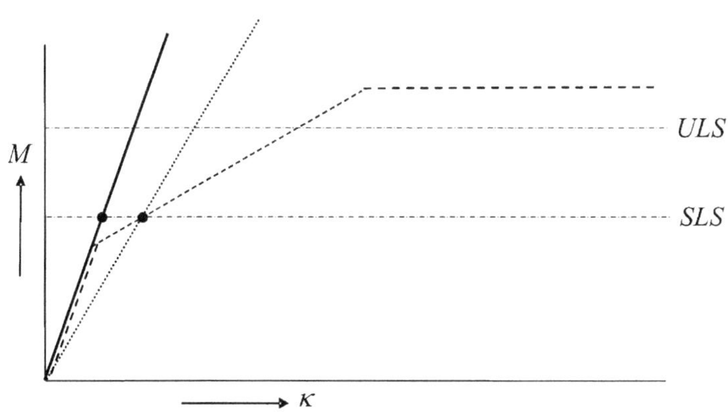

Figure 17.8 Moment-curvatures for slab with pre-tensioned floor units. Lateral reinforcement in unit.

behaviour. The reduction in lateral direction in the order of 10% is ignored again. It is justified to compute the floor as isotropic slab.

17.4.2 Ultimate Limit State

For the lateral reinforcement in the floor unit, the slab thickness is the same as for the pre-stressed direction. The only difference is now in the value of Young's modulus; a (cautious) guess is a factor of two. The ratio of stiffness becomes

$$\frac{D_{yy}}{D_{xx}} = \frac{E_l}{E} \tag{17.10}$$

where E_l is the secant modulus of elasticity for the cracked lateral direction and E for the pre-stressed longitudinal direction. Choosing $E/E_l = 2.0$ we find

$$\frac{D_{\text{yy}}}{D_{\text{xx}}} = 0.5$$

The difference between the lateral and longitudinal rigidity is much smaller than in the reinforced slab, however, still sufficient large to justify orthotropic computation. The moments in the longitudinal and lateral directions in a square simply-supported slab will be in the proportion of about 2 to 1.

17.5 Strengthened Strip Floor

A strengthened strip floor is a special case of a wide-slab floor. The floor is supported by a grid of columns. In one direction prefabricated strengthened strips are placed from one column to the other. In Figure 17.9 they are placed in the y-direction. The width of these strips is b_{o}. The strips are supports for wide-slab floor units in the other direction (x-direction). After the concrete has been poured, the thickness of the slab is h_1 at the strengthened strip, and h_{o} at the wide-slab.

If the strengthened strip and wide-slab floor units are pre-tensioned, we are justified in assuming that the wide-slab floor carries load in one direction. In that case no FE analysis is needed. If just mild steel reinforcement is applied, a FE-based analysis makes sense. Then more than one way is open for the analysis:

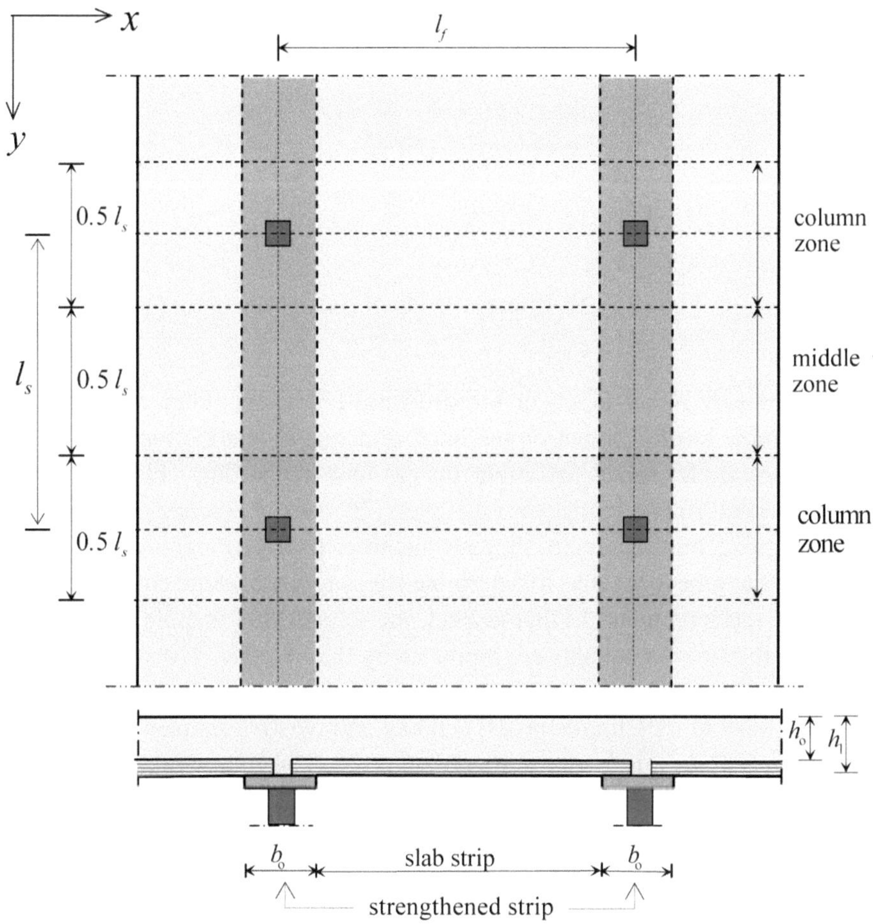

Figure 17.9 Strengthened strips over columns.

1. We may model the slab with just plate bending elements neglecting that the middle planes of the slab part with thickness h_0 and the slab part with thickness h_1 do not coincide. The output consists of moments and transverse shear forces.
2. We may model the slab with membrane-bending elements (in FE codes named shell elements). Now we automatically account for the different positions of middle planes. The output will consist of moments, transverse shear forces and membrane forces.
3. We may apply three-dimensional volume elements. Now we are able to describe the geometry most truthfully, however receive the output in terms of stresses at nodes or Gaussian points.

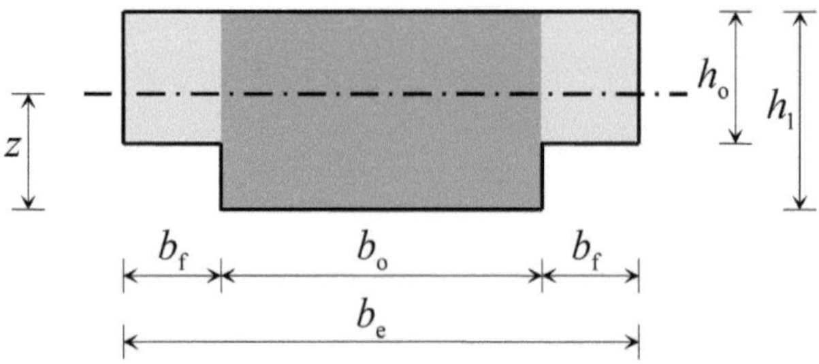

Figure 17.10 Effective width of strengthened strip.

The third way is not practical for structural engineers. They are not used to determine reinforcement on the basis of a spatial distribution of stresses, rather they prefer forces and moments per unit slab width, which are stress resultants and stress couples over thickness. If codes of practice provide rules for designing reinforcement, they always refer to forces and moments. The structural engineer first has to determine stress results, before calculating the required reinforcement. In that respect, the second way is more convenient, for then the stress resultants are produced by the FE code. The structural engineer can immediately apply the rules for designing reinforcement on the basis of a set of membrane forces, bending and twisting moments and transverse shear forces, as explained in Chapter 16. The first way of analysis is clearly the simpler one. In this chapter we restrict ourselves to this approach, because it is practical and is not expected to lead to results which are very different from the second way of analysis. Later we will justify this statement.

To some extent we can account for the different positions of the middle planes in the first way of analysis, where we just use flexural plate elements. This may be done by assigning the strengthened strip a larger thickness than h_1 or an adapted modulus of elasticity, here called h_e and E_e respectively. Because of the different positions of the middle planes, part of the slab with thickness h_o will act as a flange for the strengthened strip with thickness h_1. Let us call the width of the flange b_f. This extension occurs at both sides of the strengthened strip, see Figure 17.10. A cross-section is obtained of width $b_e = b_o + 2b_f$. National codes of practice provide rules for calculation of the effective width b_e. We call the second moment of area of the extended cross-section I_e, and if the flanges are not included, I_o. The ratio I_e/I_o determines the multiplication factor for either the thickness of the strengthened strip or its elasticity modulus.

We must take care not to duplicate the contribution of the slab parts of width b_f. If calculating I_e for the strengthened strip we must leave out the part $2b_f(h_o)^3/12$. Otherwise it occurs twice, both in the flexural rigidity of the elements in the strengthened strips and in the elements in the slab adjacent to the strengthened strip.

Example

We consider a floor and strengthened strip for which the following data hold:

$$l_s = 10\text{ m},\quad l_f = 10\text{ m},\quad E = 30{,}000\text{ N/mm}^2,$$

$$b_o = 2{,}400\text{ mm},\quad h_o = 300\text{ mm},\quad h_1 = 450\text{ mm}.$$

The assumed value b_f from the code of practice is 1,000 mm. Therefore

$$b_e = b_o + 2b_f = 2{,}400 + 2\times 1{,}000 = 4{,}400\text{ mm}.$$

The position z of the neutral line is calculated from the formula

$$z = \frac{2b_f h_o(h_1 - \frac{1}{2}h_o) + b_o h_1(\frac{1}{2}h_1)}{2b_f h_o + b_o h_1}$$
$$z = \frac{(6.0\times 10^5)(300) + (10.8\times 10^5)(225)}{6.0\times 10^5 + 10.8\times 10^5} = 251.8\text{ mm} \tag{17.11}$$

Calculation of I_e

$$I_e = 2b_f h_o\left(\left(h_1 - \frac{1}{2}h_o\right) - z\right)^2 + b_o h_1\left(\frac{1}{2}h_1 - z\right)^2 + \frac{1}{12}b_o h_1^3 \tag{17.12}$$

$$I_e = (6.00\times 10^5)\,(300 - 251.8)^2 + (10.8\times 10^5)\,(225 - 251.8)^2 + \frac{1}{12}\times 2{,}400\times 450^3$$

$$I_e = 1.39\times 10^9 + 0.78\times 10^9 + 18.23\times 10^9 = 20.40\times 10^9\text{ mm}^4$$

The last of the three terms is the second moment of area for the strengthened strip without the flanges

$$I_o = 18.23\times 10^9\text{ mm}^4$$

The ratio is of I_e and I_o is

$$\frac{I_e}{I_o} = \frac{20.40}{18.23} = 1.119$$

Therefore, we either work with

$$h_e = \sqrt[3]{1.119}\, h_1 = 1.038\, h_1 = 1.038 \times 450 = 467 \text{ mm}$$

or with equal depths and

$$E_e = 1.119\, E = 1.119 \times 30{,}000 = 33{,}570 \text{ N/mm}^2$$

Remark 1

Accounting for different positions of the middle plane implies a stiffness change of the strengthened strip in the order of 10%. This is of the same size as the margin in which we anyhow know the stiffness of cracked concrete, as said earlier in this chapter. There is not much reason to exert the extra effort for flange contribution. One more reason to neglect the difference in middle plane position is the consideration that the increase of rigidity holds only for the longitudinal direction of the strengthened strip and not for the lateral direction. If one would be precise, in fact orthotropic behaviour should be introduced. We conclude that it is not worth accounting for the different middle plane positions. Of course, the difference between the strengthened strips and the slabs between the strips is very relevant. In the example the thickness is 450 mm and 300 mm, respectively, a factor of 1.5. Then the flexural plate rigidity of the strengthened strip is 1.53 which is 3.4 times the rigidity of the slab between the strips. This difference will have a serious impact.

Remark 2

For designing the reinforcement scheme of this type of floor, we refer to the message of Chapter 14. We can plot the diagram of bending moment *mxx* in a line of constant x over the columns and at mid-span between two rows of columns. This moment distribution may be smeared over two zones, one which is called the *column zone* and the other the *middle zone*. National codes of practice may specify additional conditions with respect to redistribution and size of the compression zone in the cross-section.

17.6 Message of the Chapter

Wide-Slab Floor

- If wide-slab floor units are reinforced by mild steel only, seams can be neglected and the analysis may be done with isotropic flexural properties, both for SLS and ULS.

- If pre-stressed wide-slab floor units are used, we may distinguish between the cases when the floor unit has lateral mild steel reinforcement, and the cases that it has not. If mild steel is included in the floor unit and coupling reinforcement is placed above the seams, the FE-based analysis can be done with isotropic flexural properties for SLS. Orthotropic flexural properties should be used for ULS.

- For pre-stressed floor units without lateral mild steel, the lateral reinforcement is placed above the floor unit, which justifies computing with orthotropic flexural properties for both SLS and ULS.

- If the applied FE code is able to adapt the flexural stiffness in an iterative procedure, no choice about isotropic or orthotropic flexural behaviour needs to be made. Then we always may start with isotropic properties, and the program will gradually adapt them to orthotropic ones.

Strengthened Strip Floor

- The effect of different positions of the middle planes is negligible. It is not worth accounting for it, neither if just bending plate elements are used, nor for membrane-bending elements (called shell elements in FE codes).

- Of course, the difference in thickness between the stiffened strip and the rest of the floor is of important.

- The difference between pre-stressed and passive reinforced elements may be accounted for in the flexural rigidities, as explained for wide-slab floors.
- If the applied FE code is able to adapt the flexural stiffness in an iterative procedure, no choice about isotropic or orthotropic flexural behaviour needs to be made. Then we always may start with isotropic properties, and the program will gradually adapt them to orthotropic ones.

Chapter 18
Special Topics and Trends

We devote this chapter to four special subjects, two for membrane plates and two for plates in bending. In Section 18.1 we address the stringer-panel model for membrane calculations, and in Section 18.2 we show provisional results of membrane calculations with concrete compression stresses only. An advanced approach for orthotropic plate bending is the subject of Section 18.3. Finally, we discuss plates on soil foundations in Section 18.4.

18.1 Stringer-Panel Method

In Chapter 8 we discussed a discrete model for the analysis of membrane plates with truss elements and shear panels. There is a more accurate model with *stringers* and *panels*. In the most elementary version the stringer elements carry normal forces and the panels shear forces. This model goes back to the work of Nielsen [19]. Stringers are put in positions where we expect a compression zone or a tension band. In the tension band, we concentrate the main reinforcement. The shear panels require distributed reinforcement in two directions. This fits with practice, in which mesh reinforcement will always be applied, even when it is not required by the chosen calculation method.

The basic concept is that equilibrium is required between stringer and panel along their full boundary. The shear force in the panel is constant. This distributed shear force is the load for the stringer, therefore the normal force must vary linearly along the stringer. The method is not restricted to orthogonal grids, but is applicable to grids with quadrilateral panels as well.

J. Blaauwendraad, *Plates and FEM: Surprises and Pitfalls*, Solid Mechanics and Its Applications 171, DOI 10.1007/978-90-481-3596-7_18,

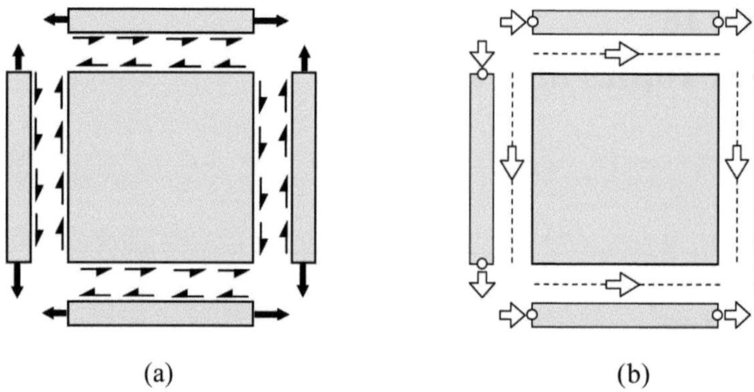

Figure 18.1 Shear panel with surrounding stringer elements.

Hoogenboom has even extended the method to cracked stringers and panels and yielding reinforcement [32]. However, then the stress state of the panels must then be enriched with distributed normal forces.

We restrict ourselves to applications with an orthogonal grid and constant shear panels. Figure 18.1 shows two exploded views of a single panel and the four stringer elements that surround the panel. The left part of the figure shows forces; the right part shows degrees of freedom. We call the boundary between stringers and panels the interface. The constant shear in the interface leads to a linear normal force in the stringer and therefore to end forces which are not equal. If a stringer lies between two panels, there is an interface at both sides of the stringer and it is loaded by two distributed shear forces, one in each interface. Stringers are pin-connected to each other in nodes at their ends. Because stringers occur in two directions, the nodes have two degrees of freedom, one in the x- and one in the y-direction. There is an additional degree of freedom at each interface between a stringer and a panel. This degree of freedom is of another type: it is the weighted average displacement in the direction of the interface. The stringer-panel model fits in the scheme of the stiffness method, but with two differences: first, the degrees of freedom are defined at uncommon positions; second, the stiffness matrices are derived on the basis of alternative work considerations [32], which falls out of the scope of this book.

For a statically determinate stringer-panel model, we can find the forces on the basis of equilibrium. For statically indeterminate models we need a program, because then compatibility starts playing a role. We show an example of each category, a beam with ends of a special shape and a shear wall, respectively.

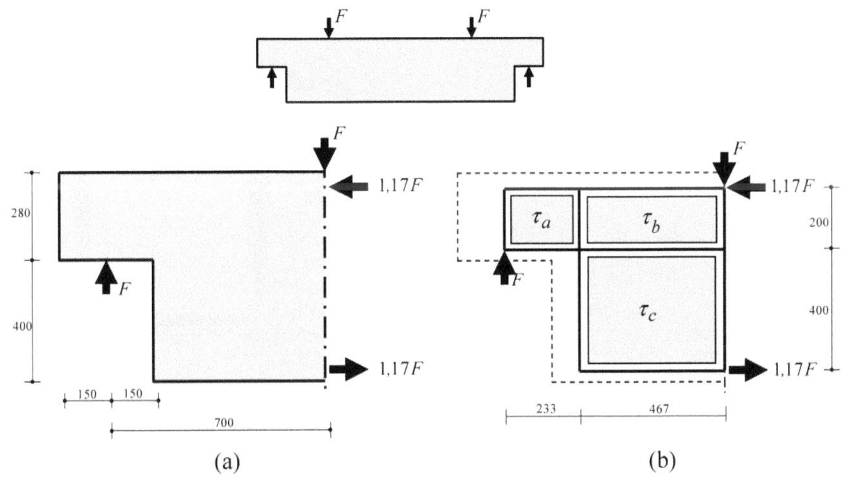

Figure 18.2 Beam with dapped ends.

18.1.1 Beam with Dapped Ends

Figure 18.2 shows a reinforced concrete beam with a special shape near the supports. The end parts of the beam have a smaller depth. We call this a structure a beam with dapped ends. It is loaded by two point loads F. There is a state of homogeneous bending moment with zero shear force between the point loads, and a disturbed stress state between the supports and the applied loads. In Figure 18.2a we show the enlarged left beam end with the forces acting on it. These forces are in equilibrium. The stringer-panel model for this structural part is shown in Figure 18.2b. We need three shear panels, and the shear forces in the panels are n_a, n_b and n_c, respectively. The model is statically determinate, so the shear forces in the panels and normal forces in the stringers can be calculated from equilibrium. Figure 18.3 shows the result of the analysis. Note that the shear forces in the figure are the total shear force along a panel edge, so they have the dimension kN. Surprisingly, the shear force n_b is zero. The transverse shear force in panel (a) is not transferred horizontally to the upper panel (b), but instead to the lower panel (c). The shear force in panel (b) is exactly zero due to the chosen geometry of the beam end. For other geometries, the panel may get a non-zero shear force, however always small compared to the shear force in panel (c).

The corresponding stringer forces are shown in Figure 18.3b. Tensile forces are red and compressive forces green. The width of the red and green lines

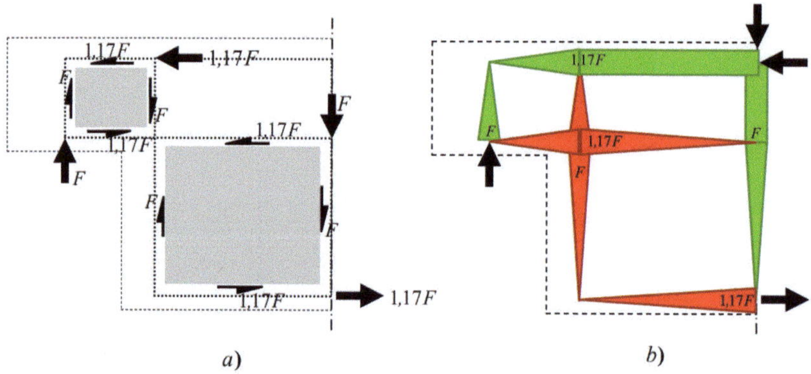

Figure 18.3 Forces in stringers and panels of beam with dapped end.

is a measure of the size of the normal force, and shows the linear variation of the normal force. The vertical stringers at the application point of the load and at the support are compressive struts. The vertical stringer that separates panel (a) from panel (c) is a tensile chord. This chord connects panel (c) to panel (a). The force in this chord has the size of the support reaction F. The horizontal tensile force above the support connects the panels (a) and (c) in the horizontal direction. The result makes clear that this reinforcement must at least be extended to the cross-section where the load is applied.

Figure 18.4a shows theoretically how the beam should be reinforced. The lower main beam reinforcement is bent up vertically and continues horizontally as an upper reinforcement in the support zone. Theoretically, distributed shear reinforcement is theoretically only required in the panels (a) and (c); in practice stirrups and hair pin reinforcement may be placed over the full depth

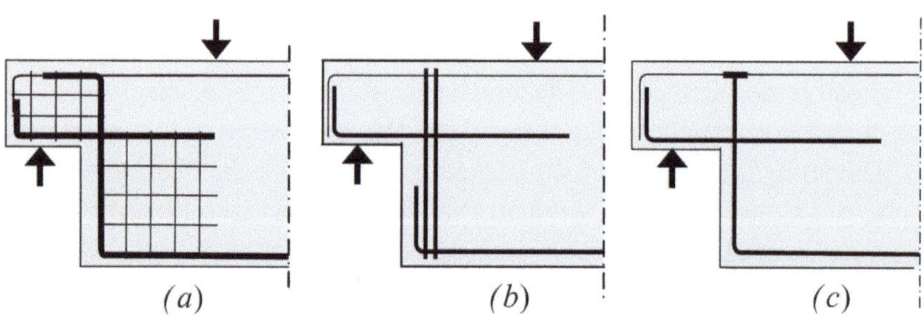

Figure 18.4 Reinforcement options for beam with dapped end.

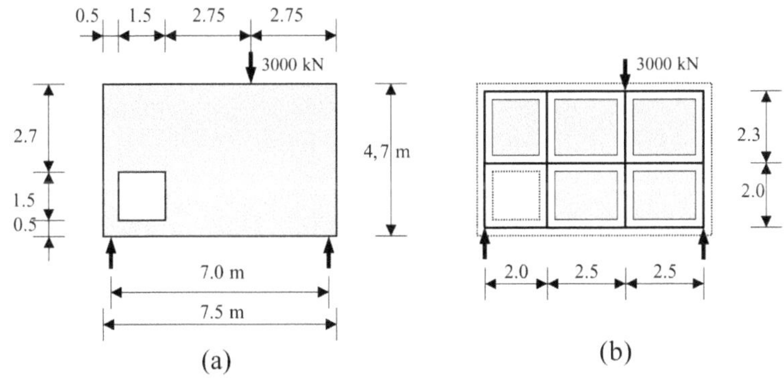

Figure 18.5 Shear wall (structure from *PCI Journal*, Vol. 32, No. 3).

of the beam in the disturbed zone. Further, Figure 18.4 shows two alternative ways of reinforcement detailing. Figure 18.4b is a solution with closed stirrups as hanging-up reinforcement. Figure 18.4c is the same as Figure 18.4a, however with a T-end anchoring. In these alternative solutions we omitted the distributed shear reinforcement for reasons of convenience.

Structural engineers who design the beam end with a strut-and-tie model will obtain similar results. In their solution compressive diagonals replace the panels (a) and (c). These diagonally struts are directed from bottom left to top right.

18.1.2 Shear Wall with Opening

In Section 16.6.1 Lourenço referred to the publication of the special report 'Toward a Consistent Design of Structural Concrete' of the *PCI Journal* [31]. We again use the shear wall with opening in Figure 18.5a, one of the examples in the report. There it was used for strut-and-tie modelling, here as an example for the stringer-panel method. The wall area is 7.0×4.7 m^2 and the thickness is 0.4 m. It is decided to apply an orthogonal reinforcement scheme in x- and y-direction so a stringer-panel model with rectangular panels is used. Figure 18.5b shows the model with the minimum possible number of panels. This model is statically indeterminate of the order one, so deformation considerations must play a role. Therefore we use a program, and we must input the cross-section areas of the stringers. We base the area of outer stringers on the distance between the wall edge and the centre of the adjacent

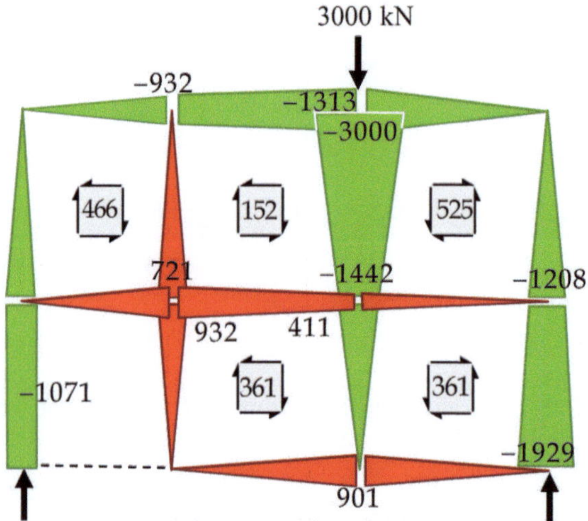

Figure 18.6 Stringer-panel results for shear wall.

panel; for inner stringers we use the centre to centre distance between panels. Young's modulus is 3.0×10^7 kN/m^2 and Poisson's ratio 0.2.

Figure 18.6 shows the results. We see similarity with the results for the beam with dapped ends. The shear force in the lower left panel is higher than in the panel on top of it. Note that shear forces in this figure have the dimension force per unit length. Two-third of the transverse shear force (size 1,071 kN) in the wall is transferred in the lower panel and one-third in the upper. Hanging-up reinforcement is again required; the maximum force in this hanger is 721 kN, two-third of the reaction force 1,071 kN. For the rest, it is instructive that the horizontal main reinforcement above the wall opening should stretch out to the very opposite edge of the wall.

18.2 Membrane Plates with Concrete Pressure Only

We have shown in this book that membrane analyses support choosing an efficient scheme in the strut-and-tie design of concrete walls and deep beams. There is an interesting development in which the program generates the strut-and-tie scheme itself. Though it is only a starting development, we pay attention to it because of its promising character. The idea is to choose the position of main reinforcement bundles and have the program generate the compressive struts. In the first run, a standard linear-elastic FE analysis is made. The principal stresses and their direction are determined in the centre of each element. If one of the principal stresses is tensile, we choose a set of local axes, such that x is the principal tensile direction, and y is the principal compressive direction. Then we omit the stiffness in the x-direction, no longer allowing for tensile stresses in that direction. Lateral contraction need no longer be considered. We replace the constitutive relationship (1.13) for isotropic behaviour by an orthotropic one:

$$\begin{Bmatrix} n_{xx} \\ n_{yy} \\ n_{xy} \end{Bmatrix} = Et \begin{bmatrix} 0 & 0 & 0 \\ 0 & 1 & 0 \\ 0 & 0 & \frac{1}{2} \end{bmatrix} \begin{Bmatrix} \varepsilon_{xx} \\ \varepsilon_{yy} \\ \gamma_{xy} \end{Bmatrix} \tag{18.1}$$

We repeat the linear-elastic analysis with these orthotropic, pressure-only, properties and find new principal stresses and directions. We recompute the rigidity matrix and make a new iteration. We do it until changes become smaller than a specified tolerance.

18.2.1 Shear Wall with Opening

Figure 18.7 shows a strong example of the program *Scia Engineer* (Nemetschek-Scia, Netherlands). It shows the shear wall with the opening used in Section 18.1. The red lines in Figure 18.7 are the inputted reinforcement bundles, and the green trajectories are the obtained compressive trajectories. These trajectories show up in narrow zones which appear as struts, so the strut-and-tie scheme emerges naturally. The lower part of Figure 18.7 shows the force distribution in the reinforcement bundles.

The shear forces in the panels are used to design distributed mesh reinforcement. In practice, structural engineers always apply minimum reinforcement as required by codes of practice. Often, this is sufficient; if not, we have to strengthen the mesh.

18.3 Advanced Orthotropy

Plates in bending often have orthotropic properties. Examples are stiffened steel bridge decks and concrete bridges composed of girders. In earlier times of hand calculations the method of Guyon–Massonnet–Bares [32] would be

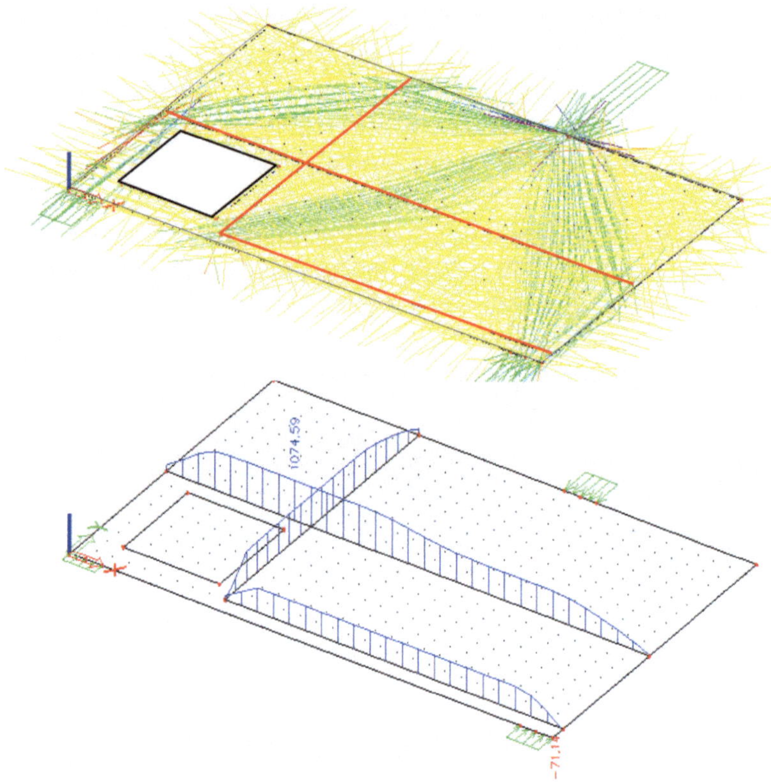

Figure 18.7 Results for pressure-only properties.

applied. Today, FE codes are available for orthotropic plate bending. Structural engineers must input bending and torsion rigidities for thin plate theory. For thick plate theory they must also input rigidities for transverse shear. It is important to remind that both theories start from the assumption that the middle planes coincide for bending in span direction, bending in lateral direction and torsion. However, this is not true in general; in fact there are not only bending and torsion moments, but also membrane forces. Bending, shearing and stretching are coupled, so the structural engineer should choose FEM software, which includes coupling of plate membrane action and bending. An example for teaching purposes is *Kola* (TU Delft, the Netherlands). We call the coupled action *advanced orthotropy* in order to distinguish it from the classical orthotropy which neglects membrane action.

The classical constitutive law for orthotropic membrane action is

$$\begin{Bmatrix} n_{xx} \\ n_{yy} \\ n_{xy} \end{Bmatrix} = \begin{bmatrix} d_{xx} & d_v & 0 \\ d_v & d_{yy} & 0 \\ 0 & 0 & d_{xy} \end{bmatrix} \begin{Bmatrix} \varepsilon_{xx} \\ \varepsilon_{yy} \\ \gamma_{xy} \end{Bmatrix} \tag{18.2}$$

and for orthotropic bending

$$\begin{Bmatrix} m_{xx} \\ m_{yy} \\ m_{av} \end{Bmatrix} = \begin{bmatrix} D_{xx} & D_v & 0 \\ D_v & D_{yy} & 0 \\ 0 & 0 & D_{av} \end{bmatrix} \begin{Bmatrix} \kappa_{xx} \\ \kappa_{yy} \\ \rho_{xy} \end{Bmatrix}. \tag{18.3}$$

The torsion moment m_{av} is the average of m_{xy} and m_{yx} and the torsion rigidity D_{av} the average of D_{xy} and D_{yx}. In classical orthotropic bending analysis the deformation due to transverse shear forces is neglected. Yet we recall the constitutive law for reasons of completeness

$$v_x = D_{sx}\,\gamma_x, \quad v_y = D_{sy}\,\gamma_y. \tag{18.4}$$

To define the constitutive law for coupled membrane action and bending we choose a common reference plane R. Because we also will account for shear deformation, we start from Mindlin theory and extend it for coupling with membrane action. We obtain

$$\begin{Bmatrix} n_{xx} \\ n_{yy} \\ n_{xy} \\ m^R_{xx} \\ m^R_{yy} \\ \hat{m}^R_{av} \\ v_x \\ v_y \end{Bmatrix} = \begin{bmatrix} d_{xx} & d_v & 0 & z_{xx}d_{xx} & z_v d_v & 0 & 0 & 0 \\ d_v & d_{yy} & 0 & z_v d_v & z_{yy}d_{yy} & 0 & 0 & 0 \\ 0 & 0 & d_{xy} & 0 & 0 & z_{xy}d_{xy} & 0 & 0 \\ z_{xx}d_{xx} & z_v d_v & 0 & D_{xx}+z_{xx}d_{xx} & D_v + z_v^2 d_v & 0 & 0 & 0 \\ z_v d_v & z_{yy}d_{yy} & 0 & D_v + z_v^2 d_v & D_{yy}+z_{yy}^2 d_{yy} & 0 & 0 & 0 \\ 0 & 0 & z_{33}g_{33} & 0 & 0 & D_{av}+z_{xy}^2 d_{xy} & 0 & 0 \\ 0 & 0 & 0 & 0 & 0 & 0 & D_{sx} & 0 \\ 0 & 0 & 0 & 0 & 0 & 0 & 0 & D_{sy} \end{bmatrix} \begin{Bmatrix} \varepsilon^R_{xx} \\ \varepsilon^R_{yy} \\ \gamma^R_{xy} \\ \kappa_{xx} \\ \kappa_{yy} \\ \rho_{xy} \\ \gamma_x \\ \gamma_y \end{Bmatrix} \tag{18.5}$$

where all z_{ij} are distances to the reference plane R. The strains and bending moments are defined with respect to the reference plane. This is indicated in Eq. (18.5) by the subscript R. If there is no difference between the reference planes for membrane action and bending then the reference plane R is chosen as the common reference plane. All z-values are zero, and membrane action, bending and transverse shear are fully uncoupled. In this limiting case Eq. (18.5) is the composition of Eqs. (18.2), (18.3) and (18.4). We refer to [33] for the working out of the advanced theory, and restrict ourselves to an example in next section. It suffices to note that the program calculates the rigidity matrix, and after that makes the FE analysis of the bridge.

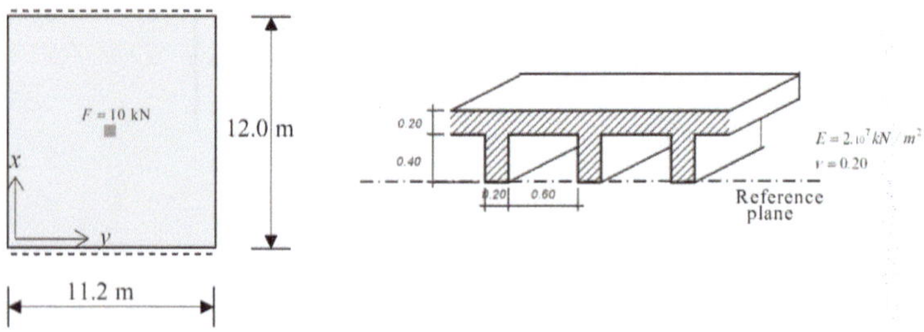

Figure 18.8 Rectangular orthotropic bridge slab.

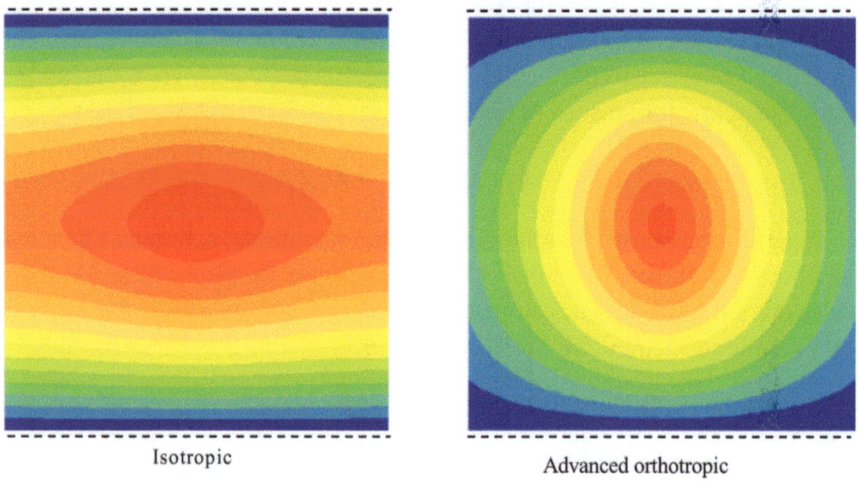

Figure 18.9 Comparison of isotropic and orthotropic deflection.

Table 18.1 Rigidity properties of T-shaped cross-section.

kN/m	kNm	m
$d_{xx} = 0.61697 \times 10^7$	$D_{xx} = 0.16217 \times 10^6$	$z_{xx} = 0.40270$
$d_v = 0.84860 \times 10^6$	$D_v = 0.29360 \times 10^4$	$z_v = 0.49804$
$d_{yy} = 0.42430 \times 10^7$	$D_{yy} = 0.14896 \times 10^5$	$z_{yy} = 0.49804$
$d_{xy} = 0.17362 \times 10^7$	$D_{av} = 0.91673 \times 10^4$	$z_{xy} = 0.49363$
$D_{sx} = 0.10873 \times 10^7$		
$D_{sy} = 0.14538 \times 10^7$		

18.3.1 Bridge with Point Load

We study the bridge in Figure 18.8 to show the effect of advanced orthotropy. The simply-supported rectangular bridge with a span of 12.0 m and width of

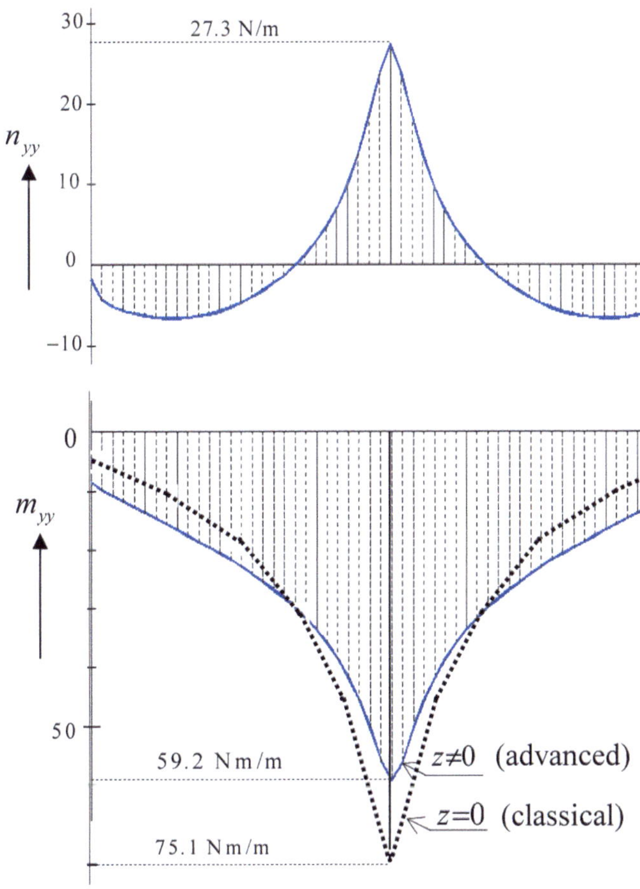

Figure 18.10 Membrane force in section at mid-span of bridge. Difference in peak moment for classical and advanced orthotropy.

11.2 m consists of 14 girders with a T-shaped cross-section. The structure is a slab with ribs. The structure is supposed to be monolithic. The bridge is subjected to a point load of 100 kN at its centre. A 25 × 25 element mesh is used, and the load is spread over the central element. The reference plane is chosen at the lower edge of the ribs. For this cross-section, the rigidity properties are assembled in Table 18.1. Rounded off, $z_{xx} = 0.4$ m and $z_{yy} = z_v = 0.5$ m. So the middle plane in the x-direction is at the boundary of the ribs and the slab, and the middle plane in the y-direction is in the middle of the slab, as we expect. The middle plane for the lateral direction is also in the middle of the slab, again agreeing with expectation. For membrane shear and torsion, the middle plane is in the lower half of the slab.

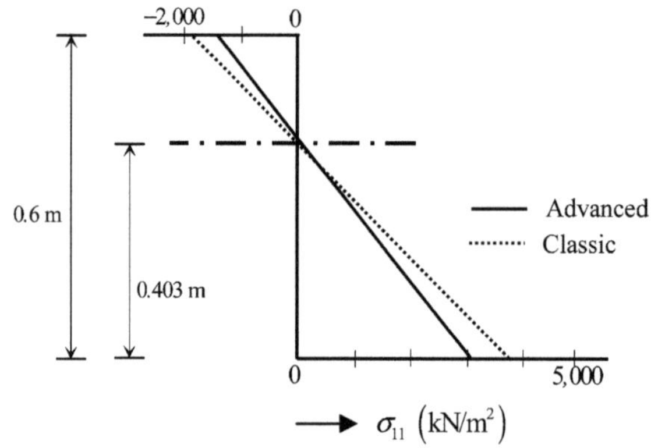

Figure 18.11 Different stresses over depth for classic and advanced orthotropy.

In Figure 18.9 we show the large difference in deflections between isotropic theory and orthotropic theory. In an isotropic plate the point load activates a wide zone of the plate; in the orthotropic plate, the deflection is localized at the plate centre. There is a large difference between isotropy and orthotropy, however not between classical orthotropy and advanced orthotropy. Therefore, just the advanced results are shown. Figure 18.10 shows distributions of the membrane force and bending moment in the section halfway across the span. The membrane force n_{yy} is tensile at the centre of the plate, and compressive at the bridge edges. The integral of n_{yy} over the section must be zero. The bending moment m_{yy} is shown for classical and advanced orthotropy. With the advanced theory a reduction from 75.1 to 59.2 kNm/m is obtained, about 20%. For classical orthotropy only the bending moment raises stresses; for advanced orthotropy we must account for both the membrane force and the bending moment. Figure 18.11 shows that the stress distributions over the depth are different. The tensile stress for the advanced theory is about 20% smaller than for the classical theory. In othe wors, the classical theory overestimates the tensile stress by about 20%.

The measure of differences between the classical and advanced theory depends on the shape of the cross-section of the bridge girders. In the example the bridge girders have a T-shape. For I-section girders, the difference between the moments for both theories is smaller, but the membrane forces are more pronounced in the advanced theory. Figure 18.12 shows the results for a square bridge with a span of 20 m and a point load of 100 kN. In this case we find relative large stress differences in the compressive zone and practically

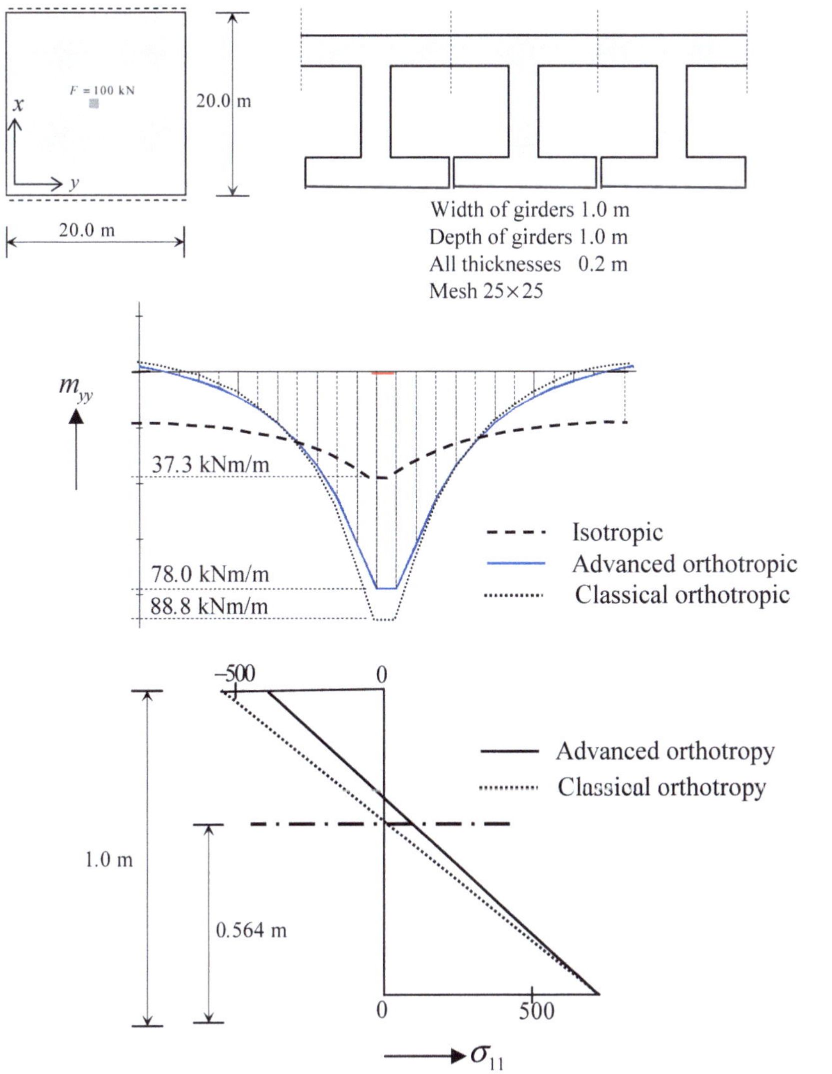

Figure 18.12 Results for bridge with I-sections.

no difference in the tensile zone. The figure includes the bending moments for isotropic properties. These are much smaller than the orthotropic ones, because more plate in the width direction is mobilized to carry the point load.

18.4 Plates on Soil Foundation

Concrete slabs on a soil bedding may be calculated as plates on an elastic foundation, which is known in literature as a Winkler foundation. In this model the soil pressure p_s is proportional to the deflection w

$$p_s = kw \tag{18.6}$$

where k is the spring constant (modulus of subgrade reaction; units kN/m^3). The model has many disadvantages. It is a difficulty that the determination of k depends on the size of the slab. Further, soil outside the slab remains stress-less, in contrast with physical reality. Due to shear stresses, surrounding soil will settle, which implies increase of the vertical soil pressure at the edge of the slab. Applying the Winkler foundation we never will find interaction between the settlements of adjacent plates, which is there in reality. Another drawback is that the model cannot account for the influence of the pore pressure of groundwater.

Many theories have been proposed to eliminate one or more omissions of the Winkler foundation. A well-known solution is the Pasternak foundation; springs cannot deflect independently on each other, but a vertical shear force between the springs comes into being when the springs have different deflections. This model is able to overcome many drawbacks of the Winkler foundation, but cannot account for the influence of groundwater, and a new soil mechanics property is introduced. Another theory introduces concentrated springs at the edge of the slab to account for surrounding soil stiffness; this again introduces a second soil mechanics parameter.

A good solution is to combine FEM software and classical knowledge from solid mechanics and geotechnics. This procedure needs a number of iterations. The analysis starts with a constant spring constant k and leads to vertical displacements w of the slab and a distribution of the soil reaction p_s. These reaction forces are lumped to point loads $P(i)$ in all N nodes i of the slab. Next, a two-step analysis is made. The solid mechanics formula of Boussinesq (or more generally Fröhlich) is used to find the change of the vertical soil stress $\sigma_{zz}(i)$ through the depth of the soil at each node $i$of the slab due to the point load $P(j)$ at node j. This is done for the point loads in all N nodes. Second, the geotechnical Terzaghi formula is applied at each node i to calculate the total settlement $s(i)$ from the distribution of stresses $\sigma_{zz}(i)$ through the depth. We stop the iteration when $s(i)$ is equal to the deflection

$w(i)$ of the slab. Otherwise we change the spring constant k. The new value is chosen so that $s(i)k(i) = p_s(i)$, and we repeat the procedure with the new set of spring constants. The procedure can incorporate the effect of pore water pressure: we use the effective stress σ_{eff} in stead of the stress σ_{zz}.

The Fröhlich formula for the change of the vertical stress at a depth z in point i is

$$\Delta\sigma_{zz}(i) = \sum_{j=1}^{N} f_{ij}\, P(j) \tag{18.7}$$

where the influence function f_{ij} is

$$f_{ij} = \frac{\mu}{2\pi z^2 \cos^{\mu+2}\theta_{ij}} \tag{18.8}$$

The angle θ_{ij} is defined by the line which connects node j with the point on a depth z at node i. The angle of this line with the vertical is θ_{ij}. The *order number* μ is defined by

$$\mu = 1/\nu + 1 \tag{18.9}$$

where ν is Poisson's ratio. For $\nu = 0.5$ the Fröhlich formula is the same as the Boussinesq formula. This Poisson's ratio is appropriate for undrained soil states. The value $\nu = 0.33$ fits normal consolidated soil with pronounced sand properties.

The Terzaghi formula for settlement is

$$s(i) = \sum_{\text{layers}} \frac{\Delta h}{C} \ln \frac{\sigma_{zz}(i) + \Delta\sigma_{zz}(i)}{\sigma_{zz}(i)} \tag{18.10}$$

This formula accounts for different soil layers of thickness Δh. The stress $\sigma_{zz}(i)$ is the existing vertical soil pressure; one may account for the reduction due to earth cutting, or do that partially. C is the dimensionless compression constant which is determined from the geotechnical cone penetration test

$$C = \alpha \frac{q_{\text{cone}}}{\sigma_{\text{eff}}} \tag{18.11}$$

where q_{cone} is the cone resistance and α is a constant between 1 and 10, dependent on the soil type of the layers. The reader is referred to Eurocode 7 *Geotechnics* [34].

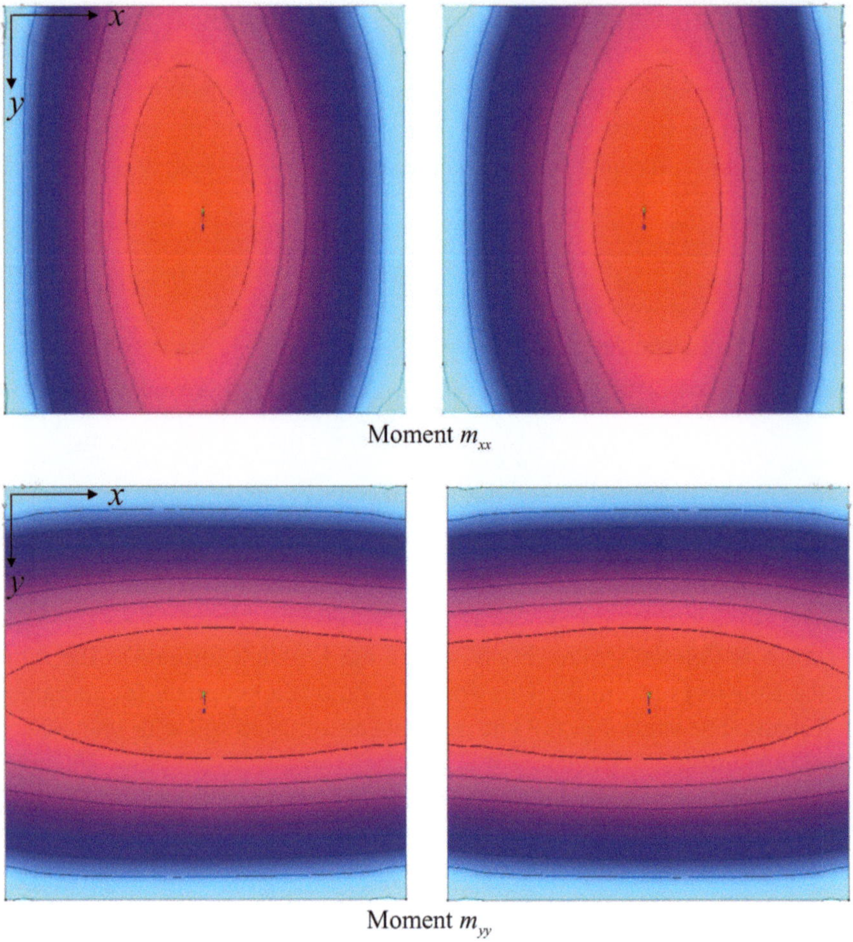

Figure 18.13 Two slabs on soil foundation on small distance of each other.

18.4.1 Two Close Slabs

We consider an application with the program *Diamonds* (Buildsoft, Belgium). Two identical square plates of 10×10 m^2 are placed at a distance of 1 m. The thickness is 0.2 m, the distributed load 25 kN/m^2, Young's modulus 30.5 kN/mm^2, Poisson's ratio 0.2 and the compression constant 10. Figure 18.13 shows the contour plot for the bending moments m_{xx} and m_{yy} in the plates. These plots would have a vertical line of symmetry per plate if there is no interaction. We see that interaction does take place. One single plate would lead to a settlement of 311 mm at the centre of the slab and

282 mm at the corners. The settlements are larger for two slabs: 334 mm in the corner closest to the other plate, 277 mm in the corner most remote from the other plate, and 350 mm halfway along the edge close to the other plate. Compared to the single plate the corner settlements are in the order of 20% larger.

The difference between the bending moments is small. The maximum bending moments in a single plate occur at the centre of the plate. They are $m_{xx} = m_{yy} = 35.1$ kNm/m. The maximum moment for two plates is in the y-direction: $m_{yy} = 35.2$ kNm/m. This is hardly different from that for the single plate. In the x-direction a noticeable reduction takes place: $m_{xx} = 25.6$ kNm/m, the order of 30%.

18.5 Message of the Chapter

- The stringer panel model for membrane analysis is a special design tool based on both equilibrium and compatibility. It helps understanding the force transfer in a beam with dapped ends and shear walls.

- A new trend in the design of concrete walls and deep beams is the iterative analysis on the basis of concrete pressure only. Then strut-and-tie schemes emerge automatically.

- Advanced orthotropy, coupling membrane action, bending and transverse shear, shows that stresses are overestimated with the classical orthotropic theory. The error is larger in the compressive zone than in the tensile zone.

- For slabs on soil foundations a combination of current FEM, classical solid mechanics and classical geotechnical knowledge leads to a convenient design tool which also accounts for the stiffness of surrounding soil.

Chapter 19
Case History of Cable-Stayed Wide-Box Bridge

In this book we focus on two-dimensional plate structures. In Chapter 10 we already stated that plate elements also may be assembled spatially. Here an example of such structure will be discussed. We consider a critical erection phase of a cable-stayed steel bridge, and will model the structure with only membrane plate elements. The case study is included in the book with the consent of Rijkswaterstaat, the national governmental agency in the Netherlands responsible for infrastructural works [35].

19.1 Introduction

In this chapter we consider a construction phase during the erection of a cable-stayed steel bridge across the river Waal, a branch in the river Rhine delta. The bridge is shown in Figure 19.1. The cross-section is a box with

Figure 19.1 Photo of the cable-stayed bridge.

J. Blaauwendraad, *Plates and FEM: Surprises and Pitfalls*, Solid Mechanics and Its Applications 171, DOI 10.1007/978-90-481-3596-7_19,

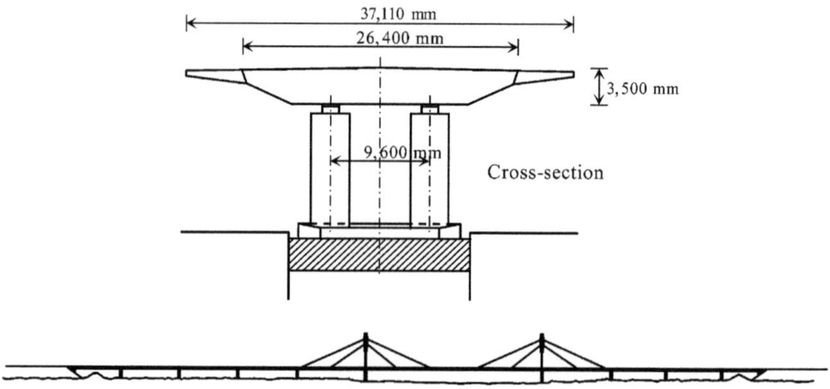

Figure 19.2 View of the bridge (dimensions in mm).

flanges. A longitudinal profile and a cross-sectional profile are shown in Figure 19.2. The pylons are in the centre of the cross-section, so the bridge has just one plane of cables. To obtain sufficient torsion stiffness the designer chose a box structure. The ratio of width and height of the box is over seven, so we consider it as a wide box. The pylon distance is 270 m. The span on the outside of the pylons at both ends is 105 m. The structural height of the box girder is 3.50 m and the width approximately 26 m. Together with the flanges, the width is about 36 meters. No longitudinal inner web is applied in the box girder, except where the stay cables are fixed. The shape of the box girder is determined partly by aerodynamic considerations. Figure 19.3 shows the box girder that is composed of diaphragms and plate panels stiffened by hat sections (prefabricated). Every 5 m there is a cross diaphragm. In Figure 19.4 the situation at an end partition above the support is drawn on a larger scale. The bridge is extended at the front. Sections of 15 m in length are added one by one using a specially developed auxiliary structure. The most dangerous erection phase occurs when the third section of 15 m is put in place. After that the first stay cable is fixed and stresses will reduce substantially. This most dangerous phase is the subject of this chapter. We study particularly the stresses in the cross-section at the pylon.

If all parts of the bridge were subdivided into membrane elements, a very large system of equations would arise. Using three degrees of freedom per node, a cross-section would have about 200 nodes when taking into account all the hat sections. This delivers about 600 degrees of freedom per cross-section. Lengthwise we can think of elements of 5 m long (we will get elon-

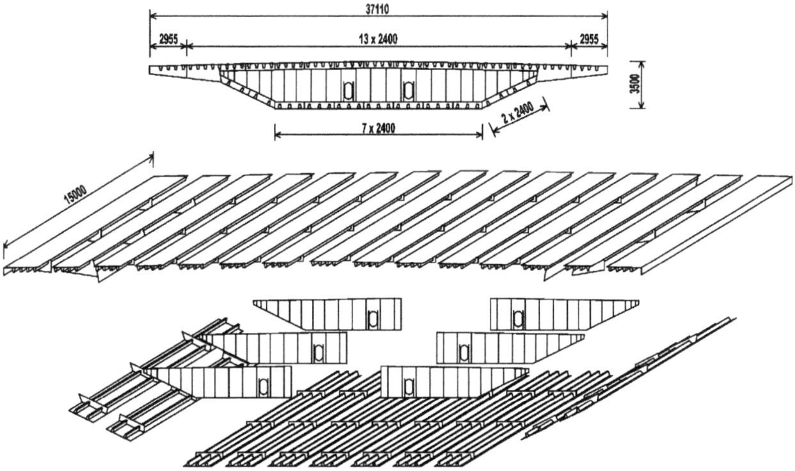

Figure 19.3 Assemblage of box girder sections (dimensions in mm).

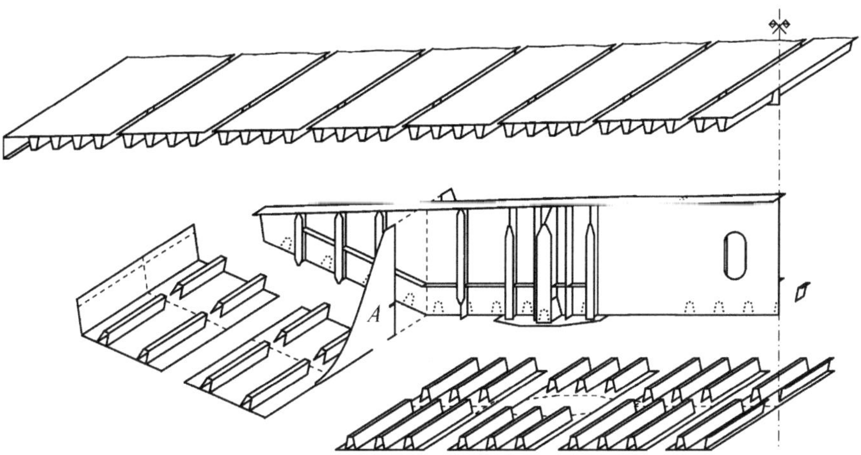

Figure 19.4 End partition and diaphragm above the support.

gated elements). The total bridge would make for approximately 100 cross-sections and therefore 50,000 degrees of freedom with half a bandwidth of about 600. In practice, this is a large system, even for high-speed computers. Such a detailed overall calculation, however, is not very useful for a bridge like this one. In the design stage a calculation with a program for 2D frame structures is sufficient. Bridge parts of a limited size can be calculated with a finite element program, for example for a critical extension phase. The panels with hat-shaped stiffeners may be considered as orthotropic plates loaded in their plane, as if the contribution of the hat sections is smeared out. Only for smaller details is a very fine mesh meaningful.

19.2 Calculation of a Construction Phase

First we state the problem in Section 19.2.1 and present a solution obtained from classical beam theory. Then, the result of the finite element calculation is presented in a number of figures in Section 19.2.2. Because surprising unexpected stress concentrations appeared and the FE code was a new tool at the time of this bridge design (1971), it was then decided to do a verifying test. So we can compare the analysis results with the outcome of a model test. An elastic model of perspex material on a scale of 1 to 50 was built by the TNO Institute for Building Research. For the sake of simplicity, the model was made of isotropic plates, after a FE analysis revealed that all the surprising outcomes appeared in that case as well. The model was subjected to strain gauge measurements and the deflection curve was measured. Hereafter the result of the isotropic FE analysis and the outcome of the test are depicted in the same figure. Such comparisons have been a substantial contribution to the acceptance of FE codes by responsible structural designers, making FE analyses a generally accepted tool. Section 19.3 is a review of the results; Section 19.4 summarizes the message of the chapter.

19.2.1 Problem Definition and Results of Beam Theory

We consider a bridge part of length 30 m which is clamped at one end. The load consists of a new bridge section of 15 m. Figure 19.5 shows how the load case is simplified to a point load causing a constant shear force. We can do this because the dangerous stresses are expected at the clamped end. Because the load is applied above the vertical box walls, a calculation with elements that are loaded only in their plane is possible. The orthotropic membrane constitutive relation reads

$$\begin{Bmatrix} n_{xx} \\ n_{yy} \\ n_{xy} \end{Bmatrix} = \begin{bmatrix} d_{xx} & d_\nu & 0 \\ d_\nu & d_{yy} & 0 \\ 0 & 0 & d_{xy} \end{bmatrix} \begin{Bmatrix} \varepsilon_{xx} \\ \varepsilon_{yy} \\ \gamma_{xy} \end{Bmatrix} \qquad (19.1)$$

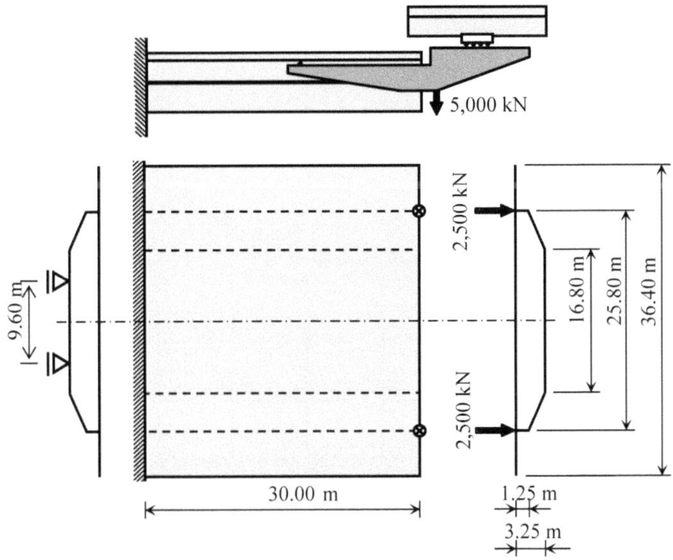

Figure 19.5 Loading in critical construction phase.

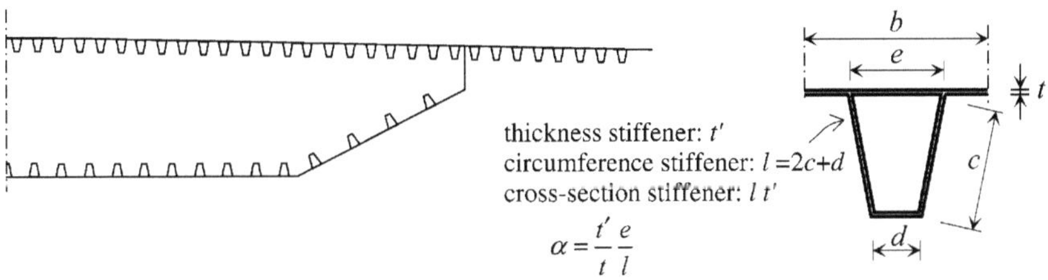

Figure 19.6 Details of hat-stiffened sections.

The calculation of the orthotropic rigidity coefficients is explained in Chapter 20. The reader is encouraged to derive these coefficients on the basis of the dimension data in Figure 19.6. The result of the derivation is

$$d_{xx} = \frac{Et}{1-\nu^2} + Et'\frac{l}{b}, \quad d_\nu = Et\frac{\nu}{1-\nu^2}$$

$$d_{yy} = \frac{E\,t}{1-\nu^2}, \quad \frac{1}{d_{xy}} = \frac{1}{G\,t}\left\{\frac{b-e}{b} + \frac{e}{b}\frac{1}{1+\alpha}\right\} \tag{19.2}$$

As stated, the model of perspex was made of isotropic plates, so from here on also the analyses regard isotropic plates. All box plates are 16 mm thick and the end diaphragm is 24 mm thick. The initial analysis is a calculation of the stresses based on classical beam theory. These stresses have been drawn in Figures 19.7 and 19.8.

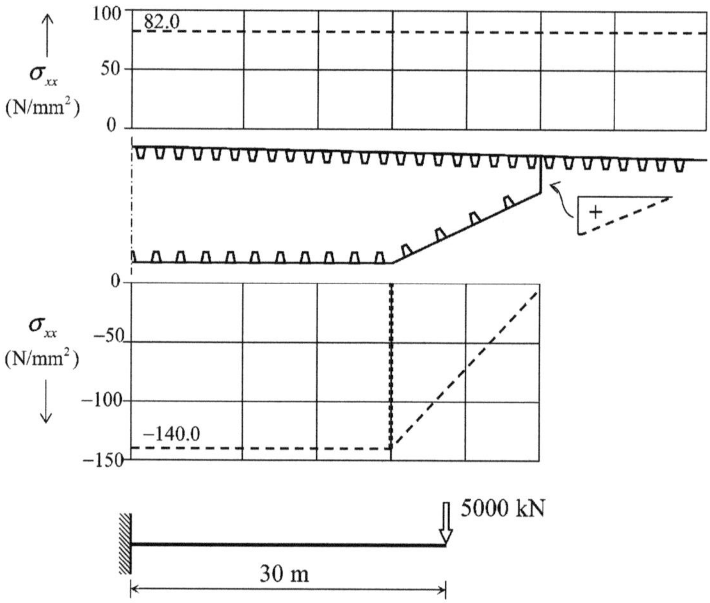

Figure 19.7 Stress σ_{xx} in top and bottom plate according to beam theory.

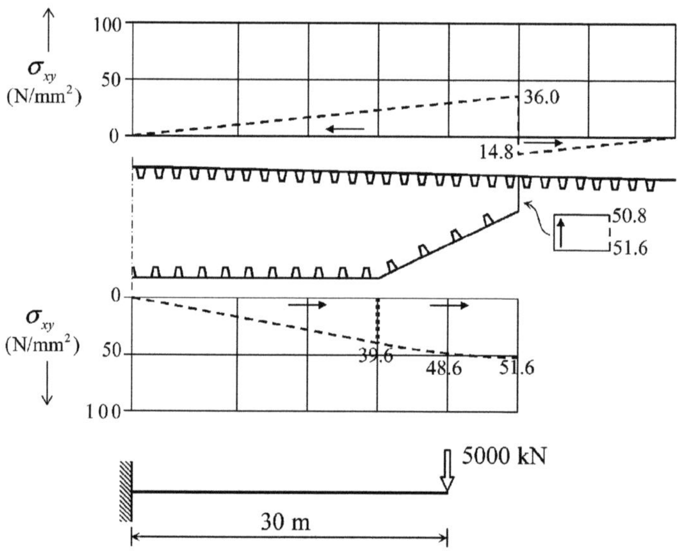

Figure 19.8 Stress σ_{xy} in top and bottom plate at clamped end according to beam theory.

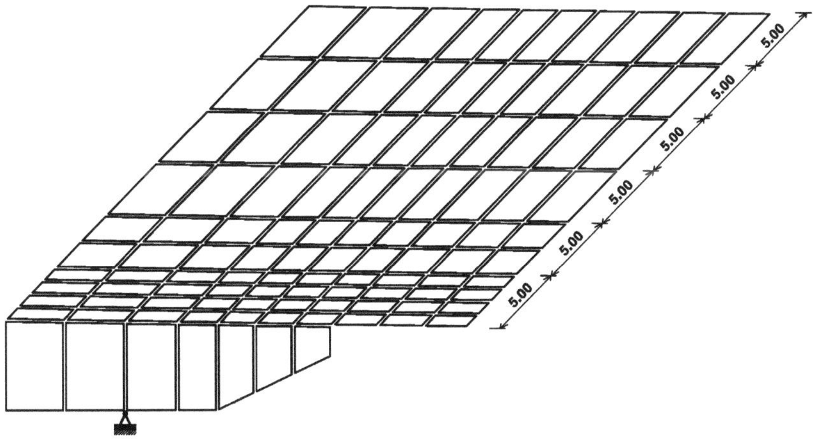

Figure 19.9 Finite element mesh displaying shrunken elements.

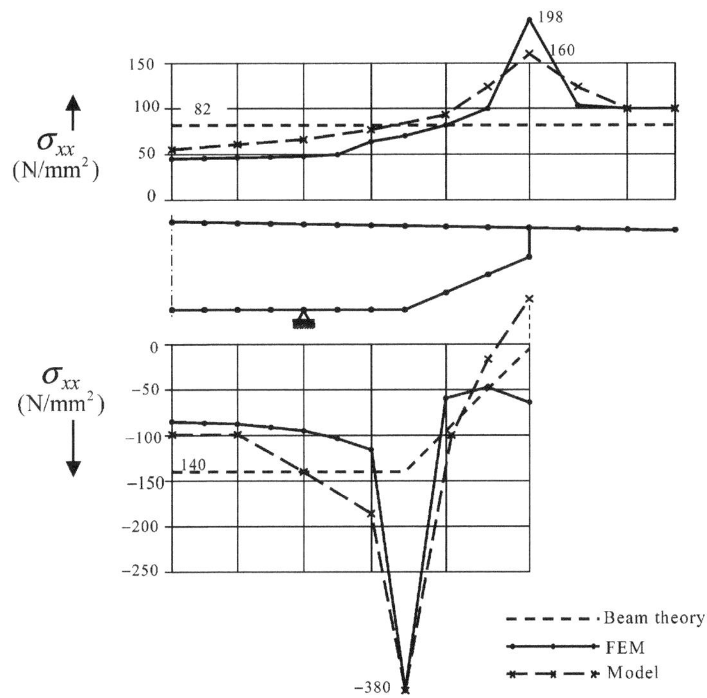

Figure 19.10 Comparison of bending stress σ_{xx} at clamped end with experiments.

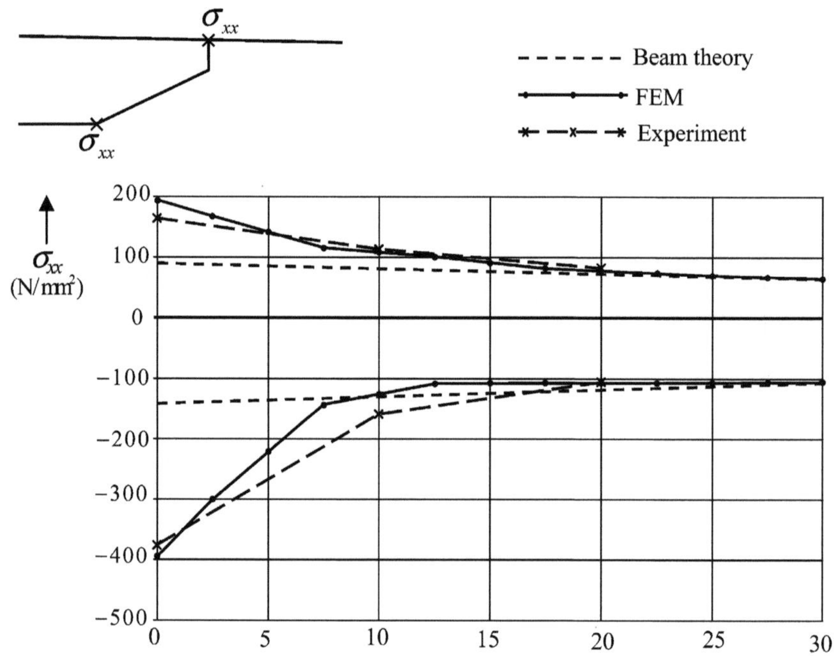

Figure 19.11 Comparison with experiments of σ_{xx} in length direction.

19.2.2 Results of the FE Analysis and the Model Test

The finite element mesh is shown in Figure 19.9. At the position where the vertical web meets the clamped end, stress concentrations were expected due to shear lag and the mesh was refined. We will review the next four figures twice. First, we discuss what FE results are found, and to what extent the model test results confirm the computation. The full lines are the FE results.

FE Results

If the box girder were subjected to a constant moment over the length, the finite element method would deliver results that correspond closely with classical beam theory. For a linearly varying moment, this is not the case. The distribution of the stress σ_{xx} at the clamped end in Figure 19.10 shows large peaks close to the connections with the vertical and the sloping webs. The amplification factor is unexpectedly large, almost 3. The effective plate width to be chosen in an analysis-as-beam is apparently much smaller than the box

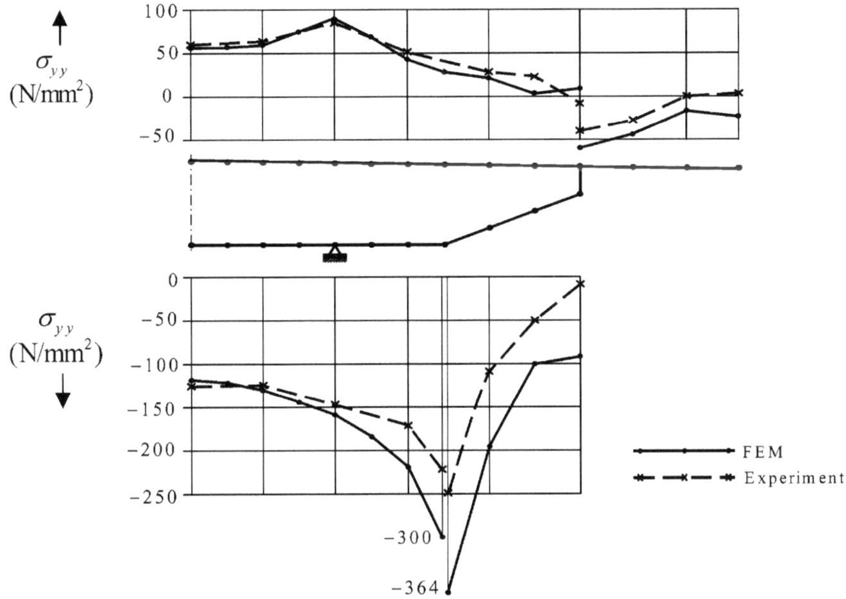

Figure 19.12 Comparison of lateral stress σ_{yy} with experiments.

girder width. Figure 19.11 displays the variation of σ_{xx} in the length direction. It shows that the peak drops off rapidly. The stress σ_{yy} in the circumferential direction is given in Figure 19.12. At the centre of the cross-section, these stresses are of the same order of magnitude as σ_{xx}. According to beam theory, they would be zero. A striking finding is the large peak close to the kink in the bottom plate. The distribution of the shear stress σ_{xy} is drawn in Figure 19.13. The deviation from the results according to beam theory is even larger now. Both in the upper plate and in the bottom plate the sign of the FE result is opposite to the sign according to beam theory. In the bottom plate the sudden change of sign and the large peak values at the kink in the bottom plate attract attention.

Model Test Results

As we said before, Figures 19.10, 19.11, 19.12, and 19.13 also include the model measurements. These are the dashed lines. We found consistent results for all the stresses. The largest difference occurs for σ_{xy} in the horizontal and sloping bottom plates, but the measured gauge values for σ_{xy} do not turn

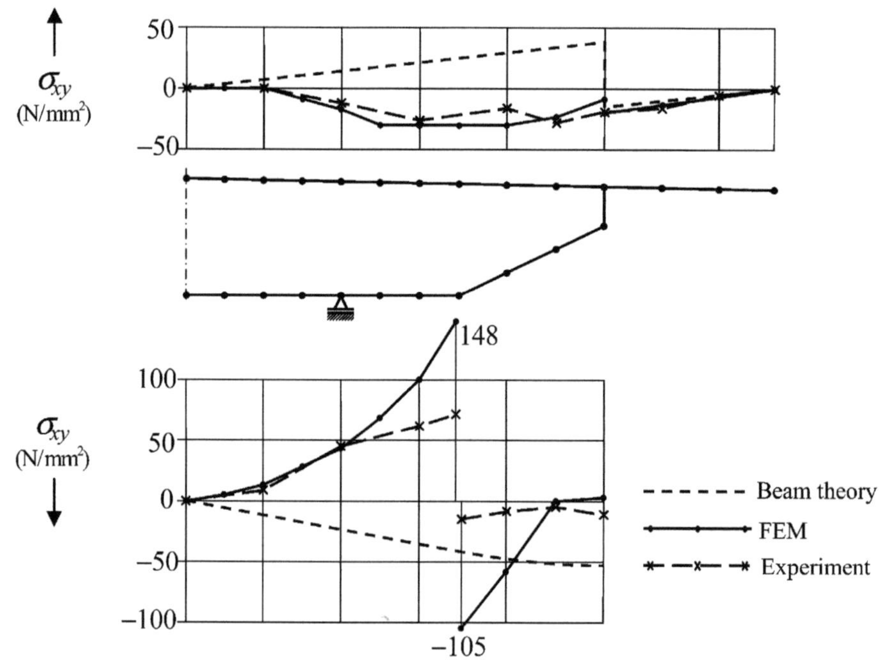

Figure 19.13 Comparison of shear stress σ_{yy} with experiments.

out to be very reliable at the kink in the plates. The vertical resultant of all measured shear stresses is not in equilibrium with the applied point load. In the finite element calculation this equilibrium is satisfied.

19.3 Review of the Results

Now that the results of the calculation have been proved reliable, it is desirable to see if the results can be explained. Perhaps it may be possible to reduce the stress peaks by making local adaptations.

19.3.1 Stress σ_{xx} in Span Direction

The stresses σ_{xx} are the bending stresses of the box-shaped cantilever beam. Their distribution does match the expectation according to the beam theory when the box girder is subjected to a constant moment, but appears not to do so when the moment varies linearly. This indicates the influence

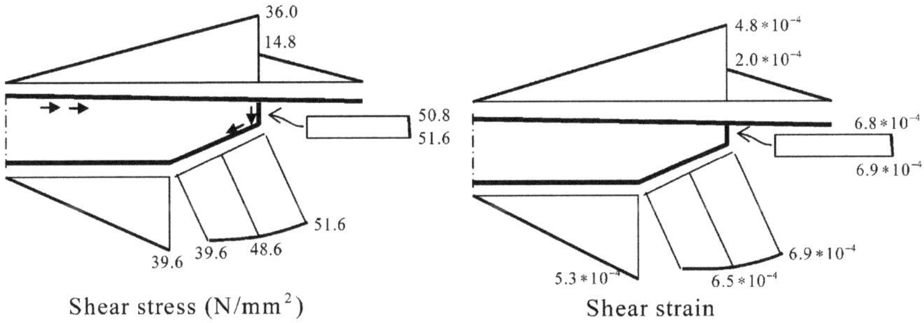

Figure 19.14 Shear stress σ_{xy} and shear strain γ_{xy} according to beam theory.

of the deformation due to the shear force. The outcomes of beam theory are possible only when the associated deformations can take place. The distribution of the shear stresses over the cross-section involves a warping of this cross-section. On the spot of the restraint above the pylon there is a vertical plane of symmetry in which warping cannot occur (clamped end in the analysis). Therefore, a local disturbance system of stresses σ_{xx} occurs to prevent the warping. In Figures 19.14, 19.15 and 19.16 we determine the warping that will occur according to beam theory for unconstrained warping. It shows how the equilibrium system of extra stresses σ_{xx} will look, if it is to bring back the warped section to a plane surface; in the side view of Figure 19.17, this surface is the straight line under angle γ.

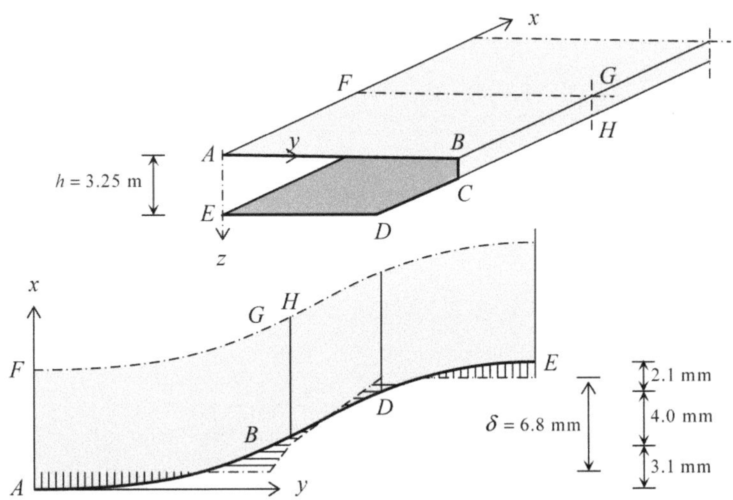

Figure 19.15 Deformation of unfolded box beam due to shear stresses in beam theory.

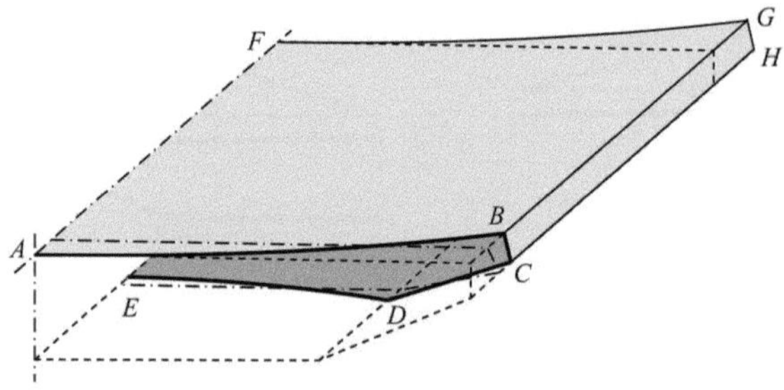

Figure 19.16 Perspective view of warping according to beam theory.

In the top horizontal plate, compression is needed in point A and tension in point B. Therefore, the constant tensile stress in the upper plate according to the beam theory will be reduced at point A and increased at point B. This corresponds with the findings for σ_{xx} in the top plate in Figure 19.10.

In the bottom plate according to Figure 19.17, tension is needed in point E and compression in point D. Compared to beam theory, the magnitude of the constant compressive force gets smaller in point E and larger in point D. This also corresponds with the calculations for the bottom plate in Figure 19.10.

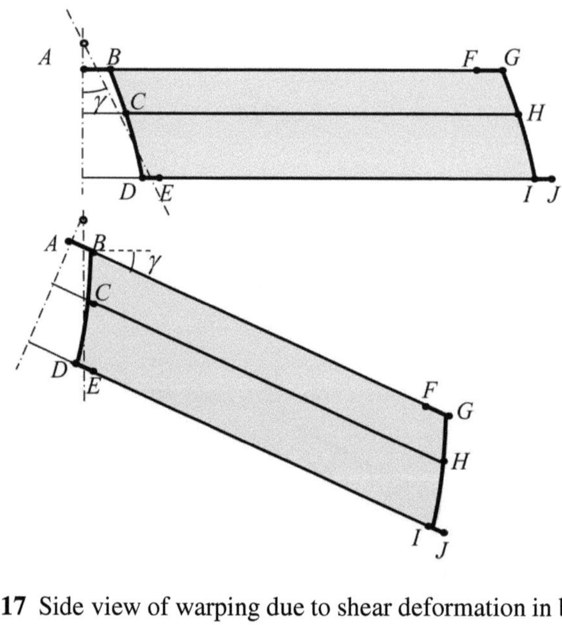

Figure 19.17 Side view of warping due to shear deformation in beam theory.

19.3.2 Stress σ_{yy} in Transverse Direction

Because the supports of the box girder are positioned relatively far inside, the end diaphragm starts to act as a cross-girder. This cross-girder has co-acting flanges, the top and bottom plates of the box. At the ends of this cross-girder, vertical forces act. These are the vertical resultants of the shear stresses in the webs of the box. The cross-girder is bent upwards between the supports. This delivers extra tensile stresses at the top, and extra compressive stresses at the bottom. This way stresses σ_{yy} can arise in the lateral direction, which are in the same order of magnitude as the longitudinal stresses.

The warping does not explain the peak for stress σ_{yy} in the bottom plate depicted in Figure 19.12. We already mentioned that the cross-girder has co-acting flange parts from the top plate and bottom plate. A kink occurs at the bottom between the horizontal flange and the sloping flange. There the flange of the cross-girder abruptly changes slope. We have already discussed this kink problem in Section 11.1.2. There we explained why the flange cannot contribute locally to the force transfer, and that the stresses σ_{yy} in the flanges have to be zero at the kink. The active width of the flange is subsequently zero, so the stresses σ_{yy} rise sharply in that section of the cross-girder. This understanding has led to the addition of a small local vertical plate partition in the longitudinal direction at this kink. In Figure 19.4, this is the part carrying the letter A. This addition halved the peak stress, a very welcome result with respect to buckling, because the bottom plate is compressed in two directions there.

19.3.3 Shear Stress σ_{xy}

If no end diaphragm were present and the stress distribution of beam theory held, the top plate would shorten in the cross direction due to lateral contraction, and the bottom plate would widen in cross direction. However, there is an end diaphragm and it is attached to the plates. So, a shear stream σ_{xy} has to act outwards along the upper plate and inwards on the lower plate. This acts against the direction of the shear stresses according to beam theory for the box bridge. In addition, this effect is amplified by the fact that the supports under the cross beam (and diaphragm) have been placed inwards. In order to get the co-acting top flange under tension and the co-acting bot-

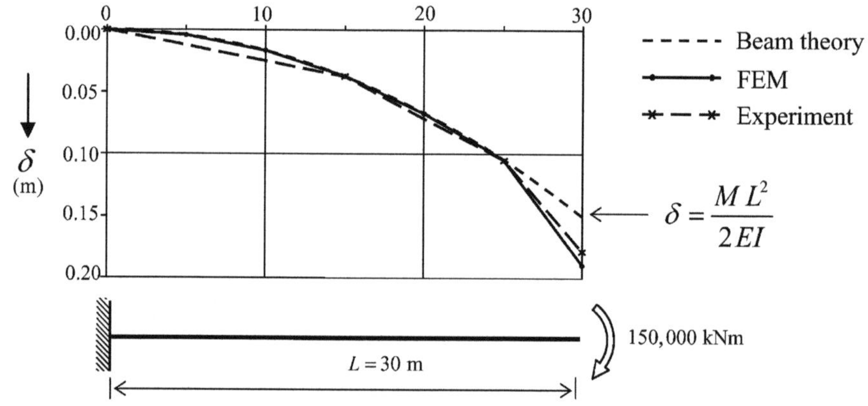

Figure 19.18 Deflection curve for a constant moment.

tom flange under compression, the shear stresses are even more opposed to the direction of beam theory for the box bridge. The final result indeed is a situation with shear stresses that have a different sign than initially found according to beam theory.

Another thing happens close to the kink in the bottom plate. There, we saw that the active flange width of the cross beam abruptly drops to zero. In Section 11.1.2 we explained that the sudden drop of stress σ_{yy} is accompanied by a large gradient in the shear stress σ_{xy}, and that the shear stress changes sign at the kink. This explains what we have seen for the shear stresses in Figure 19.13. By applying the previously mentioned local vertical plate part in the longitudinal direction at the kink, the large discontinuity in the shear stress disappears.

19.3.4 Deflection Diagram

To explain the calculated deflection curve, we first investigate the case of pure bending. If no shear force occurs, beam theory is in very good agreement with the finite element method, as appears from Figure 19.18. Only at the free end of the cantilever box do differences occur, but these can be explained by the way the constant moment is applied to the box bridge. Figure 19.19 shows the deflection curve for the constant shear force due to a point load. The model measurement is once again in good agreement with the finite element calculation.

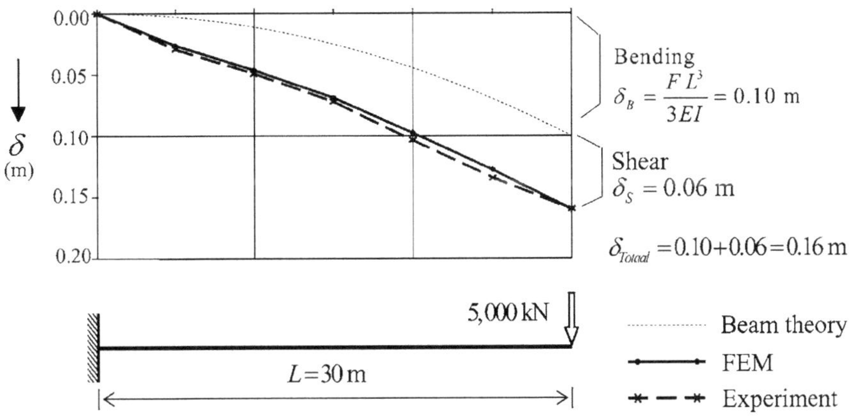

Figure 19.19 Deflection curve for an end point load.

Beam theory, taking into account only bending, delivers a deflection δ_B that is much smaller. At the end this is $FL^3/(3EI)$. This leads to 0.10 m. The difference with the finite element method result has to be due to the shear force deformation. If we use the average γ from Section 19.3.1 the δ_S can be calculated. From Figure 19.15 we know that $\gamma = 6.8/3{,}250 = 2.1 \times 10^{-3}$. Therefore, $\delta_S = \gamma L = (2.1 \times 10^{-3}) \times 30 = 0.06$ m. The total displacement is $\delta_{\text{total}} = 0.10 + 0.06 = 0.16$ m. It appears that the measured end value 0.167 m does correspond quite well with the sum of δ_B and δ_S.

Usually the average γ will be calculated in a different manner, not graphically as we did here, rather by the well-known formula $\gamma = V/GA_s$ in which G is the shear modulus and A_s is the reduced cross-sectional shear area, written as $A_s = A/\eta$, where A is the actual cross-section area and η a shape factor (6/5 for a solid rectangle). The shear area A_s can be determined by equating the expressions for the work

$$\frac{1}{2}\iint_A \frac{\sigma_{xy}^2}{G}\, dA = \frac{1}{2}\frac{V^2}{GA_s} \tag{19.3}$$

in which σ_{xy} is the shear stress according to beam theory for a shear force V. See also Figure 19.20. The procedure is to calculate the shear distribution according to classical beam theory and insert these stresses in Eq. (19.3). If done we find $A_s = 0.032$ m^2, which leads to a shear angle $\gamma = 2.1 \times 10^{-3}$, the same value which is obtained from Figure 19.15.

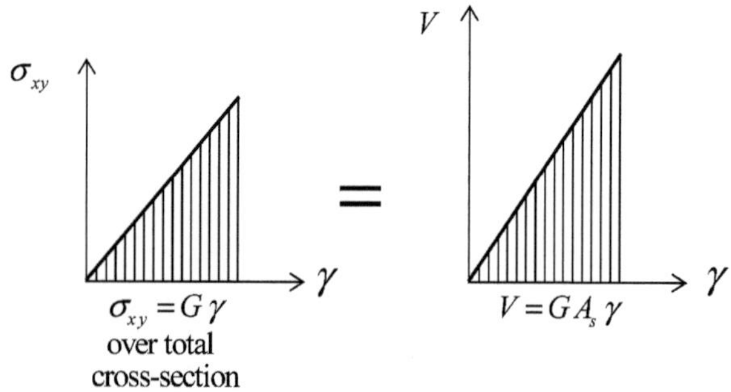

Figure 19.20 Equalizing work done by stress and force leads to the shape factor η.

19.3.5 Evaluation

We can learn two important lessons from the calculation and the interpretation. The first lesson is that the shear deformation is unexpectedly large. If a cable-stayed bridge with this cross-sectional shape is designed using a computer program for beam structures, the program must be able to take into account the shear deformation. The longitudinal profile in Figure 19.1 is deceiving in this respect. At a first glance it would seem to be a slender beam. The distance from the pylon to the first stay cable is 45 m. If the height of the structure is 3.25 m, the ratio span to height is equal to 14. For the calculation of a solid rectangular cross-section, only the bending deformation would be required. This is not true for this wide box girder.

The second lesson is that stress concentrations are important. The considered case is actually a symmetrical problem, in which the shear forces left and right of the support are of the same magnitude but opposite sign, as shown in Figure 19.21. Actually there is a shear force jump ΔV. At every place where

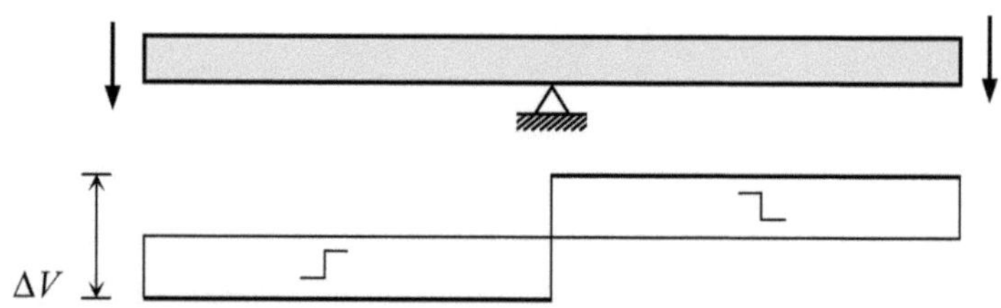

Figure 19.21 Shear force discontinuities cause shear lag disturbances.

a discontinuity in V occurs, extra stresses have to be expected. So, this takes place at each fixing point of stay cables.

19.4 Message of the Chapter

- The analysis of a wide-box bridge as a spatial assemblage of membrane elements reveals surprising stress details, which remain hidden when the structure is handled as a single beam.
- The statement, that shear deformation can be neglected in slender structures, needs refinement. The statement holds true for a derivation on the basis of a rectangular solid cross-section, of which the height is large compared to the width. In a wide-box structure (without intermediate webs in span direction) the shear deformation of the horizontal top and bottom deck can be dominant compared to the shear deformation of the vertical webs. If we want to analyze such a structure with a beam program, we must include shear deformation, even if the structure is slender.
- Restraining of warping occurs in each cross-section where the shear force is discontinuous. In the cable-stayed wide-box bridge, this is the case in each section where a cable is fixed to the bridge.
- Finally, the case study shows that 'the devil is in the detail'. Addition of a really negligibly small steel partition makes high stress melt away like snow in summer.

Part 4
Shape Orthotropy

Chapter 20
Shape-Orthotropic Membrane Rigidities

20.1 Problem Statement

Many plate structures in buildings and bridges cannot be handled as isotropic plates. Stiffeners may occur, and can be different in two orthogonal directions. The multi-cell plate in Figure 20.1 is an example. The figure shows that FE models are possible on different detailed levels. Model 1, consisting of a spatial assemblage of isotropic volume elements, explains the behaviour best. However, the input is complicated and the output massive and complex. The calculation results appear as stresses at nodes and are not easy to interpret or to translate into dimensioning of pre-stressing and reinforcement.

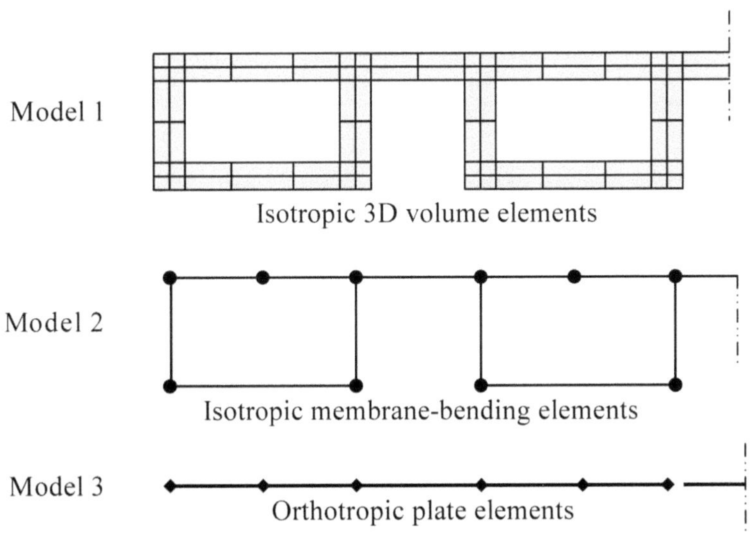

Figure 20.1 Three different levels of FE model for the same structure.

J. Blaauwendraad, *Plates and FEM: Surprises and Pitfalls*, Solid Mechanics and Its Applications 171, DOI 10.1007/978-90-481-3596-7_20,

Model 2 is less detailed. Here plane isotropic elements are used, which are appropriate for both membrane action and bending. The output consists of membrane forces and bending moments in each plate part. This output is less laborious for the structural engineer, but still awkward, because codes of practice usually refer to combinations of normal force, shear force, bending moment and/or twisting moment of total beam cross-sections (I-sections, box sections and others).

As stated earlier, FE codes refer to this element type by the name *shell* element, which suggests that the structure is curved and that coupling between the membrane state and bending state is at stake; neither is the case. We repeat: structural engineers should refuse to use the term shell element. We suggest the name *membrane-bending* element.

In model 3 of Figure 20.1 we no longer recognize the real shape of the structure. We replace it by a flat plate and model it with elements with orthotropic properties. These must be determined on the basis of the geometry of the cross-section and isotropic properties of the material used. We refer to it as *shape-orthotropy*. We can work with this model if stiffeners occur with regular spacing and there is repetition of components. We determine the rigidity properties of repeating parts of the cross-section and smear them over the width of the part. In this chapter we focus on membrane orthotropy. In Chapter 21 we discuss bending orthotropy.

20.2 Occasion of the Chapter

Commercial software with orthotropic features usually offers user-friendly formulas for a number of frequently occurring cross-sections. If the formulas do not apply for the cross-section under consideration, the client has to determine the rigidity data. To do so for shape-orthotropic plates appears to be difficult. We demonstrate this by a practical test. On our invitation three structural engineers determined the rigidity data for the two cross-sections of Figure 20.2, both for membrane action and bending. One structure consists of parallel I-sections which are connected to each other at their top flanges only, and the other of flanged box sections. All three participants are mature and experienced professionals in a consulting engineering bureau, a contractor engineering department and a state government engineering agency, respectively. Commercial FE packages require rigidity matrices for membrane action, bending and shear. Here they are defined in Eqs. (20.1), (20.2) and (20.3), respectively.

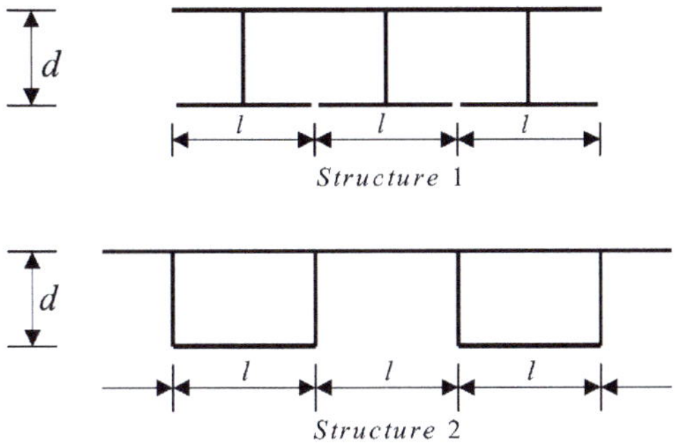

Figure 20.2 Two shape-orthotropic cross-sections.

$$\begin{Bmatrix} n_{xx} \\ n_{yy} \\ n_{xy} \end{Bmatrix} = \begin{bmatrix} d_{xx} & d_v & \\ d_v & d_{yy} & \\ & & d_{xy} \end{bmatrix} \begin{Bmatrix} \varepsilon_{xx} \\ \varepsilon_{yy} \\ \gamma_{xy} \end{Bmatrix} \tag{20.1}$$

$$\begin{Bmatrix} m_{xx} \\ m_{yy} \\ m_{xy} \end{Bmatrix} = \begin{bmatrix} D_{xx} & D_v & \\ D_v & D_{yy} & \\ & & D_{\mathrm{av}} \end{bmatrix} \begin{Bmatrix} \kappa_{xx} \\ \kappa_{yy} \\ \rho_{xy} \end{Bmatrix} \tag{20.2}$$

$$\begin{Bmatrix} v_x \\ v_y \end{Bmatrix} = \begin{bmatrix} D_{sx} & \\ & D_{sy} \end{bmatrix} \begin{Bmatrix} \gamma_x \\ \gamma_y \end{Bmatrix} \tag{20.3}$$

The submitted results for structure 1 are displayed in Figure 20.3. They are normalized by the correct value. The green colour means fine and red wrong.

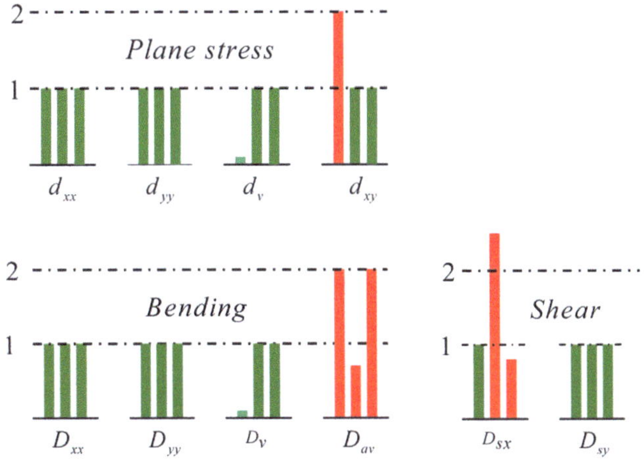

Figure 20.3 Submitted results for structure 1 (I-sections).

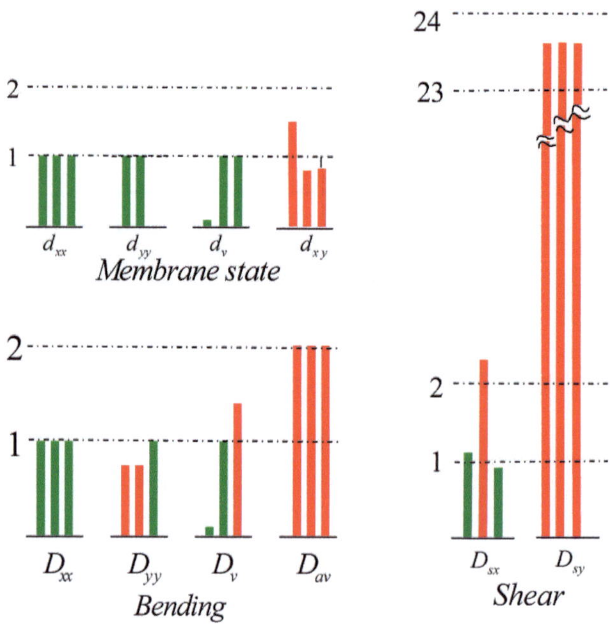

Figure 20.4 Submitted results for structure 2 (box sections).

The rigidity for normal forces and bending moments seems no problem. The rigidity for membrane shear, twisting moment and transverse shear easily is wrong by a factor two.

The results for structure 2 are displayed in Figure 20.4. Again the membrane shear and twisting rigidities are the problem. The shear rigidity for the weak direction is an order of magnitude out of the range. We conclude that it is not clear how to handle rigidities, particularly when shear is involved. We said that providers of FE software may offer user-friendly formulas for frequently occurring cross-section profiles. Experience teaches that even those formulas must be watched critically.

20.3 Membrane Plate with Stiffeners

Equation (20.1) is the constitutive law for membrane plates. The homogeneous isotropic plate is a special case for which Eq. (20.1), according to Section 1.2.2, becomes

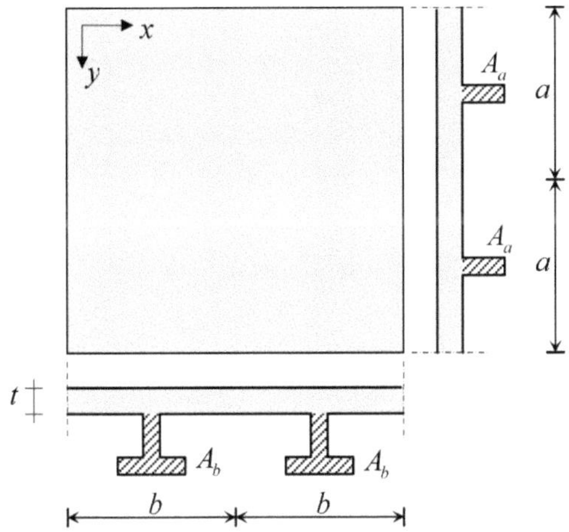

Figure 20.5 Example of orthotropic membrane plate.

$$D = \frac{Et}{1-\nu^2}\begin{bmatrix} 1 & \nu & 0 \\ \nu & 1 & 0 \\ 0 & 0 & \frac{1}{2}(1-\nu) \end{bmatrix}. \tag{20.4}$$

Figure 20.5 shows an orthotropic plate with stiffeners. The structure is an isotropic top plate with added stiffeners, ribs in the x-direction and T-sections in the y-direction. All the material is linear-elastic with Young's modulus E and Poisson's ratio ν. The top plate has thickness t. The cross-sectional area of the stiffeners in the x-direction is A_a and in the y-direction A_b. The stiffener spacings in the x- and the y-direction are a and b respectively. In reality the stiffeners are connected eccentrically, but in the analysis it is assumed that the centre line of the stiffeners coincides with the middle plane of the top plate. For this orthotropic plate we can make a simple parallel chain of the isotropic plate and the stiffeners.

$$D = \frac{Et}{1-\nu^2}\begin{bmatrix} 1 & \nu & 0 \\ \nu & 1 & 0 \\ 0 & 0 & \frac{1}{2}(1-\nu) \end{bmatrix} + \begin{bmatrix} EA_a/a & 0 & 0 \\ 0 & EA_b/b & 0 \\ 0 & 0 & 0 \end{bmatrix} \tag{20.5}$$

This expression, if summed, reads

$$D = \begin{bmatrix} \dfrac{Et}{1-\nu^2} + \dfrac{EA_a}{a} & \nu\dfrac{Et}{1-\nu^2} & 0 \\ \nu\dfrac{Et}{1-\nu^2} & \dfrac{Et}{1-\nu^2} + \dfrac{EA_b}{b} & 0 \\ 0 & 0 & Gt \end{bmatrix} \tag{20.6}$$

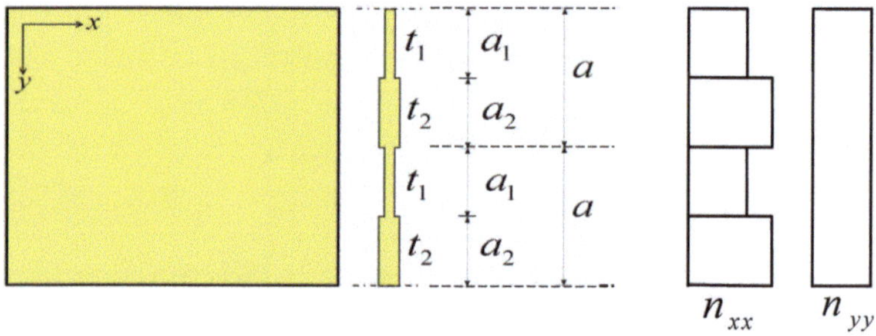

Figure 20.6 Membrane plate strips of different thickness.

After the FE analysis of the plate structure is completed, we will know the strains and can compute the membrane forces in the top plate form the first part of Eq. (20.5) and the forces in the stiffeners from the second part.

20.4 Plate Strips of Different Thickness

The plate of Figure 20.6 consists of plate strips of different thickness. Periodically the strip width is a_1 and a_2 and the thickness t_1 and t_2, respectively. Plate parts of width a repeat continuously. Admittedly, such plates do not arise in reality, but we need one as a stepping stone for membrane orthotropic plates that do occur.

20.4.1 Extensional Rigidity

First we consider the extensional rigidity for normal strains in the x- and the y-direction. The plate parts are chained in parallel in the x-direction and serially in the y-direction. Therefore, the plate parts share the strain in the x-direction, but have different strains in y-direction. At the intersection of two plate parts, $n_{yy1} = n_{yy2}$, and for the two parts together, $a\varepsilon_{yy} = a_1\varepsilon_{yy1} + a_2\varepsilon_{yy1}$. On this basis we can calculate an 'average' ε_{yy} and rigidities d_{xx}, d_{yy} and d_v. It is convenient to introduce fractions f_1 and f_2 defined as follows:

$$f_1 = \frac{a_1}{a}, \quad f_2 = \frac{a_2}{a} \tag{20.7}$$

and auxiliary thicknesses t_{serial} and t_{parallel}. The sum of the fractions is always 1. Using this, we obtain simple expressions for the rigidities

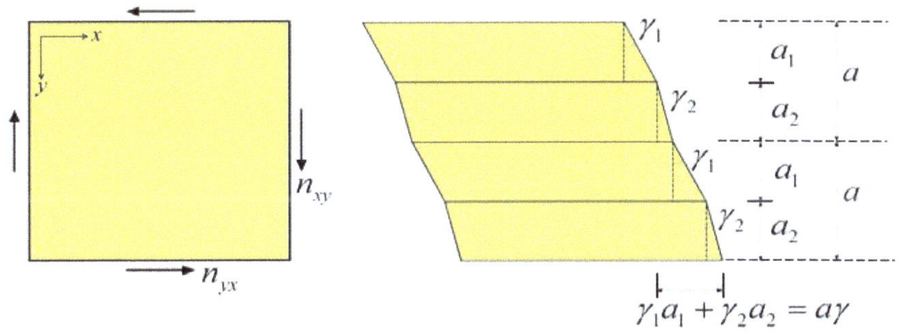

Figure 20.7 Membrane plate strips of different thickness subjected to shear.

$$f_1 = \frac{a_1}{a}, \quad f_2 = \frac{a_2}{a} \tag{20.7}$$

and auxiliary thicknesses t_{serial} and t_{parallel}. The sum of the fractions is always 1. Using this, we obtain simple expressions for the rigidities

$$d_{xx} = \frac{E t_{\text{parallel}}}{1-\nu^2}, \quad d_{yy} = \frac{E t_{\text{serial}}}{1-\nu^2}, \quad d_\nu = \nu \frac{E t_{\text{serial}}}{1-\nu^2} \tag{20.8}$$

where

$$t_{\text{parallel}} = f_1 t_1 + f_2 t_2, \quad \frac{1}{t_{\text{serial}}} = f_1 \frac{1}{t_1} + f_2 \frac{1}{t_2} \tag{20.9}$$

To obtain these formulas we just neglect a small contribution proportional to Poisson's ratio in the formula for d_{xx}.

After completion of the FE analysis, the stresses are computed in the following way. In each plate strip we know the strain ε_{xx} and the membrane force n_{yy}; the strain ε_{xx} is the same in all plate strips, and so is the membrane force n_{yy}. Then we can compute ε_{yy} and n_{xx} from Eq. (1.13). From n_{xx} and n_{yy} we find the normal stresses after division by the thickness of the plate strip.

20.4.2 Shear Rigidity in the y-Direction

We determine the shear rigidity d_{xy} on the basis of Figure 20.7. If subjected to a homogeneous field of membrane shear forces, the plate strips will experience different shear deformation γ_{xy1} and γ_{xy2}. Again serial chaining occurs and we can define an average shear deformation γ_{xy} with aid of the condition $a\,\gamma_{xy} = a_1\gamma_{xy1} + a_2\gamma_{xy2}$. The fictitious thickness t_{serial} of Eq. (2.10) plays a role again. The shear rigidity becomes

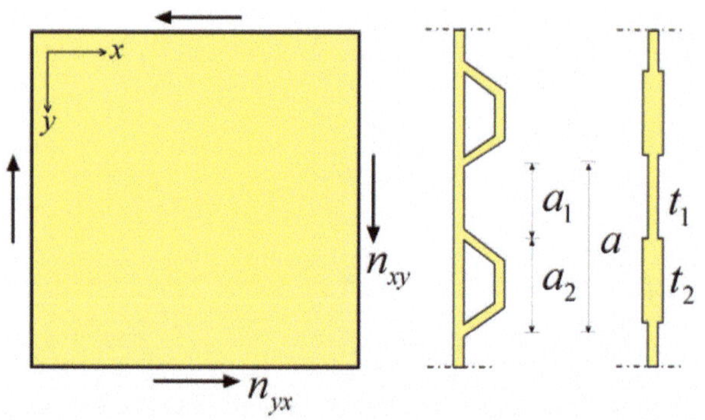

Figure 20.8 Membrane plate with hat stiffeners.

20.5 Plate with Hat Stiffeners

Decks in steel bridges may consist of a top plate stiffened by hat sections as shown in Figure 20.8. We have seen this before in Chapter 19. The extensional rigidity for normal strains is briefly addressed in Section 20.5.1. The shear rigidity needs more attention, which is paid in Section 20.5.2.

20.5.1 Extensional Rigidity

The rigidity terms d_{xx}, d_{yy} and d_ν for normal strains in the x- and the y-direction cannot cause difficulties after the discussion in Section 20.3. What holds for the rib sections there, applies for the hat sections here. The axial stiffness EA_a of the hat stiffener is smeared over the width a and added to the rigidity matrix of the top plate as in Eq. (20.6). In the y-direction the contribution of the hat section can be neglected. It has no noticeable stiffness against membrane forces in that direction.

20.5.2 Shear Rigidity

To determine the shear rigidity of the hat-stiffened plate we distinguish two plate strips of width a_1 and with a_2. The problem to be solved has something in common with the plate of Section 20.4. We can use the solution of that section if we are able to replace the strip with hat section by a single plane strip of fictitious thickness t_2.

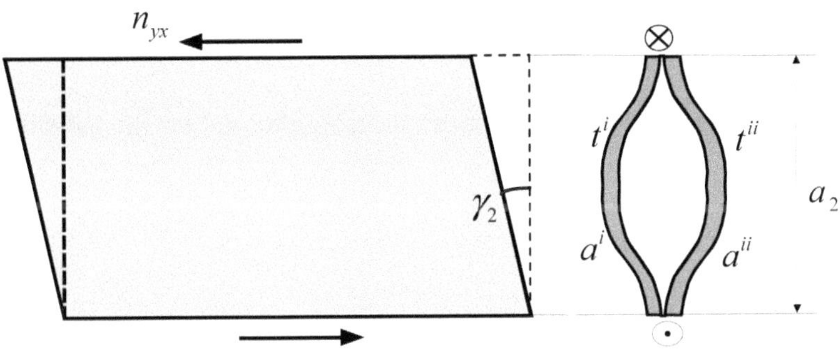

Figure 20.9 Parallel chain of plate parts under shear.

Strip 1 causes no problems; the thickness t_1 is the thickness of the top plate. The fictitious thickness t_2 of the hat-stiffened strip must be calculated from the parallel chain of two parts, the top plate and the hat section. In order to be as general as possible, we start from Figure 20.9 with two plates of developed lengths a' and a'' and thicknesses t' and t'', respectively. The two plates are welded to each other at both ends of the strip. The shear force n_{xy} spreads over the two plates in parts n' and n'', respectively. We induce a shear deformation γ_2 to the combination of the two plates; therefore the one weld will shift with reference to the other over a distance $a_2\gamma_2$. Because of compatibility both plates experience the same shift, however, they spread it over different lengths a' and a''. Therefore it holds $a'\gamma' = a_2\gamma_2$ and $a''\gamma'' = a_2\gamma_2$. From these relations we solve for the shear deformation of the two parallel plates

$$\gamma' = \frac{a_2}{a'}\gamma_2, \quad \gamma'' = \frac{a_2}{a''}\gamma_2 \tag{20.11}$$

It is helpful to introduce auxiliary quantities g' and g''

$$g' = \frac{a_2}{a'}, \quad g'' = \frac{a_2}{a''} \tag{20.12}$$

which look like fractions but will in general not sum to 1. In a parallel chain, strains are common and forces must be added. Therefore

$$n_{xy_2} = n' + n'' \tag{20.13}$$

From this equation we obtain, accounting for Eqs. (20.10) and (20.11), a simple formula for t_2:

$$t_2 = g't' + g''t'' \tag{20.14}$$

The last step in the analysis is to execute the serial chaining of plate strip 1 and the (fictitious) plate strip 2. For that purpose we make use of the formula for t_{serial} in Eq. (20.9)

$$\frac{1}{t_{\text{serial}}} = f_1 \frac{1}{t_1} + f_2 \frac{1}{t_2}, \quad f_1 = \frac{a_1}{a}, \quad f_2 = \frac{a_2}{a} \tag{20.15}$$

So we end up with the formula for the shear rigidity d_{xy} for the hat-stiffened plate

$$d_{xy} = G t_{\text{serial}} \tag{20.16}$$

20.6 Message of the Chapter

- Plates which are stiffened by ribs, T-sections or hat-sections show shape-orthotropic behaviour. They can be calculated by commercially available FE codes, offering the option of orthotropic rigidities.
- Input data for shear rigidities need special attention.
- Two main schemes apply, one for serial chaining of components and one for parallel chains.
- Formulas are derived for a number of frequently occurring plate topologies. The strategy is explained for cases that are not discussed in user manuals of commercial codes.

Chapter 21
Orthotropic Plates in Bending and Shear

21.1 Problem Statement

The constitutive laws for bending and transverse shear in isotropic homogeneous plates were discussed in Chapter 3. We refer to this chapter for definitions of moments and curvatures. There bending moments m_{xx} and m_{yy} occur, and equal twisting moments m_{xy} and m_{yx}, all defined per unit length (therefore have unit of force). The constitutive relationship between moments and curvatures is

$$\begin{Bmatrix} m_{xx} \\ m_{yy} \\ m_{xy} \end{Bmatrix} = \frac{Et^3}{12(1-\nu^2)} \begin{bmatrix} 1 & \nu & 0 \\ \nu & 1 & 0 \\ 0 & 0 & \frac{1}{2}(1-\nu) \end{bmatrix} \begin{Bmatrix} \kappa_{xx} \\ \kappa_{yy} \\ \rho_{xy} \end{Bmatrix} \tag{21.1}$$

If the structure has orthotropic properties, we again have bending moments m_{xx} and m_{yy} and twisting moments m_{xy} and m_{yx}, but now the latter two need not be equal. The twisting deformation ρ_{xy} (which is twice the twisting curvature κ_{xy}) is symmetric, but the torsion rigidities may differ in the x-direction and y-direction: $D_{xy} \neq D_{yx}$. As stated in Section 3.5 we now must replace m_{xy} in Eq. (21.1) by the average value of the two twisting moments, for which we have introduced the name m_{av}. Therefore $m_{av} = \frac{1}{2}(m_{xy} + m_{yx})$, and the constitutive relationship for the general case of bending of shape-orthotropic plates is

$$\begin{Bmatrix} m_{xx} \\ m_{yy} \\ m_{av} \end{Bmatrix} = \begin{bmatrix} D_{xx} & D_\nu & 0 \\ D_\nu & D_{yy} & 0 \\ 0 & 0 & D_{av} \end{bmatrix} \begin{Bmatrix} \kappa_{xx} \\ \kappa_{yy} \\ \rho_{xy} \end{Bmatrix} \tag{21.2}$$

The rigidity D_{xx} represents the bending stiffness in the x-direction and D_{yy} in the y-direction. The off-diagonal term D_ν is due to Poisson's ratio; it

J. Blaauwendraad, *Plates and FEM: Surprises and Pitfalls*, Solid Mechanics and Its Applications 171, DOI 10.1007/978-90-481-3596-7_21,

is zero if no lateral contraction is considered. The torsion rigidity is called D_{av} because it is related to the average moment m_{av}.

Pitfall for torsion rigidity

At this place it is necessary to draw the attention to a tricky pitfall. Many classical text books on bending of isotropic plates deal with the subject matter on the basis of the theory for thin plates and define a relation between the twisting moment m_{xy} and twisting curvature κ_{xy} instead of between the moment m_{xy} and the twisting deformation ρ_{xy}. The latter is used in FE programs and is twice κ_{xy}. The relation in the classic text books for an isotropic plate reads $m_{xy} = D_{\text{classic}}\kappa_{xy}$, where $D_{\text{classic}} = (1-\nu)D$. In the context of FEA we must use $m_{xy} = D_{\text{FEA}}\rho_{xy}$, where $D_{\text{FEA}} = \frac{1}{2}(1-\nu)D$. A factor two occurs between the two definitions of the rigidity: $D_{\text{classic}} = 2D_{\text{FEA}}$. Therefore, if one is not aware of this difference, an error of factor two is easily made. This is in fact the reason why two of three submissions in Figure 20.3 overestimated the torsion rigidity D_{33} by a factor of two.

In thick isotropic plate we must take into account transverse shear deformation. The constitutive law between shear forces v_x and v_y on the one hand and the shear angles γ_x and γ_y on the other is

$$\begin{Bmatrix} v_x \\ v_y \end{Bmatrix} = \begin{bmatrix} D_s & 0 \\ 0 & D_s \end{bmatrix} \begin{Bmatrix} \gamma_x \\ \gamma_y \end{Bmatrix} \tag{21.3}$$

where $D_s = Gt_s$ and $t_s = t/\eta$. The shape factor η accounts for the shear distribution over the cross-section, and is 6/5 for rectangular cross-sections. Note that the same rigidity D_s holds for both directions x and y. For an orthotropic plate this is no longer the case and different rigidities D_{sx} and D_{sy} are defined for the x- and y-direction, respectively. Now the rigidity relation reads

$$\begin{Bmatrix} v_x \\ v_y \end{Bmatrix} = \begin{bmatrix} D_{sx} & 0 \\ 0 & D_{sy} \end{bmatrix} \begin{Bmatrix} \gamma_x \\ \gamma_y \end{Bmatrix} \tag{21.4}$$

Special case: Mindlin theory for slender structure

We again stress that transverse shear distortion is usually taken into account for only thick plates. Then the thickness of the plate is no longer small compared to the span. Speaking about *thick* plates is correct for solid slabs, with material over the full depth. However, some slender structures also need inclusion of shear distortion and therefore application of Mindlin theory. An example is the multi-cell bridge without lateral diaphragms. Vierendeel-like cross-sections are sensitive to shear distortion and must be dealt with as thick plates, even when the structure depth is small and the structure is in fact slender.

21.2 Plate with I-Sections

The plate structure in Figure 21.1 consists of I-sections in the x-direction and a thin upper deck in the x–y plane. All material has Young's modulus E and Poisson's ratio ν. The upper deck is common for the x- and y-directions, therefore we must account for lateral contraction in this component of the cross-section.

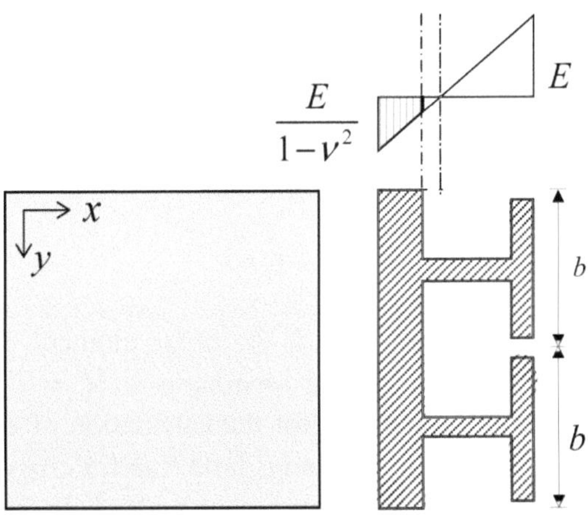

Figure 21.1 Plate with T-stiffeners (I-sections).

21.2.1 Flexural Rigidity

We start with the rigidities for bending and torsion. To determine D_{xx} we calculate the bending stiffness EI_x of the I-section of width a, and smear it over the width. Strictly speaking the I-section is composed of two different materials, because the upper flange experiences the influence of Poisson's ratio, whereas the web and bottom flange do not. So Young's moduli $E/(1-\nu^2)$ and E apply, respectively. Poisson's ratio for concrete is about 0.2, therefore the elasticity modulus of the upper plate is only 4% larger. If we keep in mind that we know the elasticity modulus E at the best with this accuracy, we may neglect the influence of ν on D_{xx}.

For D_{yy} in the lateral y-direction only the upper deck plays a role. The same applies for the off-diagonal rigidity D_ν. Because lateral contraction only occurs in the upper deck we obtain D_ν through multiplication of D_{yy} by Poisson's ratio. The result of all these considerations is

$$D_{xx} = \frac{EI_x}{b}, \quad D_{yy} = \frac{Et^3}{12(1-\nu^2)}, \quad D_\nu = \nu\frac{Et^3}{12(1-\nu^2)} \tag{21.5}$$

21.2.2 Torsional Rigidity

The torsional rigidity needs special attention, because slips are easily made. We start by considering twisting of an isotropic plate. For that purpose we first call to mind how a straight bar is deformed by a twisting moment M_t. We assume an x-axis along the centre line of the bar, call the rotation about this x-axis φ and the twisting (distortion) of the bar θ. The kinematic relation between distortion and rotation is $\theta = d\varphi/dx$. The rigidity relation between the twisting moment m_t and the twisting θ is

$$M_t = GI_t\,\theta \tag{21.6}$$

Here G is the shear modulus, and I_t the polar moment of inertia. Equation (21.6) is the constitutive law for torsion, as is $M = EI\kappa$ for bending. Figure 21.2 shows classic formulas for the calculation of I_t: for a strip, an I-section and a box section. All formulas hold true for cross-section parts in which the thickness t is small compared to the width b. In the formula for the box section, A is the area enclosed by the centre lines of the four composing walls. For equal wall thicknesses the formula becomes simpler. Here B is the circumference, the sum of all wall lengths b.

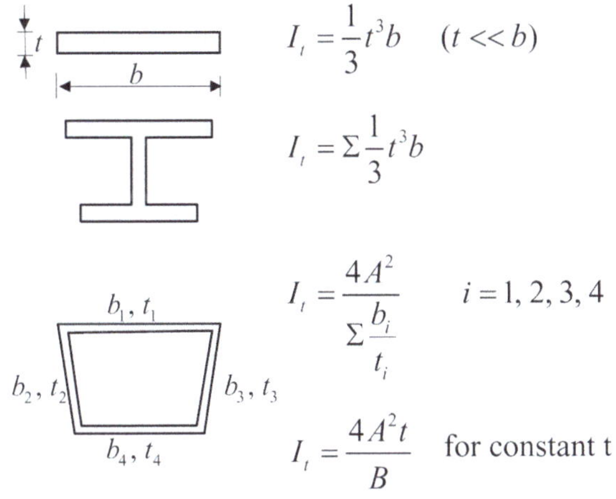

Figure 21.2 Formulas for twisted straighth bars

We repeat the strip of Figure 21.2 in Figure 21.3 and include the flow of stress. The distribution of the shear stress is linear over the thickness over almost the full width b. The resultant per unit length of these shear stresses is a moment, which we here call m_t. The stresses return at the left end with a downward shear flow and at the right end with an upward flow. The return occurs within a strip part of length which is about the strip thickness. If the thickness t is sufficient small compared to the width b then we can concentrate the resultant of all vertical stresses at both strip ends as one vertical

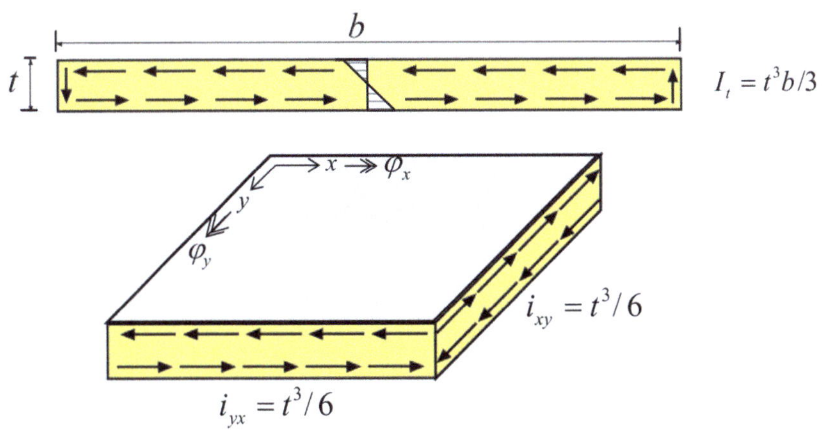

Figure 21.3 Shear stress in cross-section of strip-shaped bar.

force V at the end face. This force has the value m_t, fully comparable with that found from the theory of thick plates as discussed in Section 3.6 and thin plates as discussed in Section 4.4.2. The horizontal stresses occur over a long length, but have a very short lever arm, whereas the vertical stresses occur over a small length, but have a long lever arm. The effect is that the total moment M_t is carried half by the horizontal shear stresses and half by the vertical ones. The polar moment of inertia for the strip is $I_t = \frac{1}{3}bt^3$ and is due to both the horizontal and vertical stresses. The horizontal stresses are responsible for the half of I_t, so for the contribution $i_t = t^3/6$ per unit length.

In the bottom part of Figure 21.3 a square part is drawn, which occurs inward a homogeneous isotropic plate, subjected to a constant twisting moment m_{xy}. Each side of the part has unit length, and there are horizontal shear stresses on all four faces. If we consider this part as an element which is extracted from the strip in the top part of the figure, it immediately is understood that the torsional rigidities per unit length i_{xy} in the face normal to the x-axis and i_{yx} in the face normal to the y-axis are

$$i_{xy} = \frac{1}{6}t^3, \quad i_{yx} = \frac{1}{6}t^3 \tag{21.7}$$

Let us consider the square element as a bar in the x-direction with torsional rigidity i_{xy}. For this bar it holds $m_{xy} = Gi_{xy}\theta$. We also know $\theta = d\varphi/dx$ and $\varphi = -dw/dy$, so we arrive at the expression

$$m_{xy} = -Gi_{xy}\frac{\partial^2 w}{\partial x \partial y} \tag{21.8}$$

The mixed second derivative with negative sign is the plate torsional curvature κ_{xy} which in turn is $\rho_{xy}/2$. Therefore, Eq. (21.8) transforms into

$$m_{xy} = G\frac{i_{xy}}{2}\rho_{xy} \tag{21.9}$$

In an isotropic plate there is no difference between the x- and the y-directions, so the same equation applies for m_{yx}. Summing up for an isotropic plate, we define the rigidities by

$$m_{xy} = D_{xy}\rho_{xy}, \quad m_{yx} = D_{yx}\rho_{xy} \tag{21.10}$$

Here the rigidities are

$$D_{xy} = G\frac{i_{xy}}{2}, \quad D_{yx} = G\frac{i_{yx}}{2} \tag{21.11}$$

They are equal for isotropy, but they become different for orthotropic plates. On the basis of our discussion of isotropic plates, we can quickly clarify how to deal with orthotropic properties. It still holds true that one torsional deformation ρ_{xy} occurs, but the twisting moments and torsional moments of inertia are different: $m_{xy} \neq m_{yx}$ and $i_{xy} \neq i_{yx}$. Now we have different relations $m_{xy} = D_{xy}\rho_{xy}$ and $m_{yx} = D_{yx}\rho_{xy}$. The constitutive relation for orthotropy is $m_{\text{av}} = D_{\text{av}}\rho_{xy}$ in which $m_{\text{av}} = \frac{1}{2}(m_{xy} + m_{yx})$. From this we derive the definition for the rigidity D_{av} and the torsional moment of inertia i_{av}:

$$
\begin{aligned}
D_{\text{av}} &= G\frac{i_{\text{av}}}{2} \\
i_{\text{av}} &= \frac{1}{2}(i_{xy} + i_{yx})
\end{aligned}
\tag{21.12}
$$

We can check if this definition leads to the correct result for an isotropic plate. In that case $i_{xy} = i_{yx} = t^3/6$ and $i_{\text{av}} = \frac{1}{2}(i_{xy} + i_{yx}) = t^3/6$. Remembering $G = \frac{1}{2}E/(1 + \nu)$, we find $D_{\text{av}} = Et^3/(24(1 + \nu))$. This outcome is equivalent to the product of $Et^3/12(1 - \nu^2)$ and $\frac{1}{2}(1 - \nu)$ in Eq. (21.1). The definition in Eq. (21.12) leads to the correct torsional rigidity in an isotropic plate and is a quick guide for the determination of the rigidity in case of orthotropy.

After a FE analysis is completed, the output consists of m_{av} values, which are the average of the twisting moments m_{xy} and m_{yx}. The values m_{xy} and m_{yx} can be solved from two equations. The first equation tells us that the sum of the two moments is $2m_{\text{av}}$. The second equation tells that the quotient of the moments is i_{xy}/i_{yx}. The solution of the two equations is

$$
m_{xy} = \frac{2i_{xy}}{i_{xy} + i_{yx}} m_{\text{av}}, \quad m_{yx} = \frac{2i_{yx}}{i_{xy} + i_{yx}} m_{\text{av}} \tag{21.13}
$$

Example

We apply the derived equations to the structure with I-sections in Figure 21.1 and consider a section of width b. The continuous top deck has thickness t; the web is a strip with thickness t_w and width b_w; the bottom flange is a strip with thickness t_f and width b_f. The continuous top deck is in the state of the plate element in Figure 21.3 and therefore will contribute $t^3/6$ to both i_{xy} and i_{yx}. The web and flange contribute only to i_{xy}. Their full strip rigidities $b_w t_w^3/3$ and $b_f t_f^3/3$, respectively, are assigned to the x-direction and must be

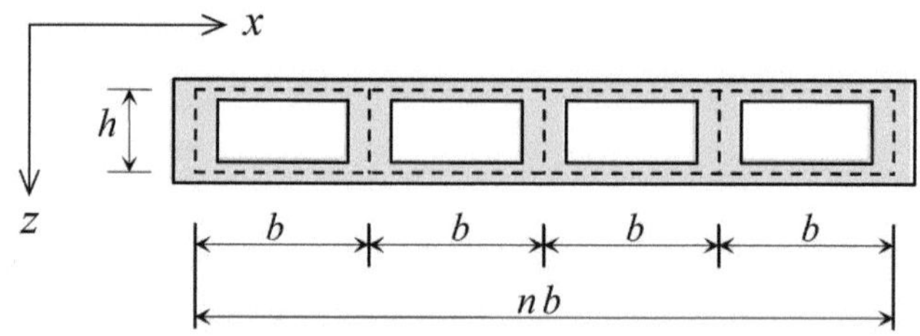

Figure 21.4 Multi-cell bridge.

smeared out over the section width b. In this way we obtain the following result:

$$i_{xy} = \left(\frac{1}{3} b_w t_w^3 + \frac{1}{3} b_f t_f^3 \right) \frac{1}{b} + \frac{1}{6} t^3, \quad i_{yx} = \frac{1}{6} t^3 \tag{21.14}$$

These must be substituted in Eq. (21.12) to find the rigidity D_{av}.

21.2.3 Shear Rigidity

A plate which consists of I-sections in the x-direction is not sensitive to transverse shear distortion. Just for completeness, we treat the shear rigidities here. If we call the shear area of the I-section A_{sx}, then the rigidities are

$$D_{sy} = \frac{5}{6} G t, \quad D_{sx} = G A_{sx} / a \tag{21.15}$$

The shear area A_{sx} is the real cross-section area A divided by the shape factor η. Engineering handbooks provide tables with values for various section shapes.

21.3 Multi-Cell Bridge

We consider the concrete plate of Figure 21.4, a chain of box cells. The x-axis is in the lateral direction of the plate. The y-axis is normal to the plane of drawing. No diaphragms have been placed in the lateral direction. Each box has the width b.

21.3.1 Flexural Rigidity

Strictly speaking we should use the elasticity modulus $E/(1-\nu^2)$ for the horizontal top and bottom walls, and E for the vertical walls. Since ν^2 is small for concrete, we will work with one elasticity modulus E when determining D_{yy}. The determination of the flexural rigidity EI_y is straightforward. The flexural rigidity Ei_x in the x-direction will be smaller than in the y-direction, because no vertical walls occur in that direction. The determination of D_ν must be based on the flexural rigidity in which the vertical webs play no role, so D_ν depends on Ei_x. We obtain

$$D_{xx} = Ei_x, \quad D_{yy} = \frac{EI_y}{b}, \quad D_\nu = \nu D_{xx} \tag{21.16}$$

21.3.2 Torsion Rigidity

Box girders have large torsional stiffness. The inner vertical webs have some influence on the stiffness but not much; the flow of shear stresses in the outer vertical walls and the two horizontal decks determines the torsional stiffness to a large extent. We make only a little error if we set the inner walls aside. So we calculate the polar moment of inertia I_t of the complete wide box of n cells with the box formula in Figure 21.2. We assign half of this polar moment of inertia to the x-direction and half to the y-direction. This leads to the result

$$i_{xy} = \frac{1}{2}\frac{I_t}{nb}, \quad i_{yx} = \frac{1}{2}\frac{I_t}{nb}, \quad i_{\text{av}} = \frac{1}{2}(i_{xy} + i_{yx}) \tag{21.17}$$

Substitution in Eq. (21.12) leads to the required torsional rigidity D_{av}. For regular structures such as the plate under consideration, we could have made a short cut and obtained the same result. We could have computed I_t of the complete wide box, divided this by the total width nb and taken half the value of it. Experienced engineers who often make this type of calculations will recognize a pattern and can make use of gained insight.

Example

A multi-cell bridge as shown in Figure 21.4 has eight rectangular cells of height $h = l$ and width $b = 2l$. All walls have thickness t. From this data it follows

$$A = (8 \times 2l)l = 16l^2, \quad B = 16 \times 2l + 2 \times l = 34l$$

$$I_t = \frac{4A^2 t}{B} = \frac{4(16l^2)^2 t}{34l} = \frac{512}{17} t l^3 \tag{21.18}$$

$$\begin{aligned} i_{xy} = i_{yx} &= \frac{I_t/2}{8\,(2l)} = \frac{16}{17} t l^2 \\ i_{\text{av}} &= \frac{1}{2}(i_{xy} + i_{yx}) = \frac{16}{17} t l^2 \end{aligned} \tag{21.19}$$

The length l is equal to $b/2$, therefore the expression transforms into

$$i_{\text{av}} = \frac{16}{17} t \left(\frac{b}{2}\right)^2 = \frac{4}{17} t b^2 \tag{21.20}$$

This must be introduced into Eq. (21.12) to find D_{av}.

21.3.3 Shear Rigidity

X-Direction

The transverse shear stiffness in the x-direction is very limited. Figure 21.5 shows the distortion due to a shear force. We want to determine D_{sx} in the shear relation

$$v_x = D_{sx}\, \gamma_x \tag{21.21}$$

We can make an analysis with a frame program to determine the deflection δ due to a given shear force v_x, from which we find the relation $v_x = K_{sx}\delta$. The shear angle γ_x over the considered box is $\gamma_x = \delta/b$. Therefore D_{sx} becomes

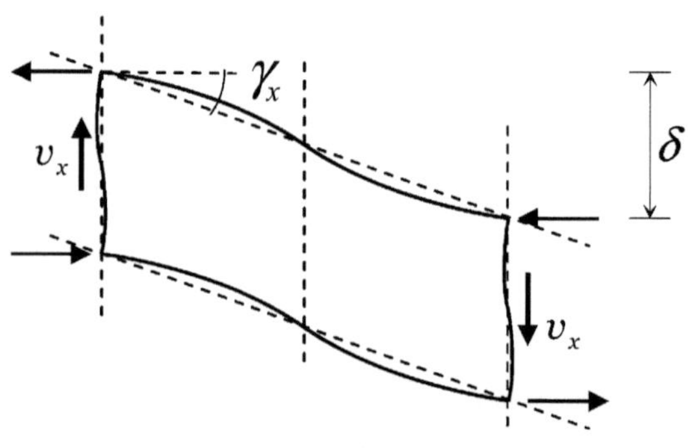

Figure 21.5 Shear distortion of one cell in a multi-cell bridge.

$$D_{sx} = bK_{sx} \tag{21.22}$$

The axial strain of the members of the box frame is formally taken into account, but its influence is in fact small. This means that the deformation due to a bending moment in the lateral x-direction of the orthotropic plate is very small compared to the distortion due to a shear force.

Usually the depth of the multi-cell plate is smaller than the length b of the cells and the thickness of the vertical walls thicker than the thickness of the horizontal walls. Therefore, the bending deformation of the vertical walls in Figure 21.5 is negligible compared to the deformation of the horizontal walls. Then we need no frame program and obtain for equal thickness t of the two horizontal walls

$$K_{sx} = 2\left\{12\frac{EI}{b^3}\right\} = 2\left\{12 \times \frac{Et^3/12}{b^3}\right\} = 2E\frac{t^3}{b^3} \tag{21.23}$$

Substitution in Eq. (21.22) leads to the shear rigidity

$$D_{sx} = bK_{sx} = 2Et\left(\frac{t}{b}\right)^2 \tag{21.24}$$

Structural engineers may just sum the shear rigidities of the top and bottom wall. For a thc top platc thcy would find $D_{sx} = Gt/\eta$ with $\eta = 1.2$. For $\nu = 0.2$ this leads to $D_{sx} = 0.35Et$, which becomes $D_{sx} = 0.7Et$ for the top and bottom plate together. This is of the order Et. In Eq. (21.24) the order is a factor $(t/b)^2$ different. Because $t/b \ll 1$, we find that Eq. (21.24) leads to a substantially smaller D_{sx} than $0.7Et$.

Pitfall for shear rigidities

Structural engineers may get in the pitfall in summing the shear rigidities of the top and bottom wall of a multi-cell bridge. Merely summing shear rigidities is very misleading. It leads to a shear rigidity which is an order of magnitude too large.

Y-Direction

The shear rigidity in the y-direction is large. It must be computed as discussed for the I-section in Section 21.2.3. The formula is

$$D_{sy} = GA_{sy}. \tag{21.25}$$

Here A_{sy} is the real cross-section area A divided by the shape factor η. If the vertical walls have thicknesses of about the size t, the value of D_{sy} will be of the order Gt. The size of η is much larger than 1, for instance 5 or more. Structural engineers may prefer to relate the shape factor to the web area because that member must in fact carry the shear force. If related to the cross-section area of only the vertical web the factor decreases to a value only little more than 1, which will appeal to structural engineers.

21.3.4 Combination of Shear Force and Twisting Moment

After completion of the FE analysis the structural engineer wants to know the shear force in the vertical webs of the multi-cell plate. Is the definition of Eq. (3.41), with reference to Figure 21.4, still valid? And can we multiply this by the spacing of the vertical webs to obtain the shear force in the web?

The shear force v_y consists of two contributions v_{yb} and v_{yt}, due to the bending moment m_{yy} and the twisting moment m_{xy}, respectively. The first contribution $v_{yb} = \partial m_{yy}/\partial y$ needs no further explanation. Therefore we will focus on the term $v_{yt} = \partial m_{xy}/\partial x$. We suppose a diagram for the twisting moment as depicted in Figure 21.6. In general, the moment has different values at the various cells. In the special case that the moment is equal in all cells, a constant shear flow n occurs in the horizontal top plate and bottom plate with size

$$n = \frac{m_{xy}}{h} \tag{21.26}$$

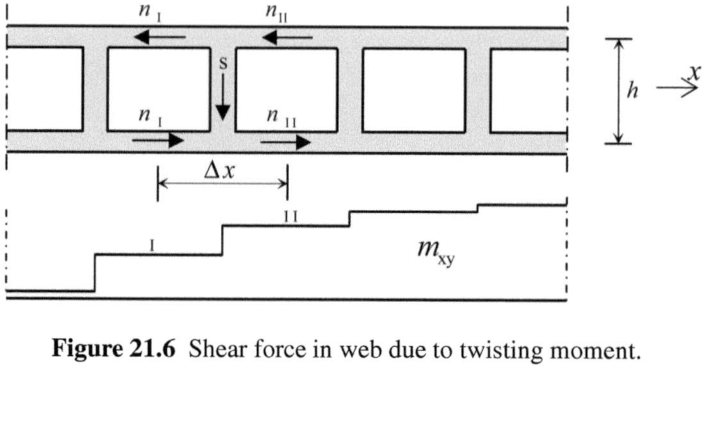

Figure 21.6 Shear force in web due to twisting moment.

where h is the distance between the middle plane of the top and bottom plate. The constant twisting moment does not raise a vertical shear force in the webs, which is in conformance with the supposition that we neglect the inner vertical webs when we determine the rigidity. In the general case of a varying twisting moment, the value of n will be different for each cell. The formula in (21.26) is now applied to each cell. The twisting moment will evoke a vertical shear force s in the inner webs of the size

$$s = n_{\mathrm{II}} - n_{\mathrm{I}} \tag{21.27}$$

This follows from equilibrium of the joint between the vertical web and the two adjacent horizontal plate parts in the top plate or bottom plate. The analogy of a fluid flow is helpful in understanding this. We call the spacing of the vertical webs Δx and will edit Eq. (21.27) by multiplying both members of the relationship by h

$$hs = hn_{\mathrm{II}} - hn_{\mathrm{I}} \tag{21.28}$$

The left-hand member is the shear force V in the web due to the twisting moment. The right-hand member is the difference of $m_{xy\,\mathrm{II}}$ and $m_{xy\,\mathrm{I}}$, so Δm_{xy}. We can write Eq. (21.28) as

$$V = \Delta m_{xy} \tag{21.29}$$

We smear out V over the width Δx and replace it by $v_{yt}\Delta x$. This transfers Eq. (21.29) into

$$v_{yt} = \frac{\Delta m_{xy}}{\Delta x} \tag{21.30}$$

Together with $v_{yb} = \partial m_{yy}/\partial y$ and if Δx is small compared to the width of the slab we obtain

$$v_y = \frac{\partial m_{yy}}{\partial y} + \frac{\partial m_{xy}}{\partial x} \tag{21.31}$$

We conclude that the equation for the shear force v_y is still fully valid for the multi-cell bridge. After completion of the FE analysis we multiply the outputted v_y by the spacing Δx to obtain the transverse shear force in the vertical web.

21.4 Plate with Separate Boxes

21.4.1 Flexural Rigidity

The cross-section shown in Figure 21.7 consists of individual flanged box sections of width b. The x-axis is in the plane of the paper and the y-axis

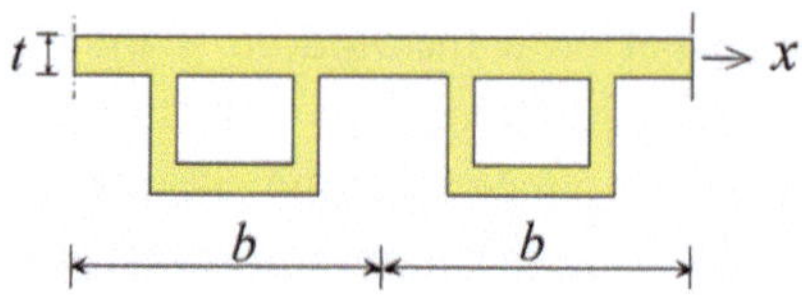

Figure 21.7 Plate of box sections with flanges.

is perpendicular to it. We determine the flexural rigidity D_{xx} by subjecting one box section to a bending moment m_{xx}, as shown in Figure 21.8, and compute the rotation φ at each end. The stiffness relation is $C = m_{xx}/2\varphi$. Because of the relation $m_{xx} = D_{xx}\kappa_{xx}$ we find

$$D_{xx} = \frac{C}{b} \tag{21.32}$$

The flexural rigidity D_{yy} in the y-direction is calculated as discussed for the plate with I-sections. We calculate EI_y of the box beam, including the flanges, and find

$$D_{yy} = \frac{EI_y}{b} \tag{21.33}$$

The off-diagonal term D_ν in the rigidity matrix is related to the stiffness of the top plate only. Application of a moment m_{xx}, as shown in Figure 21.8, leads to a smaller moment m_{xx1} in the top wall of the box. A moment $m_{yy} = \nu m_{xx}$ occurs in the flanges of the box beam and a moment $m_{yy1} = \nu m_{xx1}$ in the top wall of the box. We define a reduced rigidity $D_{xx,\text{red}}$ in order to define D_ν

$$\begin{aligned} D_\nu &= \nu D_{xx,\text{red}} \\ D_{xx,\text{red}} &= \frac{b_o m_{xx} + b_1 m_{xx,1}}{b m_{xx}} D_{xx}. \end{aligned} \tag{21.34}$$

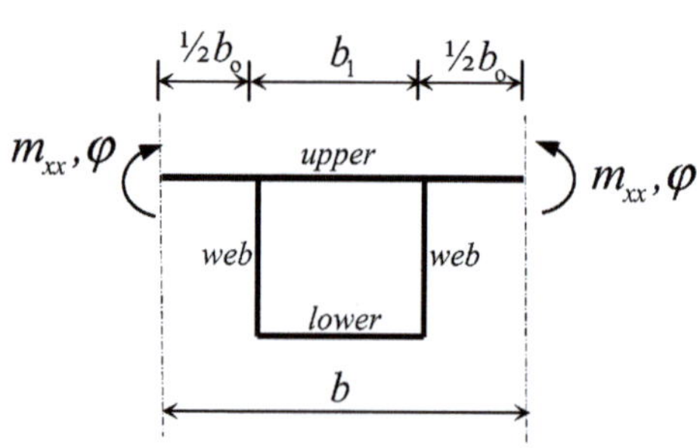

Figure 21.8 The lateral bending stiffness is determined by the relation between m_{xx} and φ.

21.4.2 Torsional Rigidity

In Figure 21.2 we have given formulas for the calculation of the torsional moment of inertia I_t of box girders with thin walls. If the wall thickness is not small compared to the width and depth of the box girder, then the contribution $bt^3/3$ or $bt^3/6$ of the individual box-sections must be added to I_t. For steel boxes this is never necessary, for boxes of reinforced concrete it may be. Defining i_{xy} for the section normal to the x-axis halfway between two boxes, and i_{yx} for the section normal to the y-axis, we find

$$i_{xy} = \frac{1}{6} t_{\text{top}}{}^3,$$

$$i_{yx} = \frac{1}{b} \left\{ I_t + t_{\text{top}}^3 b/6 + t_{\text{bottom}}^3 b_{\text{bottom}}/3 + 2 \left(t_{\text{web}}^3 b_{\text{web}}/3 \right) \right\} \quad (21.35)$$

This we must substitute in Eq. (21.12).

Example

We apply Eq. (21.35) to the structure in Figure 21.7. In this example we choose all wall lengths $l = b/2$. All thicknesses are t. Then

$$A = l^2, \quad B = 4l \rightarrow I_t = \frac{4A^2 t}{B} = \frac{4(l^2)^2 t}{4l} = tl^3 \quad (21.36)$$

Working out of Eq. (21.35) now leads to

$$i_{xy} = \frac{1}{6} t^3$$

$$i_{yx} = \frac{1}{2l} \left\{ tl^3 + (2l)\, t^3/6 + 3 \left(t^3 l/3 \right) \right\} = \frac{1}{2} tl^2 + \frac{2}{3} t^3 \quad (21.37)$$

$$= \frac{1}{2} tl^2 \left\{ 1 + \frac{3}{4} \left(\frac{t}{l} \right)^2 \right\}$$

We can assume that the ratio t/l will be about 0.2. Therefore the term between braces is practically equal to 1 and $i_{yx} = tl^2/2$. This has the consequence that i_{xy} will be negligible with respect to i_{yx}. Finally we can make the transformation from an expression in l to an expression in b, which leads to

$$i_{xy} = \frac{1}{6}t^3$$

$$i_{yx} = \frac{1}{8}tb^2 \left\{ 1 + 3\left(\frac{t}{b}\right)^2 \right\} \tag{21.38}$$

The average value of i_{xy} and i_{yx}, to be substituted in Eq. (21.12) is

$$i_{\text{av}} = \frac{1}{16}tb^2 + \frac{13}{48}t^3 \tag{21.39}$$

After the FE analysis has been done, we find one value m_{av}, from which we can calculate separate values m_{xy} and m_{yx}. The twisting moment m_{xy} will be small compared to m_{yx}, but acts in the thin flange where it must be combined with the lateral bending moment m_{xx}. This combination must be included in the consideration when designing reinforcement. In the span direction we must, strictly speaking, consider a combined state of forces and moments in each separate wall of the box. The flexural rigidity EI_y leads to both a membrane force n_{yy} and a plate bending moment m_{yy} in the top and bottom wall. Apart from this, due to Gi_{yx} we get a membrane shear force n_{yx} and a plate twisting moment m_{yx}. Reinforcement in an individual wall therefore must be designed for a combination of membrane forces and bending moments, for which we refer to Chapter 16. The structural engineer has to decide whether or not to refine the calculation to this level.

21.4.3 Shear Rigidity

X-Direction

Compared to the preceding cases of I-sections and a multi-cell bridge, the case under consideration is more complicated. In the preceding two cases the shear stiffness was either very large or very small. In the plate with I-sections, the shear rigidity is so large that shear distortion can be neglected. In the multi-cell bridge, the opposite is true; the shear distortion is so dominant that an infinitely large flexural rigidity can be assumed. The structure in Figure 21.7 is somewhat in between. It will appear that the flexural and shear deformation are of the same order. In Figure 21.8 we have already considered the case of a constant bending moment m_{xx}. A state of constant shear is not possible; it will always be accompanied of a linear distribution of bending moment, as depicted in Figure 21.9. Therefore, shear distortion in general

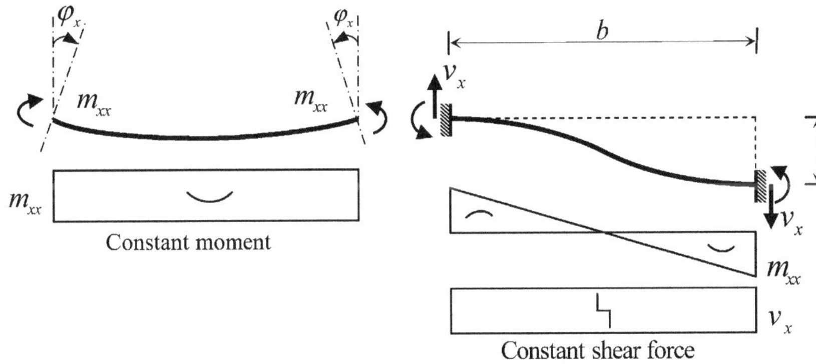

Figure 21.9 Two basic deformation states.

always interacts with flexural deformation. Returning to the box girder with flanges, we have to investigate two different cases as shown in Figure 21.10: in one case two rotations are enforced (left) and in the other a vertical shift of the one end with respect to the other (right). We search for a replacing homogeneous plate of flexural rigidity D_{xx} and shear rigidity D_{sx}. We have two unknowns and have two independent states at our disposal to determine them. The flexural rigidity has already been dealt with in Section 21.4.1 on the basis of the left deformation state of Figure 21.10. We now concentrate on the right one. We can do a frame analysis for the box girder cross-section, in which we force the right support to shift over the distance δ vertically, and calculate the evoked support reaction v_x. From the relation $v_x = K_{sx}\delta$ we calculate the stiffness K_{sx}. For the replacing homogeneous plate the relation

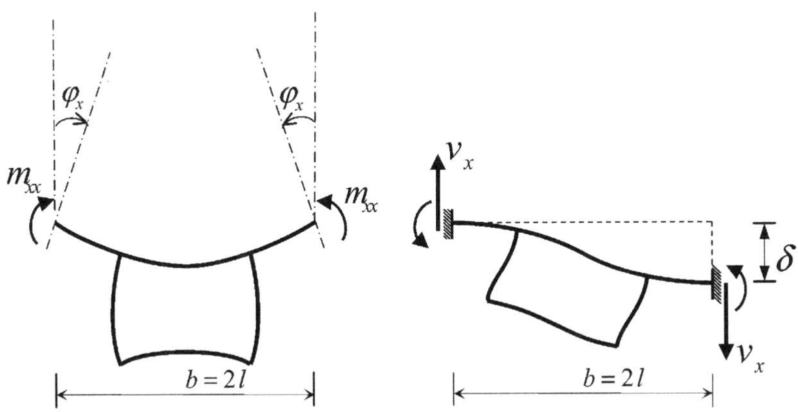

Figure 21.10 Example of square box section with flanges.

between shift and support reaction is $\delta = v_x b^3/12D_{xx} + v_x b/D_{sx}$, where the right-hand term contains both the bending deformation and the shear deformation. Equalizing this δ with v_x/K_{sx} from the frame analysis we obtain the equation

$$\frac{b^3}{12D_{xx}} + \frac{b}{D_{sx}} = \frac{1}{K_{sx}} \tag{21.40}$$

Summarizing, we must perform two frame analyses, one for the moment and one for the shear force. From the first we derive D_{xx} and from the second D_{sx}.

Example

If a structure is not too complicated the frame analyses can be done by hand calculation, and the rigidities can be obtained in closed form. In the case of Figure 21.10 with a square box of sizes l, we find from the bending analysis

$$D_{xx} = \frac{4}{33} E t^3 \tag{21.41}$$

From the shear force analysis we obtain

$$K_{sx} = \frac{8}{61} \frac{E t^3}{l^3} \tag{21.42}$$

Substitution of D_{xx} and K_{sx} in Eq. (21.40) leads to

$$\frac{(2l)^3}{12\left\{\frac{4}{33} E t^3\right\}} + \frac{2l}{D_{sx}} = \frac{1}{\left\{\frac{8}{61} \frac{E t^3}{l^3}\right\}} \tag{21.43}$$

from which we solve

$$D_{sx} = \frac{16}{17} E t \left\{\frac{t}{l}\right\}^2 \tag{21.44}$$

If we want to compare the rigidities D_{xx} and D_{sx}, the latter must be multiplied by the square width of the considered element in order to obtain quantities with the same units. Therefore, we must compare D_{xx} and $b^2 D_{sx}$. We have obtained the following order of magnitude for the three considered plates

Overview of three characteristic cases

We have obtained different orders of magnitude for three plate types

$$\text{I-section plates:} \quad \frac{D_{xx}}{D_{sx}b^2} = O\left(\frac{t}{b}\right)^2$$

$$\text{Flanged boxes:} \quad \frac{D_{xx}}{D_{sx}b^2} = O(1)$$

$$\text{Multi-cell bridge:} \quad \frac{D_{xx}}{D_{sx}b^2} = O\left(\frac{b}{t}\right)^2$$

Because $t/b \ll 1$, in the plate with I-sections the denominator is dominant (the shear rigidity), in the multi-cell bridge the numerator is dominant (the flexural rigidity) and in the plate with flanged boxes numerator and denominator are of the same order of magnitude (flexural and shear rigidity of the same order of size).

Y-direction

In the y-direction the shear rigidity is large as we have seen for the plate with I-sections, and the multi-cell plate. The structural engineer must choose the relevant value of the shape factor η and apply the formula

$$D_{sy} = G\frac{A_{sy}}{b} \tag{21.45}$$

where A_{sy} is the real cross-section area A divided by the shape factor η. This factor will be substantial larger than 1. If related to the area of only the two webs the value will be close to 1, which will appeal to structural engineers.

21.5 Message of the Chapter

- A structural engineer will sometimes severely overestimate the torsional rigidity by not recognizing the difference in definition between classical books on plate theory and Finite Element user manuals.

- In plates with longitudinal stiffeners the torsional rigidity in the span direction is different from the rigidity in the lateral direction. The contribution of the upper plate to the plate torsional rigidity must be split into a part related to the span direction and a part related to the lateral direction. Longitudinal stiffeners of the upper plate contribute only to the rigidity in the span direction.

- For the determination of the plate torsional rigidity the average rigidity of span and lateral direction must be used, and the outputted twisting moment will be the average moment for both directions. After completion of the FE analysis, we can calculate the twisting moments in the span and lateral directions on the basis of the ratio of the rigidities in both directions.

- The plate torsional rigidity of multi-cell bridges may be calculated with neglect of the intermediate vertical webs.

- After completion of the FE analysis, the shear force in vertical webs can be determined by the plate formula for the distributed shear force. This distributed shear force must be multiplied by the spacing of the webs.

- The shear rigidity in the lateral direction needs special attention. A safe method of deriving it is to consider two independent states, one constant moment and one constant shear force. This leads to two equations for the required flexural and shear rigidities.

References

1. Kurrer, K-E., *The History of the Theory of Structures, From Arch Analysis to Computational Mechanics*, Ernst & Sohn (Wiley), 2008.
2. Girkmann, K., *Flächentragwerke, Einführung in die Elastostatik der Scheiben, Platten, Schalen und Faltwerke*, Springer, Wien, 1974.
3. Timoshenko, S.P. and Goodier, J.N., *Theory of Elasticity*, second edn., McGraw-Hill Book Company, New York, 1951.
4. Timoshenko, S.P., *Schwingungsprobleme der Technik*, Verlag von Julius Springer, 1932.
5. Reissner, E. The effect of shear deformation on the bending of elastic plates. *Trans. ASME, J. Appl. Mech.* **12**, 1945, A68–77.
6. Mindlin R.D. Influence of rotary inertia and shear on flexural motions isotropic, elastic plates. *Trans. ASME, J. Appl. Mech.* **18**, 1951, 1031–1036.
7. Wang C.M., Lim G.T., Reddy J.N. and Lee K.H., Relationships between bending solutions of Reissner and Mindlin theories. *Engrg. Struct.* **23**, 2001, 838–849.
8. Todhunter, I. and Pearson, K., *History of the Theory of Elasticity*, Vol. 2, On Saint-Venant, At the University Press, Cambridge, 1893.
9. Timoshenko, S.P., *Theory of Plates and Shells*, McGraw-Hill, Auckland, 1989.
10. Szilard, R., *Theories and Applications of Plate Analysis: Classical, Numerical and Engineering Methods*, John Wiley and Sons, 2004.
11. Reddy J.N., *Theory and Analysis of Elastic Plates*, CRC Press, 2006.
12. Kirchhoff, G., Vorlesungen über Mathematische Physik, *Mechanik*, B.G. Teubner, Leipzig, 1877, 450.

13. Airy G.B., *Brit. Assoc. Advancement Sci. Rept.*, 1862.

14. Boussinesq, J., *Application des potentiels à l'étude de l'équilibre et du mouvement des solides élastiques*, Imprim. L. Danel, Lille, 1885.

15. Flaman, A.A., *Compt. Rend.* **114**, 1892, 1510.

16. Zienkiewicz, O.C. and Taylor, R.L., *The Finite Element Method*, Vol. 1, 5th edn., Butterworth-Heienmann, Oxford, 2000.

17. Hughes, T.J.R., *Finite Element Method – Linear Static and Dynamic Finite Element Analysis*, Prentice-Hall, Englewood Cliffs, 1987, 803 pp., 2000, 682 pp.

18. Rosman, R., Systeme aus durchbrochenen Wänden, *Die Bautechnik* **2**, 1964, 51–54.

19. Nielsen, M.P., *Limit Analysis and Concrete Plasticity*, 2nd ed., CRC Press, Boca Raton, London, New York, Washington DC, 1998.

20. CEB, CEB-FIB Model Code, Thomas Telford (1993).

21. Wood R.H., The reinforcement of slabs in accordance with field of moments, *Concrete Mag.* **2**, 1968, 69–75.

22. CEN, EN 1992-1-1:2004, Eurocode 2: Design of concrete structures – Part 1-1: General rules and rules for buildings, 2004.

23. CEN, EN 1992-2:2005, Eurocode 2: Design of concrete structures – Concrete bridges – Design and detailing rules, 2005.

24. Fib Bulletin 45, *Practitioner's Guide to Finite Element Modelling of Reinforced Concrete Structures*, Lausanne, 2008.

25. Lourenço, P.B. and Figueiras, J.A., Automatic design of reinforcement in concrete plates and shells, *Engrg. Comput.* **10**, 1993, 519–541.

26. Lourenço, P.B. and Figueiras, J.A. Solution for the design of reinforced concrete plates and shells, *ASCE J. Struc. Engrg.* **121**, 1995, 815–823.

27. Marti, P., Dimensioning and detailing, in *IABSE Colloquim Report*, Vol. 62, Stuttgart, Germany, 1991, pp. 411-443.

28. Kirscher, U. and Collins, M.P., Investigating the behavior of reinforced concrete shell elements, Pub. 86-09, Dept. Civ. Engrg., University of Toronto, Canada, 1986.

29. Kollegger, J., Computer programs for consistent design of surface structures, in *IABSE Colloquium Report*, Vol. 62, Stuttgart, Germany, 1991, pp. 507–512.

30. Marti, M., Leesti, P. and Khalifa, W. U., Torsion tests on reinforced concrete slabs elements, *ASCE J. Struct. Engrg.* **113**, 1987, 994–1010.

31. Schlaich, J., Schäfer, K. and Mattias, M., Toward a consistent design of structural concrete, Special Report of *PCI Journal*, Vol. 32, No. 3, 1987.

32. Bares, R. and Massonnet, C.E., *Analysis of Beam Grids and Orthotropic Plates by the Gyon–Massonnet–Bares Method*, Crosby Lockwood, London, 1968.

33. Kok, A.W.M. and Blaauwendraad, J., Shape-orthotropic stressing-bending plate model, *Engrg. Struct.* **30**, 2008, 2884–2892.

34. EN 1997, Eurocode 7 – Geotechnical design.

35. Kayser, Th.H. and Binkhorst, J., Computer calculations of a complex steel bridge verified by model investigations, *Rijkswaterstaat Communications* **23**, 1975.

Subject Index

Name Index

MIX
Papier aus verantwortungsvollen Quellen
Paper from responsible sources
FSC® C105338

If you have any concerns about our products,
you can contact us on
ProductSafety@springernature.com

In case Publisher is established outside the EU,
the EU authorized representative is:
Springer Nature Customer Service Center GmbH
Europaplatz 3, 69115 Heidelberg, Germany

Printed by Libri Plureos GmbH
in Hamburg, Germany